Lecture Notes in Computer Science 8179

Commenced Publication in 1973
Founding and Former Series Editors:
Gerhard Goos, Juris Hartmanis, and Jan van Leeuwen

T0212385

Thomas Moscibroda   Adele A. Rescigno (Eds.)

Structural Information and Communication Complexity

20th International Colloquium, SIROCCO 2013
Ischia, Italy, July 1-3, 2013
Revised Selected Papers

 Springer

Volume Editors

Thomas Moscibroda
Microsoft Research Asia
System Algorithms Research Group
No. 5, Dan Ling Street, 100080 Beijing, China
E-mail: moscitho@microsoft.com

Adele A. Rescigno
Università di Salerno
Dipartimento di Informatica
Via Ponte don Melillo, 84084 Fisciano, SA, Italy
E-mail: arescigno@unisa.it

ISSN 0302-9743   e-ISSN 1611-3349
ISBN 978-3-319-03577-2   e-ISBN 978-3-319-03578-9
DOI 10.1007/978-3-319-03578-9
Springer Cham Heidelberg New York Dordrecht London

Library of Congress Control Number: 2013952950

CR Subject Classification (1998): F.2, C.2, G.2, E.1, C.2.4

LNCS Sublibrary: SL 1 – Theoretical Computer Science and General Issues

Typesetting: Camera-ready by author, data conversion by Scientific Publishing Services, Chennai, India

Printed on acid-free paper

Springer is part of Springer Science+Business Media (www.springer.com)

Preface

The 20th International Colloquium on Structural Information and Communication Complexity (SIROCCO 2013) took place in Ischia, Italy, for three days starting July 1, 2013.

SIROCCO is devoted to the study of communication and knowledge in distributed systems from both qualitative and quantitative viewpoints. Special emphasis is given to innovative approaches and fundamental understanding in addition to efforts to optimize current designs. The typical areas include distributed computing, communication networks, game theory, parallel computing, social networks, mobile computing (including autonomous robots), peer-to-peer systems, communication complexity, fault-tolerant graph theories, and randomized/ probabilistic issues in networks.

This year, 67 papers were submitted in response to the call for papers, and each paper was evaluated by at least three reviewers. The Program Committee selected 28 papers for presentation at the colloquium and publication in this volume after in-depth discussions. The SIROCCO Prize for Innovation in Distributed Computing was awarded this year to Andrzej Pelc from the University of Quebec for his contributions to the understanding of distributed computing. A laudatio summarizing his many and important innovative achievements appears in these proceedings.

We further congratulate the recipients of the 2013 SIROCCO Best Student Paper award. This year, two papers were selected to share the award. Specifically, the 2013 SIROCCO Best Student Paper Award was given to Lakshmi Anantharamu for her paper "Broadcasting in Ad Hoc Multiple Access Channels" (with Bogdan Chlebus) and Sebastian Kniesburges and Andreas Koutsopoulos for their paper "A Deterministic Worst-Case Message Complexity Optimal Solution for Resource Discovery" (with Christian Scheideler).

The collaboration of the Program Committee members and the external reviewers enabled completion of the process of reviewing the papers and discussing them in less than four weeks. We thank them all for their devoted service to the SIROCCO community. We thank the authors of all the submitted papers; without them we could not have prepared a program of such quality. We thank Gennaro Cordasco for his assistance as publicity chair, and Yang Chen for serving as the submission chair. We also thank the keynote and invited speakers Andrea Richa, Cyril Gavoille, and Fabian Kuhn. The preparation of this event was guided by the SIROCCO Steering Committee, headed by Shay Kutten.

We are indebted to Luisa Gargano and Ugo Vaccaro for their assistance with local arrangements during the colloqium. We gratefully acknowledge the financial support of the Dipartimento di Informatica from Università di Salerno.

July 2013   Thomas Moscibroda
                                                            Adele A. Rescigno

Organization

Program Committee

Ittai Abraham	MSR Silicon Valley, USA
Chen Avin	Ben-Gurion University, Israel
Keren Censor-Hillel	Technion, Israel
Wei Chen	Microsoft Research Asia
Ajoy Datta	University of Nevada, USA
Gabriele Di Stefano	University of L'Aquila, Italy
Yuval Emek	ETH Zurich, Switzerland
Qiang-Sheng Hua	Tsinghua University, China
David Ilcinkas	Université Bordeaux, France
Toshimitsu Masuzawa	Osaka University, Japan
Friedhelm Meyer Auf der Heide	University of Paderborn, Germany
Thomas Moscibroda	Microsoft Research Asia
Calvin Newport	Georgetown University, USA
Adele Rescigno	Università di Salerno, Italy
Nicola Santoro	Carleton University, Canada
Ulrich Schmid	TU Vienna, Austria
Siddhartha Sen	Princeton, USA
Jukka Suomela	University of Helsinki, Finland
Subash Suri	UC Santa Barbara, USA

Additional Reviewers

Aspnes, James	Czyzowicz, Jurek
Auletta, Vincenzo	D'Angelo, Gianlorenzo
Biely, Martin	D'Emidio, Mattia
Borokhovich, Michael	Das, Shantanu
Bouzid, Zohir	De Marco, Gianluca
Burman, Janna	Delling, Daniel
Casteigts, Arnaud	Dereniowski, Dariusz
Chalopin, Jérémie	Devismes, Stéphane
Charron-Bost, Bernadette	Dobrev, Stefan
Cheilaris, Panagiotis	Drees, Max
Cicalese, Ferdinando	Friggstad, Zachary
Cicerone, Serafino	Függer, Matthias
Cohen, Asaf	Ganesh, Ayalvadi
Cord-Landwehr, Andreas	Gargano, Luisa
Cordasco, Gennaro	Georgiou, Chryssis

Gu, Yue
Gu, Zhaoquan
Hirvonen, Juho
Inoue, Michiko
Izumi, Taisuke
Jung, Daniel
Kakugawa, Hirotsugu
Kamei, Sayaka
Keller, Barbara
Kesselheim, Thomas
Kling, Peter
Kowalski, Dariusz
Kralovic, Rastislav
Kranakis, Evangelos
Kulkarni, Sandeep
Kutten, Shay
Köhler, Sven
Kößler, Alexander
Labourel, Arnaud
Lamani, Anissa
Larkin, Daniel
Le Merrer, Erwan
Lin, Xiao
Manne, Fredrik
Markou, Euripides
Moses, Yoram
Navarra, Alfredo
Nowak, Thomas
Ooshita, Fukuhito

Pardo Soares, Ronan
Parter, Merav
Pasquale, Francesco
Peters, Joseph
Pietrzyk, Peter
Pignolet, Yvonne-Anne
Podlipyan, Pavel
Prencipe, Giuseppe
Rajsbaum, Sergio
Robinson, Peter
Rybicki, Joel
Scalosub, Gabriel
Scheideler, Christian
Schwarz, Manfred
Suchan, Karol
Sundell, Håkan
Vaccaro, Ugo
Wang, Yajun
Weinstein, Omri
Widder, Josef
Winkler, Kyrill
Yamauchi, Yukiko
Yang, Siyu
Yu, Dongxiao
Yuan, Wen
Zhang, Jialin
Zhang, Le
Zémor, Gilles

Laudatio

2013 SIROCCO Prize for Innovation in Distributed Computing awarded to Andrzej Pelc

David Peleg
on behalf of the award committee

It is a pleasure to award the 2013 SIROCCO prize for innovation in distributed computing to Andrzej Pelc. The award is given for his significant contribution to communication paradigms for information dissemination. Let us briefly review this as well as some other major contributions and achievements of Andrzej, with a focus on those related to his many SIROCCO papers. (Notably, Andrzej coauthored 13 SIROCCO papers during the past two decades).

Much of Andrzej's research concentrated on algorithmic aspects of distributed computation and communication networks. In particular, he has contributed extensively to the study of communication paradigms designed to disseminate information in communication networks, such as broadcasting (i.e., sending a message from one source to every node in the network), gossiping (i.e., broadcasting from every node in parallel), multicasting and related paradigms. His work focused on developing algorithmic techniques for performing these tasks efficiently, by complexity measures such as time and (number and size of) messages, in different network architectures and under different assumptions, and while attempting to ensure a variety of desirable properties, most notably fault-tolerance. In addition, he studied the effects of having only partial knowledge of the network topology on the performance of broadcast algorithms in message passing networks and more recently in radio networks of different types (such as arbitrary topology, geometric graphs or unit disk graphs). Moreover, he recently published an authoritative book on these topics. Related to this line of work are Andrzej's SIROCCO papers [1–3], for example.

Another central theme of Andrzej's research involved the question of overcoming communication failures. He addressed this question in many different settings and under different assumptions on the nature of possible failures, such as random (dependent or independent) failures in nodes, links or transmissions, benign or malicious (Byzantine) faults, and globally or locally bounded faults. His studies served to broaden our knowledge and systematically organize the arsenal of tools available to us in this area. Related to this line of work are Andrzejs SIROCCO papers [4–6].

Andrzej was involved in studying fault-resilience in other contexts as well. Let me mention one area related to fault-tolerance where he made significant contributions, namely, the problem of system-level fault diagnosis in multiprocessor systems. This problem concerns a process where processors in the system can test each other for failures. It is assumed that fault-free testers correctly identify the fault status of tested processors, while faulty testers can give arbitrary test results. The goal is to develop algorithms for identifying correctly the status of all processors, assuming that the number of faults does not exceed a given upper bound. Andrzej's work explored static and adaptive solutions attempting to minimize the number of tests or the probability of error for various system topologies and under different failure models. Related to this line of work is Andrzej's SIROCCO paper [7].

Recently, Andrzej started studying algorithmic problems in systems of autonomous mobile agents or robots, which involve identical memoryless units residing in the nodes of a network or a terrain. Most notable is his work on the task of network exploration, which requires a mobile agent with small memory to explore an unknown network, i.e., traverse all its nodes and edges (possibly returning to the starting node), with no a priori knowledge of the network topology. The feasibility of this task, and the efficiency of exploration algorithms, depend on the model assumptions and on the class of allowed graph topologies. Andrzej studied this problem in different settings on trees and general graphs, paying attention mostly to the issues of memory requirements and resilience to failures. A second central problem studied extensively by Andrzej in this area is the rendezvous problem in networks, where two mobile agents, located in nodes of an unknown network, have to meet at some common location. Other problems he studied include the tasks of gathering the agents in one place and of searching for a "black hole" (namely, a destructive node) in the network. Related to this line of work are Andrzej's SIROCCO papers [8–11].

In summary, Andrzej's impressive technical achievements, his numerous influential contributions to efficient and failure-resistant algorithms in communication networks, and his leadership role in the research community in the field, make him a most highly deserving candidate for the prize.

References

1. Krzysztof Diks, Andrzej Lingas, and Andrzej Pelc. An Optimal Algorithm for Broadcasting Multiple Messages in Trees. In *Proceedings of the 4th SIROCCO*, 69–80, 1997.
2. Dariusz R. Kowalski and Andrzej Pelc. Time of Radio Broadcasting. In *Proceedings of the 10th SIROCCO*, 195-210, 2003.
3. David Ilcinkas, Dariusz R. Kowalski, and Andrzej Pelc. Fast Radio Broadcasting with Advice. In *Proceedings of the 15th SIROCCO*, 291-305, 2008.
4. Andrzej Pelc. Fast Fault-tolerant Broadcasting and Gossiping. In *Proceedings of the 2nd SIROCCO*, 159-172, 1995.
5. Andrzej Pelc. Efficient Fault Location with Small Risk. In *Proceedings of the 3rd SIROCCO*, 292-300, 1996.
6. Michel Paquette and Andrzej Pelc. Optimal Decision Strategies in Byzantine Environments. In *Proceedings of the 11th SIROCCO*, 245-254, 2004.
7. Evangelos Kranakis, Andrzej Pelc, and Anthony Spatharis. Optimal Adaptive Fault Diagnosis for Simple Multiprocessor Systems. In *Proceedings of the 5th SIROCCO*, 82-97, 1998.

8. Paola Flocchini, David Ilcinkas, Andrzej Pelc, and Nicola Santoro. Remembering without Memory: Tree Exploration by Asynchronous Oblivious Robots. In *Proceedings of the* 15th *SIROCCO*, 33-47, 2008.

9. Jurek Czyzowicz, David Ilcinkas, Arnaud Labourel, and Andrzej Pelc. Asynchronous Deterministic Rendezvous in Bounded Terrains. In *Proceedings of the* 17th *SIROCCO*, 72-85, 2010.

10. Samuel Guilbault and Andrzej Pelc. Gathering Asynchronous Oblivious Agents with Local Vision in Regular Bipartite Graphs. In *Proceedings of the* 18th *SIROCCO*, 162-173, 2011.

11. Samir Elouasbi and Andrzej Pelc. Time of Anonymous Rendezvous in Trees: Determinism vs. Randomization. In *Proceedings of the* 19th *SIROCCO*, 291-302, 2012.

Meeting in Networks
(Abstract of Award Lecture)

Andrzej Pelc

Université du Québec en Outaouais
Canada

Abstract. Two or more mobile entities, called agents or robots, starting at distinct initial positions, have to meet. This task is known in the literature as rendezvous and has many applications, both in everyday life and in computer science. Among many alternative assumptions that have been used to study the rendezvous problem, two most significantly influence the methodology appropriate for its solution. The first of these assumptions concerns the environment in which the mobile entities navigate: it can be either a terrain in the plane, or a network modeled as an undirected graph. The second assumption concerns the way in which the entities move: it can be either deterministic or randomized. In this talk we survey recent results on deterministic rendezvous in networks.

Adversarial Models for Wireless Communication
(Abstract of Keynote Lecture)

Andrea Richa

Arizona State University, U.S.A.

Abstract. In this talk, we present some recent work on adversarial modeling of wireless communication. We use an adaptive adversary to model the hard to predict physical interference, as well as other disruption in communication caused by temporary obstacles, mobility, background noise, co-existing networks, jammers, etc.

In particular, we focus on adversarial models for jamming. We present simple, local-control medium access control (MAC) protocols for wireless networks that are provably robust against adaptive adversarial jamming. Our protocols are orthogonal to physical layer protocols that rely on a broad spectrum, and can be used in conjunction with those or in networks where a broad spectrum is not available (e.g., sensor networks). We present a summary of our work in this area, going from single-hop wireless networks to multihop wireless networks modeled under SINR (signal-to-noise ratio model), and from more standard adaptive adversarial models for the jammer(s) to a more realistic adversarial model where, in addition to knowing the protocol and its entire history, the jammer also has some knowledge about the action of the nodes at the current time step. Our protocols are energy efficient, and require only very limited amount of knowledge about the jammer and the network. We also present simulation results that further validate our theoretical bounds.

We also address other recent work by the theoretical community on applications of adversarial modeling in wireless computing that focus on different paradigms (e.g., broadcasting, etc.).

The work on adversarial modeling of wireless jamming is joint work with Christian Scheideler (U. of Paderborn, Germany), Stefan Schmid (TU Berlin Telekom Labs), Jin Zhang (Google), Adrian Ogierman (U. of Paderborn) and Baruch Awerbuch (John Hopkins University).

Labeling Schemes with Forbidden-Sets
(Abstract of Invited Talk)

Cyril Gavoille

Université of Bordeaux, France

Abstract. The goal of labeling schemes is to understand how much information must be attached to the nodes of a network (formalized as labels) to solve a graph problem assuming the answer can be determined solely on the basis of the labels of the nodes invoked in the query. In this talk, I give a short survey on an extension of labeling schemes that can answer graph problems where some of the nodes may be turn off (or forbidden).

Distributed Computation in Directed and Dynamic Networks
(Abstract of Invited Talk)

Fabian Kuhn University of Freiburg, Germany

Abstract. We consider simple distributed data aggregation and information dissemination problems such as computing the minimum or the sum of a bunch of values or broadcasting multiple messages to all nodes in a network. In standard, undirected networks, these tasks are well studied and can be solved by simple distributed algorithms in time proportional to the diameter of the network. In my talk, I will discuss the complexity of such fundamental problems in networks with unidirectional links and in networks with dynamic topology. We will see that in absence of stable, bidirectional links, also the most basic distributed computation and information dissemination tasks become challenging, leading to a number of fascinating new research questions.

Table of Contents

Dynamic Networks Algorithms

Algorithms 1

Online Algorithms

Social Networks Systems

Distributed Algorithms

Robots

Wireless Networks

Algorithms 2

Algorithms 3

Distributed Community Detection
in Dynamic Graphs*
(Extended Abstract)

Andrea Clementi[1], Miriam Di Ianni[1], Giorgio Gambosi[1], Emanuele Natale[1],
and Riccardo Silvestri[2]

[1] Università Tor Vergata di Roma
{clementi,diianni,gambosi}@mat.uniroma2.it, emanatale@gmail.com
[2] Sapienza Università di Roma
silvestri@di.uniroma1.it

Abstract. Inspired by the increasing interest in self-organizing social opportunistic networks, we investigate the problem of distributed detection of unknown communities in dynamic random graphs. As a formal framework, we consider the dynamic version of the well-studied *Planted Bisection Model* dyn-$\mathcal{G}(n, p, q)$ where the node set $[n]$ of the network is partitioned into two unknown communities and, at every time step, each possible edge (u, v) is active with probability p if both nodes belong to the same community, while it is active with probability q (with $q << p$) otherwise. We also consider a time-Markovian generalization of this model.

We propose a distributed protocol based on the popular *Label Propagation Algorithm* and prove that, when the ratio p/q is larger than n^b (for an arbitrarily small constant $b > 0$), the protocol finds the right "planted" partition in $O(\log n)$ time even when the snapshots of the dynamic graph are sparse and disconnected (i.e. in the case $p = \Theta(1/n)$).

Keywords: Distributed Computing, Dynamic Graphs, Social Opportunistic Networks.

1 Introduction

Community detection in complex networks has recently attracted wide attention in several research areas such as social networks, communication networks, biological systems [14]. Understanding the community structure of a complex network is a challenging crucial issue in several applications (good surveys on this topic can be found in [2,12,20]). A modern application scenario (the one this paper is inspired from) is that of *Opportunistic Networks* where recent studies show that *social-aware* protocols provides efficient solutions for basic communication tasks [5].

---

* Partially supported by Italian MIUR under the PRIN 2010-11 Project ARS TechnoMedia.

T. Moscibroda and A.A. Rescigno (Eds.): SIROCCO 2013, LNCS 8179, pp. 1–12, 2013.

The static *Planted Bisection Model* [3,4,13] (or *Stochastic Blockmodel*, as it is known in the statistics community [15]) is a popular framework to formalize the problem of detecting communities in random graphs.

The (Static) Planted Bisection Model: Centralized Algorithms. The (static) Planted Bisection Model is defined as a static random graph $\mathcal{G}(n, p, q)$ (with $p, q \in (0, 1)$ such that $q << p$) where the node subset $[n] = \{1, 2, \ldots, n\}$ is partitioned into two equal-sized *unknown communities* V_1 and V_2 and each possible edge (u, v) is included with probability p if u and v both belong to the same community while it is included with probability q otherwise[1]. The goal here is to identify the unknown partition.

In [13] and, successively, in [10,19], some efficient centralized algorithms have been presented for the above problem. Such algorithms are based on centralized, expensive procedures such as simulated annealing and spectral-graph computations: all of them require the full knowledge of the graph adjacency matrix and, moreover, they work on static graphs only.

Community Detection in Opportunistic Networks. Recent studies in opportunistic networks focus on the impact of the *agent social behavior* on some basic communication tasks such as routing and broadcasting [5]. Recently, this issue has been investigated in an emerging class of opportunistic networks called *Intermittently-Connected Mobile Networks (ICMNs)* [23]: such networks are characterized by wireless links, representing opportunities for exchanging data, that sporadically appear among network nodes (usually mobile radio devices). So-called *social-aware* communication protocols rely on the reasonable intuition that, since mobile devices are carried by people who tend to form *communities*, members (i.e. nodes) of the same community are used to communicate with each other much more often than nodes from different communities. Experiments on real-data sets have widely shown that identifying communities can strongly help in improving the protocol performances [5]. It thus follows that community detection in ICMNs is a crucial issue.

As observed above, several centralized community-detection methods have been proposed in the literature that may result useful for offline data analysis of mobile traces. However, it is a common belief that next-future technologies will yield a dramatic growth of *self-organizing* ICMNs where the network protocols work without relying on any centralized server. In this new communication paradigm, it is required that community detection is performed in a fully distributed way. To this aim, in this paper we consider an algorithmic solution to community detection in ICMNs that relies on the epidemic mechanism known as *Label Propagation Algorithms* [18,21].

The Dynamic Planted Bisection Model. In order to capture the high dynamicity of ICMNs, we consider the natural dynamic version of the $\mathcal{G}(n, p, q)$ model. A *dynamic graph* is a probabilistic process that describes a graph whose topology changes with time: so it can be represented by a sequence $\mathcal{G} = \{G_t =$

---

[1] Observe that when $p = q$ the random graph model is the well-known Erdös-Rényi model.

$([n], E_t) : t \in \mathbb{N}\}$ of graphs with the same set $V = [n]$ of nodes, where G_t is the *snapshot* of the dynamic graph at time step t.

The dynamic version of the Planted Bisection Model, denoted as dyn-$\mathcal{G}(n, p, q)$, consists of a dynamic graph where n is the number of nodes while $p = p(n)$ and $q = q(n)$ are the edge-probability functions. At every time step t, each edge (u, v) is included in E_t with probability p if both u and v belong to the same community V_i $(i = 1, 2)$ while it is included with probability q otherwise (this model can also be seen as a non-homogeneous version of the *dynamic Erdös–Rényi graph* model [5]). So, the dynamic state (on/off) of an edge over the time is a random variable having Bernoully distribution with parameter p or q, respectively. Of course, relevant simplifications have been assumed in this model. However, in [6], experimental validations have shown that some real ICMNs exhibit some crucial connectivity properties (such as hop-diameter) which are well-approximated by *sparse* dynamic Erdös–Rényi graphs.

A strong simplifying assumption in the dynamic Erdös–Rényi graph model is *time independence*: the graph topology at time t is fully independent from the topology at time $t - 1$. *Edge Markovian Evolving Graphs* (in short *edge-MEG*) were first introduced in [7] as a generalization of the dynamic Erdös–Rényi graph model that captures the strong dependence between the existence of an edge at a given time step and its existence at the previous time step. An edge-MEG is a dynamic random graph $\mathcal{G}(n, p_\uparrow, p_\downarrow, E_0) = \{G_t = ([n], E_t) : t \in \mathbb{N}\}$ defined as follows. Starting from an initial random edge set E_0, at every time step, every edge changes its state (existing or not) according to a two-state Markovian process with probabilities $p_\uparrow$ and $p_\downarrow$. If an edge exists at time t, then, at time $t + 1$, it disappears with probability $p_\downarrow$. If instead the edge does not exist at time t, then it will come up at time $t + 1$ with probability $p_\uparrow$. We observe that the setting $p_\downarrow = 1 - p_\uparrow$ yields a sequence of independent Erdös–Rényi random graphs, i.e., *dynamic Erdös–Rényi graphs*, with edge probability $p = p_\uparrow$. Edge-MEGs have been adopted as concrete models for several real dynamic networks such as faulty networks [8], peer-to-peer systems [22], mobile ad-hoc networks [22]. Furthermore, Edge-MEGs have been considered by Whitbeck et al [23] as a concrete model for analyzing the performance of epidemic routing on sparse ICMNs. In this paper, we consider the Edge-MEG as a mathematical model for ICMNs. The dynamic Planted Bisection Model can be easily generalized in order to include edge-MEGs: here, we have two edge-probability parameter pairs $(p_\uparrow, p_\downarrow)$ and $(q_\uparrow, q_\downarrow)$ between two nodes u and v depending on whether they both belong to the same community or not. So, if both u and v belongs to the same community then the edge (u, v) is governed by the 2-state Markov chain with parameters $(p_\uparrow, p_\downarrow)$ otherwise the edge is governed by the 2-state Markov chain with parameter $(q_\uparrow, q_\downarrow)$. We assume that $q_\uparrow \ll p_\uparrow$ and, according to the parameter tuning performed in [23], it turns out that the best fitting to real scenarios is achieved by setting $p_\downarrow$, (and $q_\downarrow$) as absolute constants.

The algorithmic goal in the dyn-$\mathcal{G}(n, p, q)$ model is to design a fully-distributed protocol that computes a *good (node) labeling* for the dynamic graph $\mathcal{G}$: we say that a function $Z : V \rightarrow \{1, 2\}$ is a *good labeling* for $\mathcal{G}$ if Z labels each community with a different label: $\forall i, k \in \{1, 2\} \, \forall u \in V_i \, \forall v \in V_k \, : \, Z(u) = Z(v) \, \leftrightarrow \, i = k$. Nodes are entities that share a global clock and know (a good approximation of) the number n of nodes but it is not required they have distinct IDs. Initially, each node does not know anything about communities while, at every time step, it can exchange information with its current neighbors.

In [16], some greedy protocols are tested on specific sets of real mobility-trace datas. By running such protocols, every node constructs and updates its own community-list according to the length and the rate of the contacts observed so far by itself and by the nodes it meets. So, the protocol exploits the intuition that communities are formed by nodes that use to meet each other often and for a long time. However, no analytical result is given for such heuristics that, moreover, require nodes to often update and transmit relatively large lists of node-IDs.

Label Propagation Algorithms. A well-studied community-detection strategy is the one known as *Label Propagation Algorithms (LPA)* [21]. This strategy is based on a simple epidemic mechanism which can be efficiently implemented in a fully-distributed fashion since it requires easy local computations: it is thus very suitable for opportunistic networks such as ICMNs. In its basic version, some distinct labels are initially assigned to a subset of nodes; at every step, each node updates its label (if any) by choosing the label which most of its (current) neighbors have (the *majority* label); if there are multiple majority labels, one label is randomly chosen. Several versions of LPA-based protocols have been tested on a wide range of social networks [11,18,21]: such works experimentally show that LPA-based protocols work quite efficiently and are effective in providing *almost* good labeling. Based on extensive simulations, [21] empirically shows that the average convergence time of the (synchronous) LPA-based protocols is bounded by some logarithmic function on n. Clearly, the goal of the protocol is to converge to a good labeling for dyn-$\mathcal{G}(n, p, q)$. Despite the simplicity of LPA-based protocols, very few analytical results are known on their performance over relevant classes of graphs. As observed in [17], it seems hard to derive, from empirical results, any fundamental conclusions about LPA behavior, even on specific families of graphs. Recently, [11] provided a semi-synchronous version of the LPA-based protocol and formally prove that it guarantees finite convergence time on any static graph. In [17], an LPA-based protocol has been analyzed on the Planted Partition Model for highly-dense topologies. In particular, their analysis considers the static model $\mathcal{G}(n, p, q)$ with $p = \Omega(1/n^{1/4-\epsilon})$ and $q = O(p^2)$. In this restricted static case, it is shown that the protocol converges in constant expected time and conjectured a logarithmic bound for sparse topologies. In general, providing analytical bounds on the convergence time of LPA-based protocols over relevant classes of networks is an important open question that has been proposed in several papers arising from different areas [11,17,21].

Our Algorithmic Contribution. We provide an efficient distributed LPA-based protocol on the dynamic Planted Bisection Model dyn-$\mathcal{G}(n, p, q)$ with arbitrary $p > 0$ and $q = O(p/n^b)$ where $b > 0$ is any *arbitrarily small* constant.

Our protocol yields with high probability[2] (in short *w.h.p.*) a good labeling in $O\left(\max\{\log n, \frac{\log n}{pn}\}\right)$ time. The bound is tight for any $p = O(1/n)$ while it is only a logarithmic factor larger than the optimum for the rest of the parameter range (i.e. for more dense topologies). For the first time, we thus formally prove a logarithmic bound on the convergence time of an LPA-based protocol on a class of sparse and disconnected dynamic random graphs (i.e. for $p = \Theta(1/n)$). The local labeling rule adopted by the protocol is simple and requires no node IDs: the only exchanged informations are the labels. Our protocol can be easily adapted in order to construct a good labeling in the presence of a larger number of equal-sized communities (provided that this number is an absolute constant) and, more importantly, it also works for the Edge-MEG model $\mathcal{G}(n, p_\uparrow, p_\downarrow, q_\uparrow, q_\downarrow, E_0)$ in the parameter range $q_\uparrow = O(p_\uparrow/n^b)$, where b is *any* positive constant. In the latter model, the completion time is w.h.p. bounded by $O\left(M \cdot \max\left\{\log n, \frac{\log n}{p_\uparrow n}\right\}\right)$ where M is a bound on the *mixing time* of the two 2-state Markov chains governing the edges of the dynamic graph. It is known that (see for example [7])

$$M = O\left(\max\left\{\frac{1}{p_\uparrow + p_\downarrow}, \frac{1}{q_\uparrow + q_\downarrow}, \log n\right\}\right)$$

Observe that, when $p_\downarrow$ and $q_\downarrow$ are some arbitrary positive constants and $p_\uparrow = \Omega(1/n)$ (this case includes the "realistic" range derived in [23]), then $M = O(\log n)$ and the bound on the completion time becomes $O(\log^2 n)$. This bound is only a logarithmic factor larger than the optimal labeling time in the case of sparse topologies, i.e., when $p_\uparrow = \Theta(1/n)$. We run our protocol over hundreds of random instances according to the dyn-$\mathcal{G}(n, p, q)$ model with n varying from 10^3 to 10^6. Besides a good validation of our asymptotical analysis, the experiments show further positive features of the protocol. Our protocol is indeed *tolerant* to non-homogeneous edge-probability functions. In particular, the protocol almost-always returns a good labeling in *Bernoullian* graphs where the edge probability is not uniform, i.e., for each pair (u, v) of nodes in the same community, the parameter $p_{u,v}$ is suitably chosen in order to yield irregular sparse graphs. A detailed description of the experimental results together with all technical proofs can be found in the full version of the paper [9]

2 The Protocol and Its Analysis

In this extended abstract, we consider the dynamic graph dyn-$\mathcal{G}(n, p, q)$ under the following restrictions: the parameter p is known by every node; there are only 2 communities V_1 and V_2, each of size $n/2$ (n is an even number); the labeling process starts with (exactly) two *source nodes*, $s_1 \in V_1$ that is labeled by z_1 and $s_2 \in V_2$ that is labeled by z_2 with $z_1 \neq z_2$. The parameters p and q belong to the following ranges

$$\frac{1}{n} \leqslant p \leqslant \frac{d \log n}{n} \text{ and } q = O\left(\frac{p}{n^b}\right), \text{ for some constants } d > 0 \text{ and } b > 0. \quad (1)$$

---

[2] As usual, we say an event holds with high probability if it holds with probability at least $1 - \frac{1}{n^{\Theta(1)}}$.

Such restricions make the description easier, thus allowing us to focus on the main ideas of our protocol and of its analysis. In the full version [9], we show how to obtain the general result stated in the Introduction.

The protocol relies on the simple and natural properties of LPA. Starting from two source nodes (one in each community), each one having a different label, the protocol performs a label spreading by adopting a simple labeling/broadcasting rule (for instance, every node gets the label it sees most frequently in its neighbors). Since links between nodes of the same community are much more frequent than the other ones, we can argue that the *good-labeling* will be faster than the *bad-labeling* (in each community, the good labeling is the one from the source of the community while the bad labeling is the one coming from the other source).

However, providing a rigorous analysis of the above process requires to cope with some non-trivial probabilistic issues that have not been considered in the analysis of information spreading in dynamic graphs made in previous papers [1,7,8]. Let us consider any local labeling rule that depends on the label configuration of the (dynamic) neighborhood of the node only. At a given time step, there is a subset $I_c \subseteq [n]$ of labeled nodes and we need to evaluate the probabilities P_g (P_b) that a non-labeled node gets a good (bad) label in the next step. After an initial phase, there is a non-negligible probability that some nodes will get the bad label. Then, such nodes will start a spreading of the bad labeling at the same rate of the good one. Observe also that good-labeled nodes may (wrongly) change their state as well, so, differently from a standard single-source broadcast, the epidemic process is not monotone with respect to good-labeling.

It turns out that the probabilities P_g and P_b strongly depend on the *label-balance* between the sizes of the subsets of well-labeled nodes and of the badly-labeled ones in the two communities. Keeping a tight balance between such values during all the process is the main technical goal of the protocol. In arbitrary label configurations over *sparse* graph snapshots, getting "high-probability" bounds on the rate of new (well/badly) labeled nodes is a non-trivial issue: indeed, it is not hard to show that, given any two nodes $v, w \in [n] \setminus I_c$, the events "$v$ will be (well/badly-)labeled" and "w will be (well/badly-)labeled" are not independent. As we will see, such issues are already present in this "restricted" case. A first important step of our approach is to describe the combination between the labeling process and the dynamic graph as a finite-state Markovian process. Then, we perform a step-by-step analysis, focusing on the probability that the Markovian Process visits a sequence of states having "good-balance" properties.

Our protocol applies local rules depending on the current node's neighborhood and on the current time step only. The protocol execution over the dynamic graph can be represented by the following *Markovian Process*: for any time step t, we denote as $\left(k_1^{(t)}, k_2^{(t)}, h_1^{(t)}, h_2^{(t)}; E_t \right)$ the *state* reached by the Markovian Process where $k_i^{(t)}$ denotes the number of nodes in the i-th community labeled by label z_i at time step t and $h_i^{(t)}$ denotes the number of nodes in the i-th community labeled by label z_j at time step t, for $i, j = 1, 2$ and $j \neq i$. In particular, the Markovian Process works as follows

$$\cdots \rightarrow \left(k_1^{(t)}, k_2^{(t)}, h_1^{(t)}, h_2^{(t)}; E_t\right) \xrightarrow{\text{dyn-}\mathcal{G}(n, p, q)} \left(k_1^{(t)}, k_2^{(t)}, h_1^{(t)}, h_2^{(t)}; E_{t+1}\right) \xrightarrow{\text{protocol}}$$

$$\xrightarrow{\text{protocol}} \left(k_1^{(t+1)}, k_2^{(t+1)}, h_1^{(t+1)}, h_2^{(t+1)}; E_{t+1}\right) \xrightarrow{\text{dyn-}\mathcal{G}(n, p, q)} \cdots$$

The main advantage of this description is the following: observe the process in any fixed state and consider the set of nodes U still having no label. Then it is not hard to verify that, in the next time step, the events { "node v gets a good/bad label", $v \in U$}, are mutually independent. This will allow us to prove strong-concentration bounds on the label-balance discussed above for a sufficiently-long sequence of states visited by the Markovian Process, thus getting a large fraction of well-labeled nodes in each community within a short time; this corresponds to a first protocol stage called *fast spreading* of the good labels.

Unfortunately, this independence property does not hold among labeled nodes of the same community, let's see why in the next simple scenario. Assume that the rule is the majority one, consider two nodes u and v having the same label z at time t, and assume the event $\mathcal{E}$ = "node u will keep label z at time $t + 1$" holds. Then the event "$(u, v) \in E_t \mid \mathcal{E}$" is more likely and, thus, according to the majority rule, the event "v gets label $z \mid \mathcal{E}$" is more likely as well. This clearly shows a key-depencence in the label spreading.

In order to overcome this issue, our protocol allows every node to change its first label-updating rule only after a *spreading stage* of suitable length (we will see later this stage is in fact formed by 3 consecutive phases): we can thus analyze the spreading of the good labeling (only) on the current set of unlabeled nodes (where stochastic independence holds) and prove that the process reaches a state with a large number of well-labeled nodes. After this spreading stage, labeled nodes (have to) start to update their labels according to some simple rule that will be discussed later. In the full version [9], we prove this *saturation phase* has logarithmic convergence time by providing a simple and efficient method to cope with the above discussed stochastic dependence.

2.1   A Restricted Setting: Formal Description

The protocol works in 5 consecutive temporal phases: the goal of this phase partition is to control the rate of new labeled nodes as function of the expected values reached by the random variables (r.v.s) $k_i^{(t)}, h_i^{(t)}$ (at the end of each phase). Indeed, when such expected values reach some specific thresholds, the protocol and/or its analysis must change accordingly in order to keep the label configuration well-balanced in the two communities during all the process and to manage the stochastic depencence described above.

At any time step t, we denote, for each node $v \in V_i$, the number of z_i-labeled neighbors of v as $N_i^v(t)$, for $i = 1, 2$. Given a node $v \in V$, the set of its neighbors at time t will be denoted as $\Gamma_t(v)$. For the sake of brevity, whenever possible we will omit the parameter t in the above variables and, in the proofs, we will only analyze the labeling in V_1, the analysis for V_2 being the same.

Stage I: Spreading

Phase 1: Source Labeling. The phase runs for $\tau_1 = c_1 \log n$ time steps, where $c_1 > 0$ is an explicit constant that will be fixed later. In this phase, only the neighbors of the sources will decide their label. The goal is to reach a state such that w.h.p. $k_i = \Theta(\log n)$ and $h_i = 0$ $(i = 1, 2)$. For any non-source node v, the labeling rule is the following.

- Let $i \in \{1, 2\}$; v gets label z_i if there is a time step $t \leq \tau_1$ such that $s_i \in \Gamma_t(v)$ and, for $j \neq i$ and for all t such that $1 \leqslant t \leqslant \tau_1$, it holds that $s_j \notin \Gamma_t(v)$;
- In all other cases, v remains unlabeled.

At the end of this phase, it is possibile to prove that a node gets the good label with probability $\Theta(p\tau_1)$ and, w.h.p., no node will get the bad label. From this fact, we can prove the following

Theorem 1. *Let $d_1 > 0$ be any (sufficiently large) constant. Then, a constant $c_1 > 0$ can be fixed so that, at time step $\tau_1 = c_1 \log n$ the Markovian Process w.h.p. reaches a state such that*

$$k_1^{(\tau_1)}, k_2^{(\tau_1)} \in \left[\frac{d_1}{16} pn \log n, 4d_1 pn \log n \right] \quad \text{and} \quad h_1^{(\tau_1)}, h_2^{(\tau_1)} = 0 \qquad (2)$$

Phase 2: Fast Labeling I. This phase of the Protocol aims to get an exponential rate of the good-labeling inside every community in order to reach, in $\tau_2 = O(\log n)$ steps, a state such that the number of well-labeled nodes is bounded by some root of n and the number of badly-labeled ones is still 0. Differently from Phase 1, unlabeled nodes can get a label at every time step according to the following rule: for $\tau_1 < t \leq \tau_1 + \tau_2$, at time step t of Phase 2 every *unlabeled* node v

- gets label z_1 at time $t + 1$ iff $N_1^v(t) > 0$ and $N_2^v(t) = 0$,
- gets label z_2 at time $t + 1$ iff $N_2^v(t) > 0$ and $N_1^v(t) = 0$,
- remains unlabeled at time $t + 1$ otherwise.

In the next theorem, we assume that, at time step τ_1 (i.e. at the end of Phase 1), the Markovian Process reaches a state satisfying Cond. (2). In particular, we assume that $k_i^{\tau_1} \geqslant \frac{d_1}{16} pn \log n$. Thanks to Theorem 1, this event holds w.h.p. In what follows, we will make use of the following function

$$F(n, k) = 2 \max \left\{ \sqrt{\frac{\log n}{k}}, \frac{\text{polylog} \, n}{n^{1-a}} \right\}$$

At the end of Phase 2, we can prove the Process w.h.p. satisfies the following properties.

Theorem 2. *For any $\eta > 0$, constants a and ϕ can be fixed so that, at the final step of Phase 2*

$$\tau_2 = \frac{1}{\log \left(1 + \left(\frac{np}{2} \right) \right)} \log \left(\frac{n^a}{\phi \log^3 n} \right) + \log^{-1} \left[\left(1 + \frac{np}{2} \right) \left(1 - F(n, \underline{k}_1^{(\tau_1)}) \right) \right] \log \left(\frac{\log^3 n}{\underline{k}_1^{(\tau_1)}} \right) + \tau_1,$$

it holds w.h.p. that

$$\text{for } i = 1, 2, \ n^a \ \leqslant \ k_i^{(\tau_2)} \ \leqslant \ n^a \log^\eta n, \ \text{ and } \ h_i^{(\tau_2)} \ = \ 0. \tag{3}$$

Phase 3: Fast Labeling II. In this phase nodes apply the same rule of Phase 2 but we need to separate the analysis from the previous one since, when the "well-labeled" subset gets size larger than some root of n, we cannot anymore exploit the fact that the bad labeling is w.h.p. not started yet (i.e. $h = 0$). However, we will show that when the well-labeled sets get size $\Theta(n/\text{polylog}\,n)$, the bad-labeled sets have still size bounded by some root of n. We assume that, at the end of Phase 2, the Markovian Process reaches a state satisfying Cond. (3) of Theorem 2.

Theorem 3. *For any constant $\eta > 0$, constants $a_1 < 1$ and $\gamma > 0$ can be fixed so that at the final time step of Phase 3*

$$\tau_3 = \frac{1}{\log\left(1 + \left(\frac{np}{2}\right)\right)} \log\left(\frac{n^{1-a}}{\gamma \log^3 n}\right) + \tau_2$$

for $i = 1, 2$, it holds w.h.p. that

$$\frac{n}{\log^3 n} \leqslant k_i^{(\tau_3)} \leqslant \frac{n}{\log^{3-\eta} n} \ \text{ and } \ h_i^{(\tau_3)} \leqslant n^{a_1} \tag{4}$$

Theorems 2 and 3 guarantee a very tight range for the r.v. k_1 and k_2 at the final step of Phase 2 and 3, respectively. This tight balance is crucial for removing the hypothesis on the existence of the two leaders (see [9]).

Stage II: Saturation
Phase 4: Controlled Saturation. At the end of Phase 3, the Markovian Process w.h.p. reaches a state that satisfies the properties stated in Theorem 3. The goal of Phase 4 is to obtain a (large) constant fraction α (say, $\alpha = 3/4$) of the nodes of each community that get the good label and, at the same time, to ensure that the number of bad-labeled nodes is still bounded by some root of n. We cannot guarantee this goal by applying the same labeling rule of the previous phase: the number of bad-labeled nodes would increase too fast. The protocol thus performs a much "weaker" labeling rule that is enough for the good labeling while keeping the final number of bad-labeled nodes bounded by some root of n. The fourth phase consists of three consecutive identical time-windows during which *every* (labeled or not) node $v \in V$ applies the following simple rule:

Time Window of Phase 4.
For any $t \in [1, T_4 = c_4 \log n]$, v looks at the labels of its neighbors at time t and:

- If v sees only one label (say, z) for *all* the window time steps, then v gets label z;
- In all the other cases (either v sees more labels or v does not see any label), v either keeps its label (if any) or it remains unlabeled.

Remark. Observe that, departing from the previous phases, we now need to analyze the label-spreading of the above rule over nodes having previously-assigned
labels. This rises the stochastic dependence described earlier in this section. In
order to solve this issue, we first observe that if the graph has random *oriented*
edges, then the r.v.s describing the label of every node become mutually independent. In order to make our graph oriented, the nodes thus run a simple
procedure at the very beginning of every step. This procedure simulates a virtual
dyn-$\mathcal{G}(n, p, q)$ where the edges inside each community $\mathcal{V}$ are generated according
to a directed $G_{n/2, \tilde{p}}$ model, where $\tilde{p} = 1 - \sqrt{1 - p}$. Moreover, the procedure
makes the resulting probability of the edges between communities still bounded
by $O(q)$: it thus preserves the polynomial gap between p and q. The detailed
description of the procedure and its properties are available in [9]

The Protocol-Window of Phase 4 is repeated 3 times for a specific setting of
the constant c_4 that will be determined in (the proof of) Theorem 4. Thanks to
Theorem 3, we can assume that the Markovian Process w.h.p. terminates Phase
3 reaching a state that satisfies Eq. (4) and show the following

Theorem 4. *Let α be any constant such that $0 < \alpha < 1$. Then, constants c_4
and $a_1 < 1$ can be fixed so that, at time step $\tau_4 = \tau_3 + 3T_4$, the Markovian
Process w.h.p reaches a state such that, for $i = 1, 2$,*

$$k_i^{\tau_4} \geqslant \alpha n , \text{ and } h_i^{\tau_4} \leqslant n^{a_1} \text{polylog} \, n. \tag{5}$$

Phase 5: Majority Rule. Theorem 4 states that, at the end of Phase 4, the Markovian Process w.h.p. reaches a state where a (large) constant fraction of the nodes
(say, $3/4$) in both communities is well-labeled while only $O(n^{a_1} \text{polylog} \, n)$ nodes
are bad-labeled. We now show that a further final phase, where nodes apply a
simple majority rule, yields the good labeling, w.h.p.. Remind that every node
also applies the "orienting-edge" Procedure discussed in the previous phase. *Every* node $v \in V$ applies the following labeling rule:

 - For every $t \in [1, T_5 = c_5 \log n]$, *every* node v observes the labels of its
   neighbors at time t and, for every label z_i $(i = 1, 2)$, v computes the number
   f_i^t of its neighbors labeled with z_i.
 - Then, node v gets label z_1 if $\sum_{t \in [1, ..., \tau_5]} f_1^t > \sum_{t \in [1, ..., \tau_5]} f_2^t$, otherwise v gets
   label z_2 (break ties arbitrarily).

Let us assume the Markovian Process starts Phase 5 from a state satisfying Eq.
(5) (say with constant $\alpha = 3/4$).

Theorem 5. *A constant $c_5 > 0$ can be fixed so that, at time $\tau_5 = \tau_4 + c_5 \log n$,
every node of each community is well-labeled, w.h.p.*

2.2   Overall Completion Time of the Protocol and Its Optimality

When p and q satisfy Cond. (1), we have shown that every phase has length
$O(\log n)$: the Protocol has thus an overall completion time $O(\log n)$. In the full

version of the paper [9], we will show that for $p = o(1/n)$ the length of each phase must be stretched to $\Theta\left(\frac{\log n}{pn}\right)$. It is easy to verify that, if $p = O(1/n)$, starting from the initial random snapshot, there is a non-negligible probability that some node will be isolated for $\tau(n)$ time steps where $\tau(n)$ is any increasing function such that $\tau = o\left(\frac{\log n}{pn}\right)$: this implies that, in the above range, our protocol has optimal completion time.

3 Conclusions

This paper introduces a framework that allows an analytical study of the distributed community-detection problem in dynamic graphs. Then, it shows an efficient algorithmic solution in two classes of such graphs that model some features of opportunistic networks such as ICMNs. We believe that the problem deserves to be studied in other classes of dynamic graphs that may capture further relevant features of social opportunistic networks such as geometric constraints.

Acknowledgements. We thank Stefano Leucci for its help in getting an efficient protocol simulation over large random graphs.

References

1. Baumann, H., Crescenzi, P., Fraigniaud, P.: Parsimonious flooding in dynamic graphs. In: Proceedings of the 28th ACM Symposium on Principles of Distributed Computing, PODC 2009, pp. 260–269. ACM, New York (2009)
2. Boccaletti, S., Latora, V., Moreno, Y., Chavez, M., Hwang, D.-U.: Complex networks: Structure and dynamics. Physics Reports 424(4-5), 175–308 (2006)
3. Boppana, R.B.: Eigenvalues and graph bisection: An average-case analysis. In: Proceedings of the 28th Annual Symposium on Foundations of Computer Science, SFCS 1987, pp. 280–285. IEEE Computer Society, Washington, DC (1987)
4. Bui, T.N., Leighton, F.T., Chaudhuri, S., Sipser, M.: Graph bisection algorithms with good average case behavior. Combinatorica 7(2), 171–191 (1987)
5. Chaintreau, A., Hui, P., Crowcroft, J., Diot, C., Gass, R., Scott, J.: Impact of human mobility on opportunistic forwarding algorithms. IEEE Transactions on Mobile Computing 6(6), 606–620 (2007)
6. Chaintreau, A., Mtibaa, A., Massoulie, L., Diot, C.: The diameter of opportunistic mobile networks. In: Proceedings of the 2007 ACM CoNEXT Conference, CoNEXT 2007, pp. 12:1–12:12. ACM, New York (2007)
7. Clementi, A.E., Macci, C., Monti, A., Pasquale, F., Silvestri, R.: Flooding time in edge-markovian dynamic graphs. In: Proceedings of the Twenty-seventh ACM Symposium on Principles of Distributed Computing, PODC 2008, pp. 213–222. ACM, New York (2008)
8. Clementi, A.E., Monti, A., Pasquale, F., Silvestri, R.: Information spreading in stationary markovian evolving graphs. In: IEEE International Symposium on Parallel & Distributed Processing, IPDPS 2009, pp. 1–12. IEEE (2009)
9. Clementi, A.E.F., Ianni, M.D., Gambosi, G., Natale, E., Silvestri, R.: Distributed community detection in dynamic graphs. Technical report (2013)

10. Condon, A., Karp, R.M.: Algorithms for graph partitioning on the planted partition model. Random Structures and Algorithms 18(2), 116–140 (2001)
11. Cordasco, G., Gargano, L.: Label propagation algorithm: a semi–synchronous approach. International Journal of Social Network Mining 1(1), 3–26 (2012)
12. Danon, L., Diaz-Guilera, A., Duch, J., Arenas, A.: Comparing community structure identification. Journal of Statistical Mechanics: Theory and Experiment 2005 (09), P09008 (2005)
13. Dyer, M., Frieze, A.: The solution of some random np-hard problems in polynomial expected time. Journal of Algorithms 10(4), 451–489 (1989)
14. Girvan, M., Newman, M.E.J.: Community structure in social and biological networks. Proceedings of the National Academy of Sciences 99(12), 7821–7826 (2002)
15. Holland, P.W., Laskey, K.B., Leinhardt, S.: Stochastic blockmodels: First steps. Social Networks 5(2), 109–137 (1983)
16. Hui, P., Yoneki, E., Chan, S.Y., Crowcroft, J.: Distributed community detection in delay tolerant networks. In: Proceedings of 2nd ACM/IEEE International Workshop on Mobility in the Evolving Internet Architecture, page 7. ACM (2007)
17. Kothapalli, K., Pemmaraju, S.V., Sardeshmukh, V.: On the analysis of a label propagation algorithm for community detection. In: Frey, D., Raynal, M., Sarkar, S., Shyamasundar, R.K., Sinha, P. (eds.) ICDCN 2013. LNCS, vol. 7730, pp. 255–269. Springer, Heidelberg (2013)
18. Liu, X., Murata, T.: Advanced modularity-specialized label propagation algorithm for detecting communities in networks. Physica A: Statistical Mechanics and its Applications 389(7), 1493–1500 (2010)
19. Mossel, E., Neeman, J., Sly, A.: Stochastic Block Models and Reconstruction. ArXiv e-prints (February 2012)
20. Newman, M.E.J., Girvan, M.: Finding and evaluating community structure in networks. Physical Review E 69(2), 026113 (2004)
21. Raghavan, U.N., Albert, R., Kumara, S.: Near linear time algorithm to detect community structures in large-scale networks. Phys. Rev. E 76, 036106(2007)
22. Vojnovic, M., Proutiere, A.: Hop limited flooding over dynamic networks. In: 2011 Proceedings IEEE INFOCOM, pp. 685–693. IEEE (2011)
23. Whitbeck, J., Conan, V., de Amorim, M.D.: Performance of opportunistic epidemic routing on edge-markovian dynamic graphs. IEEE Transactions on Communications 59(5), 1259–1263 (2011)

Exploration of the T-Interval-Connected Dynamic Graphs: The Case of the Ring

David Ilcinkas* and Ahmed Mouhamadou Wade*

LaBRI, CNRS & Bordeaux University
{ilcinkas,wade}@labri.fr

Abstract. In this paper, we study the T-interval-connected dynamic graphs from the point of view of the time necessary and sufficient for their exploration by a mobile entity (agent). A dynamic graph (more precisely, an evolving graph) is T-interval-connected ($T \geq 1$) if, for every window of T consecutive time steps, there exists a connected spanning subgraph that is stable (always present) during this period. This property of connection stability over time was introduced by Kuhn, Lynch and Oshman [6] (STOC 2010). We focus on the case when the underlying graph is a ring of size n, and we show that the worst-case time complexity for the exploration problem is $2n - T - \Theta(1)$ time units if the agent knows the dynamics of the graph, and $n + \frac{n}{\max\{1, T-1\}}(\delta - 1) \pm \Theta(\delta)$ time units otherwise, where δ is the maximum time between two successive appearances of an edge.

Keywords: Exploration, Dynamic graphs, Mobile agent, T-interval-connectivity.

1 Introduction

Partly due to the very important increase of the number of communicating objects that we observe today, the distributed computing systems are becoming more and more dynamic. The computational models for static networks are clearly not sufficient anymore to capture the behavior of these new communication networks. In fact, even the computational models that take into account a certain degree of fault tolerance become insufficient for some very dynamic networks. Indeed, the classical models of fault tolerance either assume that the frequency of fault occurrences is small, which gives enough time to the algorithm to adapt to the changes, or that the system stabilizes after a certain amount of time (as in the self-stabilizing systems for example). Therefore, in the last decade or so, many more or less equivalent models have been developed that take into account the extreme dynamism of some communication networks. An interested reader will find in [1] a very complete overview of the different models and studies of dynamic graphs (see also [7]).

---

* Partially supported by the ANR project DISPLEXITY (ANR-11-BS02-014), the INRIA project CEPAGE, and the European project EULER.

T. Moscibroda and A.A. Rescigno (Eds.): SIROCCO 2013, LNCS 8179, pp. 13–23, 2013.
© Springer International Publishing Switzerland 2013

One of the first developed models, and also one of the most standard, is the model of evolving graphs [3]. To simplify, given a static graph G, called the underlying graph, an evolving graph based on G is a (possibly infinite) sequence of spanning but not necessarily connected subgraphs of G (see Section 2 for precise definitions). This model is particularly well adapted for modeling dynamic synchronous networks.

In all its generality, the model of evolving graphs allows to consider an extremely varied set of dynamic networks. Therefore, to obtain interesting results, it is often required to make assumptions that reduce the possibilities of dynamic graphs generated by the model. One example is the assumption of connectivity over time, which states that there is a journey (path over time) from any vertex to any other vertex. Another example is the assumption of constant connectivity, for which the graph must be connected at all times. This latter assumption, which is very usual, has been recently generalized in a paper by Kuhn, Lynch and Oshman [6] by the notion of T-interval-connectivity. Roughly speaking, given an integer $T \geq 1$, a dynamic graph is T-interval-connected if, for any window of T consecutive time steps, there exists a connected spanning subgraph which is stable throughout the period. (The notion of constant connectivity is thus equivalent to the notion of 1-interval-connectivity). This new notion, which captures the connection stability over time, allows the finding of interesting results: the T-interval-connectivity allows to reduce by a factor of about $\Theta(T)$ the number of messages that is necessary and sufficient to perform a complete exchange of information between all the vertices [2,6] (gossip problem).

In this paper, we carry on the study of these T-interval-connected dynamic graphs by considering the problem of exploration. A mobile entity (called agent), moving from node to node along the edges of a dynamic graph, must traverse/visit each of its vertices at least once (the traversal of an edge takes one time unit). This fundamental problem in distributed computing by mobile agents has been widely studied in static graphs since the seminal paper by Claude Shannon [8]. As far as highly dynamic graphs are concerned, only the case of periodically-varying graphs has been studied [4,5]. We focus here on the (worst-case) time complexity of this problem, namely the number of time units used by the agent to solve the problem in the T-interval-connected dynamic graphs. The problem of exploration, in addition to its theoretical interests, can be applied for instance to the network maintenance, where a mobile agent has to control the proper functioning of each vertex of the graph.

We consider the problem in two scenarios. In the first one, the agent knows entirely and exactly the dynamic graph it has to explore. This situation corresponds to predictable dynamic networks such as transportation networks for example. In the second scenario, the agent does not know the dynamics of the graph, that is the times of appearance and disappearance of the edges. This case typically corresponds to networks whose changes are related to frequent and unpredictable failures. In this second scenario, Kuhn, Lynch and Oshman [6] noted that the exploration problem is impossible to solve under the single assumption of 1-interval-connectivity. In fact, it is quite easy to convince oneself that by adding the assumption that each edge of the underlying graph must appear infinitely often, the exploration problem becomes possible, but the time complexity remains unbounded. In this article, and only for the second scenario, we therefore add the assumption of δ-recurrence, for some integer $\delta \geq 1$: each edge of the underlying graph appears at least once every δ time units.

It turns out that the problem of exploration is much more complex in dynamic graphs than in static graphs. Indeed, let us consider for example the first scenario (known dynamic graph). The worst-case exploration time of n-node static graphs is clearly in $\Theta(n)$ (worst case $2n - 3$). On the other hand, the worst-case exploration time of n-node (1-interval-connected) dynamic graphs remains largely unknown. No lower bound better than the static bound is known, while the best known upper bound is quadratic, and directly follows from the fact that the temporal diameter of these graphs is bounded by n. Therefore, we focus here on the study of T-interval-connected dynamic graphs whose underlying graph is a ring. Note that, in this particular case, the T-interval-connectivity property, for $T \geq 1$, implies that at most one edge can be absent at a given time.

Our results. We determine in this paper the exact time complexity of the exploration problem for the n-node T-interval-connected dynamic graphs based on the ring, when the agent knows the dynamics of the graph. This is essentially $2n - T - 1$ time units (see Section 3 for details). When the agent does not know the dynamics of the graph, we add the assumption of δ-recurrence, and we show that the complexity increases to $n + \frac{n}{\max\{1, T-1\}}(\delta - 1) \pm \Theta(\delta)$ time units (see Section 4 for details).

2   Model and Definitions

This section gives the precise definitions of the concepts and models informally mentioned in the introduction. Some definitions are similar or even identical to the definitions given in [6].

Definition 1 (Dynamic graph). *A dynamic graph is a pair $\mathcal{G} = (V, \mathcal{E})$, where V is a static set of n vertices, and $\mathcal{E}$ is a function which maps to every integer $i \geq 0$ a set $\mathcal{E}(i)$ of undirected edges on V.*

Definition 2 (Underlying graph). *Given a dynamic graph $\mathcal{G} = (V, \mathcal{E})$, the static graph $G = (V, \bigcup_{i=0}^{\infty} \mathcal{E}(i))$ is called the underlying graph of $\mathcal{G}$. Conversely, the dynamic graph $\mathcal{G}$ is said to be based on the static graph G.*

In this article, we consider the dynamic graphs based on the n-node ring, denoted C_n.

Definition 3 (T-interval-connectivity). *A dynamic graph $\mathcal{G} = (V, \mathcal{E})$ is T-interval-connected, for an integer $T \geq 1$, if for every integer $i \geq 0$, the static graph $G_{[i, i+T[} = (V, \bigcap_{j=i}^{i+T-1} \mathcal{E}(j))$ is connected.*

Definition 4 (δ-recurrence). *A dynamic graph is δ-recurrent if every edge of the underlying graph is present at least once every δ time steps.*

A mobile entity, called *agent*, operates on these dynamic graphs. The agent can traverse at most one edge per time unit. It may also stay at the current node (typically to wait for an incident edge to appear). We say that an agent *explores* the dynamic graph if and only if it visits all the nodes.

3   The Agent Knows the Dynamics of the Graph

In this section, we assume that the agent perfectly knows the dynamic graph to be explored.

3.1   Upper Bound

The following theorem shows that the worst-case exploration time is actually small, bounded by $2n$, when the underlying graph is a ring. Furthermore, it shows that the agent can benefit from the T-interval-connectivity to spare an additive term T. Note that our upper bound is constructive.

Before proceeding with the formal theorem and its proof, let us informally describe the key ingredients of the proof of the most general case.

We consider two algorithms, being the algorithms always going in the clockwise, resp. counter-clockwise, direction, traversing edges as soon as the dynamic graph allows it. At the beginning of the process, the agents try to traverse distinct edges and thus, at each time step, at least one of them progresses. During this phase, the average speed of the two agents is thus $1/2$ (edge traversals per time unit). However, when the agents are about to meet each other (thus after time at most n), their progression can be stopped by the absence of a unique edge e.

If this edge e is absent for at least $n-1$ time steps, then any agent has enough time to change its direction and to explore all the nodes of the graph in the other direction, hence completing exploration within $2n$ steps.

If the edge e does not stay absent long enough and reappears at time t, we modify the two algorithms as follows. The agent previously progressing in the clockwise, resp. counter-clockwise, direction, starts now by exploring the ring in the opposite direction, before going back in the usual direction the latest possible so that it reaches the edge e at time at most t. At time t, the two modified algorithms cross each other, and then continue their progression in their usual direction until one of them terminates the exploration. Note that, after time t, we have again the property that, at each time step, at least one agent progresses.

Globally, except during the period when e is absent, the average speed of the two agents is $1/2$. Besides, the modification of the algorithms allows each of the agent to explore an additional part of the ring. Unfortunately, these parts of the ring are traversed twice instead of once. Nevertheless, intuitively, the speed of both the modified agents is 1 during the period when e is absent. This compensates the loss induced by traversing twice some parts of the ring. Overall, the average speed is thus globally of at least $1/2$, which implies that at least one of the two modified agents performs exploration within time $2n$.

When the dynamic graph is T-interval connected, all edges must be present during $T - 1$ steps between the removal of two different edges. This fact is used to gain an additive term of $T - 1$ on the exploration time, yielding to a time of roughly $2n - T$. A much more precise analysis of the modified algorithms allows us to obtain the exact claimed bounds.

Theorem 1. *For every integers $n \geq 3$ and $T \geq 1$, and for every T-interval-connected dynamic graph based on C_n, there exists an agent (algorithm) exploring this dynamic graph in time at most*

$$\begin{cases} 2n - 3 & \text{if } T = 1 \\ 2n - T - 1 & \text{if } 2 \leq T < (n+1)/2 \\ \lfloor \frac{3(n-1)}{2} \rfloor & \text{if } T \geq (n+1)/2 \end{cases}$$

Proof. Fix $n \geq 3$ and an arbitrary dynamic graph based on the ring C_n. Let $v_0, v_1, \cdots, v_{n-1}$ be the vertices of C_n in clockwise order. Assume that the agent starts exploration from v_0 at time 0. In order to prove this theorem, we will describe various algorithms, and we will show that at least one of them will allow the agent to perform exploration within the claimed time bound. Let $\mathcal{T}$ be this bound.

First assume that at most one edge e is absent during the time interval $[0, \mathcal{T})$. Then, an agent going to the closest extremity of e and then changing direction will explore all nodes of the ring in time at most $3(n - 1)/2 \leq \mathcal{T}$. So let us assume from now on that at least two different edges are absent at least once each during the time interval $[0, \mathcal{T})$.

Before proceeding with the rest of the proof, we introduce the following notations. Given a time interval I and two algorithms A and B, let d_A^I be the number of edge traversals performed by agent A during the time interval I, let α_A^I, resp. $\alpha_{A,B}^I$, be the number of time steps in I for which agent A, resp. both agents A and B, do(es) not move. Note that it never helps to wait at a node when all its incident edges are present. Hence, without loss of generality, an agent always stays at a node because of the absence of an incident edge. Finally, let β^I be the number of time steps in I for which no edges are absent.

Let us now consider two simple algorithms. L, respectively R, is the algorithm always going in the clockwise, resp. counter-clockwise, direction, traversing edges as soon as the dynamic graph allows it. Now consider the sum of the number of edges traversed by each of the two algorithms until some time t. Since only one edge can be absent at a given time, this sum increases by at least one (and obviously by at most two) at each time step, until this sum is larger or equal to $n - 1$. So let e be the unique unexplored edge when this sum reaches $n - 1$. If the sum jumps directly from $n - 2$ to n, then fix e to be any of the last two unexplored edges. In both cases, let t_1 be the first time one of the two agents reaches one extremity of e. We consider two cases.

Case 1. The edge e is absent during the whole interval $[t_1, t_1 + n - 1)$.

In this case, the first agent to reach an extremity of e, at time t_1, goes back in the opposite direction and explores the ring in $n - 1$ further steps. This gives an exploration time of at most $t_1 + n - 1$. Let $I_1 = [0, t_1)$. We have

$$t_1 = \begin{cases} d_L^{I_1} + \alpha_L^{I_1} & \text{(1)} \\ d_R^{I_1} + \alpha_R^{I_1} & \text{(2)} \end{cases}$$

and, since L and R are always trying to traverse distinct edges during I_1 and at most one edge may be removed at any time, we also have

$$\alpha_L^{I_1} + \alpha_R^{I_1} + \beta^{I_1} \leq t_1 \tag{3}$$

Besides, we have $d_L^{I_1} + d_R^{I_1} \leq n - 1$ (4)

and since there are at least two removed different edges during the whole interval $[0, t_1 + n - 1)$, we have

$$\beta^{I_1} \geq T - 1 \tag{5}$$

(1)+(2)+(3)+(4)+(5) $\to t_1 + n - 1 \leq 2n - T - 1$.

For $T = 1$, this bound is one unit larger than the claimed bound. If the inequality (4) is in fact strict, then the correct bound is obtained. Otherwise, it means that at time $t_1 - 1$, both agents were free to move. This implies that either $\beta^{I_1} \geq 1$ or that the inequality (3) is strict. In both cases, this also gives the correct bound.

Case 2. The edge e is not absent during the whole interval $[t_1, t_1 + n - 1)$.

Then let t_2 be the smallest time $t \geq t_1$ such that the edge e is present at time t. We define two new algorithms, one of which will explore the dynamic graph within $\mathcal{T}$.

Let L' be the algorithm that is equal to L until some time t, at which L' goes back in the other direction forever. More precisely, L' is the algorithm for which t is the largest possible value such that L' arrives at the extremity of e at time at most t_2. Similarly, let R' be the algorithm that is equal to R until some time t, at which R' goes back in the other direction forever. More precisely, R' is the algorithm for which t is the largest possible value such that R' arrives at the extremity of e at time at most t_2. Let $\mathcal{T}_{exp}$ be the exploration time of the first between L' and G' exploring the dynamic graph.

In order to analyze the algorithms L' and R', we introduce two other algorithms. Let L'', respectively R'' be the algorithm defined as L', resp. R', but turning back exactly one time unit later than L', resp. R'.

Let $I_1 = [0, t_1)$, $I_2 = [t_1, t_2)$, $I_{1,2} = [0, t_2)$, $I_3 = [t_2, \mathcal{T}_{exp})$, and $I = [0, \mathcal{T}_{exp})$. On I_1, we have

$$t_1 \geq \alpha_{L''}^{I_1} + \alpha_{R''}^{I_1} - \alpha_{L'',R''}^{I_1} + \beta^{I_1} \tag{1}$$

$$t_1 \geq \alpha_L^{I_1} + \alpha_R^{I_1} + \alpha_{L'',R''}^{I_1} + \beta^{I_1} \tag{2}$$

As in the first case, we have

$$t_1 = \begin{cases} d_L^{I_1} + \alpha_L^{I_1} & \text{(3)} \\ d_R^{I_1} + \alpha_R^{I_1} & \text{(4)} \end{cases}$$

(1)+(2)+(3)+(4) $\to \alpha_{L''}^{I_1} + \alpha_{R''}^{I_1} + 2\beta^{I_1} \leq d_L^{I_1} + d_R^{I_1}$ (5)

On $I_{1,2}$, we have

$$t_1 + t_2 = \begin{cases} d_{L''}^{I_{1,2}} + \alpha_{L''}^{I_{1,2}} & (6) \\ d_{R''}^{I_{1,2}} + \alpha_{R''}^{I_{1,2}} & (7) \end{cases}$$

Note that, by definition of L'' and R''

$$d_{L''}^{I_{1,2}} \leq d_{L'}^{I_{1,2}} + 1 \tag{8}$$

$$d_{R''}^{I_{1,2}} \leq d_{R'}^{I_{1,2}} + 1 \tag{9}$$

$(6)+(7)+(8)+(9) \rightarrow 2(t_1 + t_2) \leq d_{L'}^{I_{1,2}} + d_{R'}^{I_{1,2}} + \alpha_{L'}^{I_{1,2}} + \alpha_{R'}^{I_{1,2}} + 2 \tag{10}$

Note that on I_2 L'' and R'' are not blocked because the edge e is absent during this interval. Hence

$$\alpha_{L''}^{I_{1,2}} = \alpha_{L''}^{I_1} \tag{11}$$

$$\alpha_{R''}^{I_{1,2}} = \alpha_{R''}^{I_1} \tag{12}$$

$(5)+(10)+(11)+(12) \rightarrow 2(t_1 + t_2) + 2\beta^{I_1} \leq d_L^{I_1} + d_R^{I_1} + d_L^{I_{1,2}} + d_R^{I_{1,2}} + 2 \tag{13}$

On I_3, we have

$$\mathcal{T}_{exp} - (t_1 + t_2) \geq \alpha_{L'}^{I_3} + \alpha_{R'}^{I_3} + \beta^{I_3} \tag{14}$$

and

$$\mathcal{T}_{exp} - (t_1 + t_2) = d_{L'}^{I_3} + \alpha_{L'}^{I_3} \tag{15}$$

$$\mathcal{T}_{exp} - (t_1 + t_2) = d_{R'}^{I_3} + \alpha_{R'}^{I_3} \tag{16}$$

$(14)+(15)+(16) \rightarrow \mathcal{T}_{exp} - (t_1 + t_2) + \beta^{I_3} \leq d_{L'}^{I_3} + d_{R'}^{I_3} \tag{17}$

$(17)+\frac{1}{2}(13) \rightarrow \mathcal{T}_{exp} + \beta^{I_1} + \beta^{I_3} \leq \frac{1}{2}(d_L^{I_1} + d_R^{I_1} + d_{L'}^{I_{1,2}} + d_{R'}^{I_{1,2}}) + d_{L'}^{I_3} + d_{R'}^{I_3} + 1$
(18)

Note that $\beta^{I_1} + \beta^{I_3} = \beta^I$

Let x, resp. y, be the number of edges traversed by L', resp. R', before turning back. Then

$$d_{R'}^{I_{1,2}} = 2x + d_L^{I_1} \tag{19}$$

$$d_{L'}^{I_{1,2}} = 2y + d_R^{I_1} \tag{20}$$

$$d_{R'}^{I_3} = d_L^{I_1} - x \tag{21}$$

$$d_{L'}^{I_3} = d_R^{I_1} - x \tag{22}$$

$$d_L^{I_1} + d_R^{I_1} \leq n - 1 \tag{23}$$

and since there are at least two removed different edges during the interval

$$\beta^I \geq T - 1 \tag{24}$$

Finally, we get the sought result

$(18)+(19)+(20)+(21)+(22)+(23)+(24) \rightarrow \mathcal{T}_{exp} \leq 2n - T - 1$.

One can again argue similarly as in the first case to gain one time unit in the case $T = 1$, which concludes the proof. $\qquad\square$

3.2   Lower Bound

We now prove that the precise bound given in Section 3.1 is actually the exact worst-case time complexity of the exploration problem.

Theorem 2. *For every integers $n \geq 3$ and $T \geq 1$, there exists a T-interval-connected dynamic graph based on C_n such that any agent (algorithm) needs at least*

$$\begin{cases} 2n - 3 & \text{if } T = 1 \\ 2n - T - 1 & \text{if } 2 \leq T < (n+1)/2 \\ \lfloor \frac{3(n-1)}{2} \rfloor & \text{if } T \geq (n+1)/2 \end{cases}$$

time units to explore it.

Proof. For any integers $n \geq 3$, and $2 \leq T \leq \lceil (n+1)/2 \rceil$, we define a T-interval-connected dynamic graph $\mathcal{G}_{n,T}$ based on C_n. Let $v_0, v_1, \cdots, v_{n-1}$ be the vertices of C_n in clockwise order. Assume that the exploration starts from v_0 at time 0. In $\mathcal{G}_{n,T}$, the edge $\{v_0, v_1\}$, respectively $\{v_{T-1}, v_T\}$, is absent in the time interval $[0, n - 2T + 1)$, respectively $[n - T, 2n)$. Note that this dynamic graph is indeed T-interval-connected.

Consider any agent (algorithm). We will now prove that the time it uses to explore $\mathcal{G}_{n,T}$ is at least $2n - T - 1$. Since the agent must explore all vertices, it must in particular explore both v_{T-1} and v_T. We consider two cases.

Case 1. v_{T-1} is explored before v_T.

To visit v_{T-1} without going through v_T, the agent must traverse the edge $\{v_0, v_1\}$. By construction, this edge is absent until time $n - 2T + 1$. Moreover, the length of the path between v_0 and v_{T-1} without going through v_T is $T - 1$. Thus the agent needs at least $n - T$ time units to reach v_{T-1} for the first time. Since the edge $\{v_{T-1}, v_T\}$ is absent in the time interval $[n - T, 2n)$, the fastest way of reaching v_T is to traverse the whole ring through v_0, inducing $n - 1$ additional time units. So in this first case, the agent needs at least $2n - T - 1$ time units to explore $\mathcal{G}_{n,T}$.

Case 2. v_T is explored before v_{T-1}.

To visit v_T without going through v_{T-1}, the agent must use the path v_0, v_{n-1}, up to v_T, which is of length $n - T$. When at node v_T, and since the edge $\{v_{T-1}, v_T\}$ is absent in the time interval $[n - T, 2n)$, the fastest way of reaching v_{T-1} is to traverse the whole ring through v_0, inducing $n - 1$ additional time units. Thus also in the second case, the agent needs at least $2n - T - 1$ time units to explore $\mathcal{G}_{n,T}$.

This proves the theorem for values of T in $[2, \lceil (n+1)/2 \rceil]$. In fact, this also proves the theorem for $T = 1$ because $\mathcal{G}_{n,2}$ is obviously also 1-interval-connected, and the claimed bound is the same for $T = 1$ and $T = 2$. Besides, note that only one edge is ever removed in $\mathcal{G}_{n, \lceil (n+1)/2 \rceil}$. This dynamic graph is therefore 1-interval-connected for any T, and thus the theorem is also proved for values of T larger than $(n+1)/2$. □

4   The Agent Does Not Know the Dynamics of the Graph

In this section, we assume that the agent does not know the dynamics of the graph, i.e., it does not know the times of appearance and disappearance of the edges. As explained in the introduction, we assume here the δ-recurrence property, for a given $\delta \geq 1$, in order for the problem to be solvable in bounded time.

4.1   Upper Bound

We first prove that there exists a very simple algorithm that is able to explore all the δ-recurrent T-interval-connected dynamic graphs based on the ring. This algorithm consists in moving as much and as soon as possible in a fixed arbitrary direction, see Algorithm 1.

Algorithm 1. STUBBORN-TRAVERSAL(dir)

Input: a direction dir
   for each time step **do**
      if   the edge in the dir direction is present **then**
         traverse it
      else
         wait
      end if
   end for

Theorem 3. *For every integers $n \geq 3$, $T \geq 1$ and $\delta \geq 1$, and for any direction* dir*, Algorithm* STUBBORN-TRAVERSAL*(*dir*) explores any δ-recurrent T-interval-connected dynamic graph based on C_n in time at most*

$$n - 1 + \left\lceil \frac{n-1}{\max\{1, T-1\}} \right\rceil (\delta - 1).$$

Proof. Fix an arbitrary direction dir and let us analyze the algorithm STUBBORN-TRAVERSAL(dir). Note first that it will complete exploration after traversing exactly $n - 1$ edges. To bound its exploration time, it thus remains to bound the number of time steps when the agent cannot move.

Since the dynamic graph is δ-recurrent, an edge cannot be absent for more than $\delta - 1$ consecutive time steps. Furthermore, since the dynamic graph is T-interval-connected, two time steps in which two different edges are absent must be separated by at least $T - 1$ time steps in which all edges are present. Therefore, the agent can traverse at least $\max\{1, T-1\}$ edges between two consecutive blocks at different nodes. To summarize, the agent can be blocked at most $\left\lceil \frac{n-1}{\max\{1, T-1\}} \right\rceil$ times during at most $\delta - 1$ time steps.

Putting everything together, the agent will perform edge traversals for $n - 1$ time steps and will wait for at most $\left\lceil \frac{n-1}{\max\{1, T-1\}} \right\rceil (\delta - 1)$ time steps, which gives the claimed bound. $\square$

4.2   Lower Bound

It turns out that the simple and natural Algorithm 1, described and analyzed in Section 4.1, is almost optimal, up to an additive term proportional to δ.

Theorem 4. *For every integers $n \geq 3$, $T \geq 1$, and $\delta \geq 1$, and for every agent (algorithm), there exists a δ-recurrent T-interval-connected dynamic graph based on C_n such that this agent needs at least*

$$n - 1 + \left\lfloor \frac{n-3}{\max\{1, T-1\}} \right\rfloor (\delta - 1)$$

time units to explore it.
This result holds even if the agent knows n, T and δ.

Proof. Let $n \geq 3$, $T \geq 1$, and $\delta \geq 1$. Fix an arbitrary agent (algorithm) A. We construct as follows the δ-recurrent T-interval-connected dynamic graph $\mathcal{G}_{n,T,\delta}(A)$ based on C_n that this agent will fail to explore in less than the claimed bound.

Let $v_0, v_1, \cdots, v_{n-1}$ be the vertices of C_n in clockwise order. Assume that the agent starts exploration from v_0 at time 0. For any integer $1 \leq i \leq n-1$, if the node v_i is explored by going from v_0 in the counter-clockwise direction, then node v_i is denoted v_{i-n}. Finally, let $\tilde{T} = \max\{1, T-1\}$.

In the dynamic graph $\mathcal{G}_{n,T,\delta}(A)$, only the edges $\{v_{\tilde{T}+1}, v_{\tilde{T}+2}\}$, $\{v_{2\tilde{T}+1}, v_{2\tilde{T}+2}\}$, and so on, and $\{v_0, v_{-1}\}$, $\{v_{-\tilde{T}}, v_{-\tilde{T}-1}\}$, $\{v_{-2\tilde{T}}, v_{-2\tilde{T}-1}\}$, and so on, may be absent. The actual times of appearance and disappearance of these edges depend on the algorithm A. For any integer $i \geq 0$, each time the agent arrives at node $v_{-i\tilde{T}}$ in the counter-clockwise direction, the edge $\{v_{-i\tilde{T}}, v_{-i\tilde{T}-1}\}$ is removed until either the δ-recurrence forces the edge to reappear or the agent leaves the node $v_{-i\tilde{T}}$ to go on $v_{-i\tilde{T}+1}$. Similarly, for any integer $i \geq 1$, each time the agent arrives at node $v_{i\tilde{T}+1}$ in the clockwise direction, the edge $\{v_{i\tilde{T}+1}, v_{i\tilde{T}+2}\}$ is removed until either the δ-recurrence forces the edge to reappear or the agent leaves the node $v_{i\tilde{T}+1}$ to go on $v_{i\tilde{T}}$. Note that between two time steps with two different absent edges, there are at least $T-1$ time steps for which no edges are absent. The dynamic graph is therefore T-interval-connected. It is also δ-recurrent by construction.

By definition of the dynamics of the graph, the agent needs to wait $\delta - 1$ time units to go from $v_{-i\tilde{T}}$ to $v_{-i\tilde{T}-1}$, for $i \geq 0$, or to go from $v_{i\tilde{T}+1}$ to $v_{i\tilde{T}+2}$, for $i \geq 1$. To explore all the vertices, the agent needs to perform at least $\left\lfloor \frac{n-3}{\max\{1, T-1\}} \right\rfloor$ such traversals. The waiting time of the agent is thus at least $\left\lfloor \frac{n-3}{\max\{1, T-1\}} \right\rfloor (\delta - 1)$. Since the agent needs also at least $n - 1$ time units to traverse enough edges so that all vertices are explored, we obtain the claimed bound. $\square$

5   Conclusion

We studied in this paper the problem of exploration of the T-interval-connected dynamic graphs based on the ring in two scenarios, when the agent is specific to the dynamic graph, and when the agent does not know the dynamics of the graph. The next objective is obviously to extend these results to larger families of underlying graphs. Unfortunately, this problem is much more difficult than it seems: proving that any dynamic graph based on a tree of cycles (a cactus) can be explored in time $O(n)$ is already a challenging open problem.

References

1. Casteigts, A., Flocchini, P., Quattrociocchi, W., Santoro, N.: Time-varying graphs and dynamic networks. International Journal of Parallel, Emergent and Distributed Systems 27(5) (2012)
2. Dutta, C., Pandurangan, G., Rajaraman, R., Sun, Z.: Information spreading in dynamic networks. CoRR, abs/1112.0384 (2011)
3. Ferreira, A.: Building a reference combinatorial model for MANETs. IEEE Network 18(5), 24–29 (2004)
4. Flocchini, P., Mans, B., Santoro, N.: On the exploration of time-varying networks. Theoretical Computer Science 469, 53–68 (2013)
5. Ilcinkas, D., Wade, A.M.: On the Power of Waiting when Exploring Public Transportation Systems. In: Fernàndez Anta, A., Lipari, G., Roy, M. (eds.) OPODIS 2011. LNCS, vol. 7109, pp. 451–464. Springer, Heidelberg (2011)
6. Kuhn, F., Lynch, N.A., Oshman, R.: Distributed computation in dynamic networks. In: 42nd ACM Symposium on Theory of Computing (STOC), pp. 513–522 (2010)
7. Kuhn, F., Oshman, R.: Dynamic networks: models and algorithms. ACM SIGACT News 42(1), 82–96 (2011)
8. Shannon, C.E.: Presentation of a maze-solving machine. In: 8th Conf. of the Josiah Macy Jr. Found (Cybernetics), pp. 173–180 (1951)

A Characterization of Dynamic Networks
Where Consensus Is Solvable

Étienne Coulouma* and Emmanuel Godard

LIF, Université Aix-Marseille and CNRS

Abstract. We consider the Consensus problem in arbitrary dynamic networks. A dynamic network is a communication network whose topology evolves from round to round. We make no assumptions on the possible topologies. We give the first complete necessary and sufficient condition for dynamic networks where it is possible to solve Consensus.

We show that we can complement the necessary condition for solvability of Consensus given, in the context of omission faults, in [GP11] in the context of dynamic networks. We prove that this condition is actually sufficient by presenting a new Consensus algorithm. This algorithm is based upon reconstructing a partial, but significant, view of the actual communications that occurred during the execution.

1 Introduction

Designing algorithms for static networks is an area that has been studied with numerous approach (distributed / centralized, online / offline, ...). This is one of the main themes of distributed computing. Designing algorithms for dynamic networks, where the network structure can be modified during the computation is less understood. Numerous research projects exist about systems where the origin of the dynamicity is from faults (consequences being deleting or adding nodes or edges to the network). Indeed, fault-tolerance is probably one of the main endeavors in distributed computing. However, faults are in general of limited scope, in limited number and, above all, are considered as anomalies with respect to the normal and correct behavior of the system. So, here we consider systems that are never stable, where the number of changes is not bounded and changes are continuously occurring, where these changes are not considered anomalies but are an integral part of the system at hand. Such highly dynamic systems do exist, and they are actually quite common, and they are becoming pervasive.

We consider communication networks in which the topology can evolve from round to round. A specific link can dynamically disappear and then appear again after an unpredictable number of rounds, and it can continue to alternate between being present and absent in an unpredictable way. This model is more general than other models, such as *component failure* models, whose evolutions, once they appear somewhere, are located there permanently. Interestingly, we

---

* Research supported by ANR PANDA while being at LIF.

T. Moscibroda and A.A. Rescigno (Eds.): SIROCCO 2013, LNCS 8179, pp. 24–35, 2013.

will show, using the notion of communication events, that this model is also closely related to a failure model that was introduced in [SW89], and was called the *mobile faults* or *dynamic faults* model. This synchronous presentation of the communications has also been shown to be good for *layered analysis* [MR02].

An important property of the systems we are studying here is that the set of possible simultaneous communication is the same for each round. In some sense, the system has no "memory" of the previous evolutions. Real systems often exhibit such memory-less behaviour. Moreover, we do not restrict ourselves to network with complete connectivity as it is usually done. In this paper we consider the *most general case* of such systems, i.e. systems in which the set of possible simultaneous communications is arbitrary. This allows the modelling of any system in which communication can happen intermittently, in any arbitrary pattern, including systems in which the communications are not symmetric.

We investigate the Consensus problem in these networks. While it has long been known that solvability of the Broadcast problem implies solvability of the Consensus problem, we show here the precise relationship between those two. In [GP11], an impossibility proof for Consensus was presented in the context of omission faults. In this paper, we show that the necessary condition for solvability of Consensus of [GP11] is actually sufficient by presenting a new Consensus algorithm. Theorem 4.7 is the first complete characterization of solvability of Consensus in dynamic networks with arbitrary set of possible topologies.

The Consensus and Broadcast Problems. The Consensus problem is a very well studied problem in the area of Distributed Algorithms. It is defined as follows. Each node of the network starts with an initial value, and all nodes of the network have to agree on a common value, which is one of the initial values. Many versions of the problem concern the design of algorithms for systems that are unreliable. Two of the most widely studied patterns of information propagation in communication networks are *broadcasting* and *gossiping*. A *broadcast* is the distribution of an initial value from one node of a network to every other node of the network. A *gossip* is a simultaneous broadcast from every node of the network. The Broadcast problem that we study in this paper is to find a node from which a broadcast can be successfully completed.

Our Contribution. In this paper, we investigate dynamic networks where the topology evolves arbitrarily from round to round and nodes do not know their neighbours at a given instant [KLO10]. The topology of the network at a given instant is called a *communication event*. We give a necessary and sufficient condition about the dynamic networks with given set of possible communication events for which Consensus is solvable. In [GP11], the necessary condition was given in the context of omissions faults. Here we show that the technique extends to dynamic behaviours as communication events capture the right notion of communication common to both models and we prove the reciprocal. Having the full characterization is important from a theoretical point of view, but this is also very interesting as both sides illustrate the problem at hand. Furthermore, we give a more constructive presentation of our tools, especially the relation β (to be defined later), see Definition 5.1.

A node from which it is possible to broadcast if the system is restricted to a given communication event is called a *source* for the communication event. We define an equivalence relation β on the communication events that is based on the collective local observations of the events by the sources. The characterization is as follows: *it is possible to solve Consensus if and only if for all C equivalence class of β, there is a node that can broadcast when only events from C can occur*. It is very simple to characterize Broadcastability (see Theorem 4.3), so we get very simple and efficient conditions about solving Consensus on dynamic networks or systems with omissions faults.

Related Work. Our model of intermittent communications can be considered as a very general model of communication networks. In this paper, it corresponds to dynamic networks where the sending primitive is a broadcast to neighbours as introduced in [KLO10]. In [KLO10, KMO11], the network has a T−interval connectivity which is a stronger requirement than in our study since it implies that all nodes are sources for every communication event; it means that Consensus is obviously solvable and therefore a variant of Consensus, the *Coordinated Consensus* is studied in [KMO11]. A survey of dynamic networks with general behaviour has been done in [CFQS12].

The Consensus problem has been widely studied in the context of shared memory systems and message passing systems in which any node can communicate with any other node. Surprisingly, there have been few studies in the context of communication networks, where the communication graph is not a complete graph. In one of the first such studies [SW07] (after [SW89]), the Consensus and related Agreement problems are investigated for networks in which there are at most f omissions during any given round. It is proved that it is impossible to solve Consensus if f is at least the minimum degree of the graph. A Consensus algorithm is presented for the case where f is strictly smaller than the connectivity of the network. In [FG11, GP11], these results are generalized, showing that exact limits for Consensus can be derived from exact limits for Broadcast. In particular when the number of omissions is bounded (in any way) the Consensus problem is exactly the Broadcast problem. In [SWK09], Schmid *et al* investigate Consensus for communication networks with locally bounded number of faulty links. Following this work, in [BRS12], some necessary conditions and some sufficient conditions are given for solving Consensus in dynamic networks with unidirectional links.

In [CBS09], Charron-Bost and Schiper present a model that can describe benign faults. This model is called the "Heard-Of" model. It is a round-based model for an omission-prone environment in which the set of possible communication events is not necessarily the same for each round.

2   Definitions and Notations for Dynamic Networks

We model a communication network by a digraph $\mathcal{G} = (V, E)$ which does not have to be symmetric. We always assume that nodes have unique identities. This digraph $\mathcal{G}$ will be fixed throughout this paper. This is the *underlying graph*.

2.1 Dynamic Networks

In this section, we introduce our model and the associated notation. Communication in our model is reliable, and is performed in rounds, but with changing topology from round to round. Communication with a given topology is described by a spanning subgraph G of $\mathcal{G}$. Intuitively, if an arc is missing in G, then, the (unidirectionnal) link has disappeared. The precise semantics are defined more formally in Section 2.3.

We define the set $\Sigma = \{(V, E') \mid E' \subseteq E\}$. This set represents all possible simultaneous communications given the underlying graph $\mathcal{G}$. For ease of notation, we will always identify a spanning subgraph in Σ with its set of arcs.

Definition 2.1. *An element of Σ is called a* communication event *(or* event *for short). A* communication scenario *(or* scenario *for short) is an infinite sequence of communication events. A* dynamic network *with support $\mathcal{G}$ is a set of communication events.*

A natural way to describe communications is to consider Σ to be an alphabet, with communication events as letters of the alphabet, and scenarios as infinite words. We will use standard concatenation notation when describing sequences. If σ and σ' are two sequences, with σ a finite sequence, then $\sigma\sigma'$ is the sequence that starts with the ordered sequence of events σ followed by the ordered sequence of events σ'. Given an event H, and $k \in \mathbb{N}$, H^k is the finite sequence constituted of k consecutive H. We denote ε the empty sequence.

We describe a dynamic network with $\mathbf{G} \subseteq \Sigma$. Given $\mathbf{G}$ we denote by $\rho(\mathbf{G})$ the set of infinite sequences of elements of $\mathbf{G}$. The set of finite sequences is denoted by $\rho_f(\mathbf{G})$. These sequences describe exactly all the possible evolutions for the dynamic network.

2.2 Examples of Dynamic Networks

We present examples for systems with two processes but they can be easily extended to any arbitrary graph. The set $\Sigma = \{\circ\leftrightarrows\bullet, \circ\leftarrow\bullet, \circ\rightarrow\bullet, \circ\ \ \bullet\}$ is the set of directed graphs with two nodes $\circ$ and $\bullet$. Note that these schemes can also be interpeted as systems with intermittent omission faults.

Example 2.2. The dynamic network $\{\circ\leftrightarrows\bullet\}$ corresponds to a static system. The dynamic network $\mathcal{O}_1 = \{\circ\leftrightarrows\bullet, \circ\leftarrow\bullet, \circ\rightarrow\bullet\}$ is well understood and corresponds to the situation in which there is at least one unidirectional link in each round.

Example 2.3. The dynamic network $\mathcal{H} = \{\circ\leftarrow\bullet, \circ\rightarrow\bullet\}$ describes a system in which at most one message can be successfully received in any round, since the networks is possibly alternating around a unidirectional link.

2.3 Execution of a Distributed Algorithm

Given a dynamic network $\mathbf{G}$, we define what is an execution of a given algorithm $\mathcal{A}$ with a given initial configuration ι. Every process can execute the following communication primitives [KLO10]:

– $send(msg)$ to send the same message msg to all out-neighbours,
– $recv()$ to get the messages from all in-neighbours.

An *execution*, or *run*, of algorithm $\mathcal{A}$ *subject to* scenario $\sigma \in \rho(\mathbf{G})$ is the following. Consider process u and one of its out-neighbours v in $\mathcal{G}$. During round $r \in \mathbb{N}$, a message msg is sent from u to all its neighbours according to the instructions in algorithm $\mathcal{A}$. The node v will receive the corresponding message msg only if H, the r-th element of σ, is such that $(u, v) \in H$. All messages sent in a round can only be received in the same round. After sending and receiving messages, all processes update their states according to $\mathcal{A}$ and the messages they received. Given all nodes have unique identities, when a message is received, it is known from which neighbour it is received. A *configuration* corresponds to the set of local states at the end of a given round.

Given $w \in \rho_f(\mathbf{G})$, and an initial configuration ι, let $s_\iota^x(w)$ denote the state of process x at the end of the $|w|$-th round of algorithm $\mathcal{A}$ subject to scenario w, with initial configuration ι. The initial state of x is therefore $\iota(x) = s_\iota^x(\varepsilon)$. When ι is clear from the context, we might omit it and note simply $s^x(w)$. An *execution* of $\mathcal{A}$ subject to scenario $\sigma \in \rho(\mathbf{G})$ is the (possibly infinite) sequence of such message exchanges and corresponding configurations.

Definition 2.4. *An algorithm $\mathcal{A}$ solves a problem $\mathcal{P}$ in $\mathbf{G}$ if, for any scenario $\sigma \in \rho(\mathbf{G})$, for any initial configuation ι, there exists a finite prefix w of σ such that the state $s_\iota^x(w)$ of each process $x \in V$ satisfies the specifications of $\mathcal{P}$ for initial configuration ι. In such a case, $\mathcal{P}$ is said to be* solvable *on* $\mathbf{G}$.

3   The Problems

3.1   The Binary Consensus Problem

A set of synchronous processes wishes to agree about a binary value. This problem was first identified and formalized by Pease, Shostak and Lamport [PSL80]. Given a set of processes, a consensus protocol must satisfy the following properties for any combination of initial values [Lyn96]:

Termination : every process decides some value;

Validity : if all processes have the same initial value v, then every process decides v;

Agreement : if a process decides v, then every process decides v.

Consensus with these termination and decision requirements is more precisely referred to as *Uniform Consensus* (see [Ray02] for a discussion). Given a network environment, the natural questions are: is Consensus solvable, and if it is solvable, what is the minimum number of rounds to solve it? In this paper, we describe exactly for which dynamic networks $\mathbf{G}$ Consensus is solvable on $\mathbf{G}$.

3.2   The Broadcast Problem

The *Broadcast problem for $u \in V$ on scheme* $\mathbf{G}$ is to find a distributed algorithm $\mathcal{A}$ such that any value stored in u is successfully transmitted to all nodes of $\mathcal{G}$ in all scenarios of $\rho(\mathbf{G})$.

By extension, we say that Broadcast is solvable on **G**, if the Broadcast problem for some $u \in V$ is solvable on **G**. If Broadcast is solvable on **G**, we also say **G** is *broadcastable*. The next proposition is from folklore but leads to very interesting questions. If a node u can broadcast in any execution then, it can be used to solve Consensus since the initial value of u can always be used as a correct decision value for all nodes.

Proposition 3.1. *Let* **G** *be a dynamic network. If* **G** *is broadcastable, then Consensus is solvable on* **G**.

However the converse proposition is not necessarily true. Example 2.3 is a case where it is not possible to broadcast (because if you want to broadcast from ○, then ○←● can occur as the only event, and symmetrically if you want to broadcast from ● then ○→● can occur) but where Consensus is solvable. Indeed, the following is a Consensus algorithm. Each node sends its initial value. If a value is received, then this value is decided. Otherwise, the initial value is decided. From this remark, it shall be noted that, in general dynamic networks, the Consensus and Broadcast problems are not equivalent.

4 Broadcastability

4.1 Characterizations of Broadcastability with Arbitrary Dynamic Networks

We start with a basic definition and lemma.

Definition 4.1. *Given a dynamic network* **G**, *consider a digraph G in* **G** *and a node $u \in V$. A node $v \in V$ is* reachable *from u in G if there is a directed path from u to v in G. Node u is a* source *for G if every $v \in V$ is reachable from u in G. The set of sources of G is denoted $\mathcal{S}(G)$.*

Definition 4.2. *Consider the dynamic network* **G** $= \{G_1, \ldots, G_q\}$. *It is said to be* source-incompatible *if* $\bigcap_{1 \leq i \leq q} \mathcal{S}(G_i) = \emptyset$.

If one event has no source then the dynamic network is source-incompatible.

Theorem 4.3 ([GP11]). *Let* **G** *be a dynamic network. Then* **G** *is broadcastable if and only if* **G** *is not source-incompatible.*

4.2 Towards a Converse Reduction

We consider **G** a given dynamic network. We define now a precise relations about indistinguishability. Given an equivalence relation γ on **G**, we denote $[G]_\gamma$ the equivalence class of $G \in$ **G**. Such a set is called a γ-class.

Let G be a digraph. Given a subset X of vertices, we denote $In_X(G) = \{(v, u) \in G \mid u \in X\}$.

Definition 4.4. *Given three digraphs $G, H, K \in$* **G**, *we define the following relation denoted by $G\alpha_K H$ if $In_{\mathcal{S}(K)}(G) = In_{\mathcal{S}(K)}(H)$. The relation α^* is the transitive closure of α_K relations for any $K \in$* **G**.

The relation α_K describes how some communication events are indistinguishable to all the nodes of $\mathcal{S}(K)$. The relation α^* is the transitive closure of such indistinguishability.

Definition 4.5. *We denote β the coarsest equivalence relation included in α^* such that for all graphs G, H*
(Closure Property) $G\beta H \implies \exists H_0, \ldots, H_q$ and $K_1, \ldots, K_q$ such that
  (i) $G = H_0, H = H_q$,
  (ii) $\forall i \geq 1, H_i \beta G, K_i \beta G$,
  (iii) $\forall i \geq 1, H_{i-1}\alpha_{K_i} H_i$.

The relation β is well defined as the equality relation satisfies such a closure property. And for any two relations R_1 and R_2 that satisfy the property, we have that the transitive closure of $R_1 \cup R_2$ that satisfies the Closure property. It can be constructed incrementally, see section 5.1. The relation β is expressing, as will be shown later, the maximal indistinguishability for events from **G**. A key point is that, in β, indistinguishability is within the equivalence classes, contrary to α^* where property (ii) of the closure property is not satisfied.

Example 4.6. In $\mathcal{O}_1$ from Example 2.2, there is only one equivalence class. Let's see why. First, the sets of sources to consider are: $\mathcal{S}(\circ\!\leftarrow\!\bullet) = \{\bullet\}$; $\mathcal{S}(\circ\!\rightarrow\!\bullet) = \{\circ\}$; $\mathcal{S}(\circ\!\leftrightarrows\!\bullet) = \{\circ, \bullet\}$. We have $In_{\{\circ\}}(\circ\!\leftarrow\!\bullet) = In_{\{\circ\}}(\circ\!\leftrightarrows\!\bullet)$. Hence $\circ\!\leftarrow\!\bullet \; \alpha_{\circ\rightarrow\bullet} \; \circ\!\leftrightarrows\!\bullet$. Similarly, $\circ\!\rightarrow\!\bullet \; \alpha_{\circ\leftarrow\bullet} \; \circ\!\leftrightarrows\!\bullet$. Therefore, there is only one α^*−class and all communication events are β-equivalent.

In Example 2.3, α^*, therefore β, has two singleton equivalence classes. Every node can compute immediately which communication event happened.

Finally, we can now state the main theorem.

Theorem 4.7. *Let* **G** *be a dynamic network. Consensus is solvable on* **G** *if and only if all β-classes of* **G** *are broadcastable.*

Note that this result can also be extended to dynamic networks where the nodes can sense the presence or the absence of the links by changing the definition of α to also take into account the outgoing arcs.

4.3   Proof of Necessary Condition

The proof of the necessary condition uses a bivalency approach that is similar to the adjacency and continuity techniques of [SW07]. We show that even restricting the graphs on which the properties apply, it is still possible to derive the impossibility of Consensus. In some sense, we describe the minimal set of subgraphs to which these properties need to apply. This proof can be found in the updated version of [GP11, section 5].

5   An Algorithm for Solving Consensus

5.1   Notations and First Properties

We show here that when the β-classes are broadcastable, then the following algorithm solves Consensus. The algorithm is presented as a full information protocol. Every node receives the value of the states of the neighbours, append this to its local state and then re-transmits its entire state to all neighbours. This generates messages of exponential size. As it is long known, see eg [FHMV95, chap 6, Section 6.6.2], it is possible to encode everything in messages of polynomial size. However we keep the full information protocol presentation for simplicity, as we are here mainly interested in the computability aspect of the problem.

The specific part of a full information protocol is to describe the halting criterion and the computation of the decided value. The algorithm uses the following technique. The number of rounds is known a priori by a (huge) upper bound on the number of necessary rounds. To each β-class C, we associate s_C, one of the common sources for this class. To get the decided value, we will compute (select) a common β-class C on all node and use the initial value of s_C as decision value. Hence, the crucial part is to compute the β-classes of the events that occurred during the computation and select the first one that occurs sufficiently often (to be formally precised later). The reader should be aware that it is not always possible to compute (even with some delay) the β-classes of all the previous events. However, as will be shown, it is possible to do it for the α^*-classes. From these α^*-classes, we will construct a refinement in such a way that some relevant β-classes can actually be computed and in such a way that it is possible to select the same β-class on each node. We define the following equivalence relations β_i by refinement, starting with $\beta_0 = \alpha^*$.

Definition 5.1. Let $i \geq 0$, let $G, H \in \mathbf{G}$. We say that $G\beta_{i+1}H$ if $G\beta_i H$ and $\exists G_0, \ldots, G_s \in [G]_{\beta_i}$, and $\exists K_1, \ldots, K_s \in [G]_{\beta_i}$ such that
- $G_0 = G, G_s = H,$
- $\forall j, G_{j-1}\alpha_{K_j}G_j.$

Because they are defined by refining the previous relation, a $\beta_{i+1}-$class is always included in some β_i-class. Looking at the structure that is induced from the set inclusion relation we get a sub-lattice of the boolean lattice where the α^*-classes are at the top and the β-classes are at the bottom. We present now the following lemmas to explain the structure of this "lattice of indistinguishability". This will be used to compute which received initial value should be decided. These relations are also an equivalent way to define the β relation.

Lemma 5.2. There exists p such that $\beta_p = \beta$.

Proof. The number of possible refinements is finite (the number of events is indeed finite). Therefore there is $p \in \mathbb{N}$ such that $\beta_p = \beta_{p+1}$. By construction the relation β_p satisfies the Closure Property of Def. 4.5 and is maximal. Hence it is exactly the relation β. □

Given two set of arcs F and G, we define the partial order relation $\Subset$ by $F \Subset G$ if $F \subseteq G$ and for all $(v, u) \in F$, for all $(w, u) \in G$, we have $(w, u) \in F$.

Given a set F of arcs, we define $\mathrm{Comp}(F) = \{G \in \mathbf{G} \mid F \Subset G\}$. If F is a set of radius 1 balls, this can be interpreted as the set of events that are "compatible" with F. And now given two digraphs G and K, we set $\mathrm{Comp}_K(G) = \mathrm{Comp}(In_{S(K)}(G))$. It is the set of events that are "compatible" with event G, when this event is seen from the sources of K. Or equivalently, it is the set of events that are indistinguishable from G for the set of sources of K.

Lemma 5.3. *For all events* $G, K \in \mathbf{G}$, $\mathrm{Comp}_K(G) \subseteq [G]_{\alpha^*}$.

Proof. Consider $H \in \mathrm{Comp}_K(G) = \mathrm{Comp}(In_{\mathcal{S}(K)}(G))$. We have $In_{\mathcal{S}(K)}(G) \in In_{\mathcal{S}(K)}(H)$, so $\forall u \in \mathcal{S}(K)$, $In_u(H) = In_u(G)$ by definition. This yields exactly $G\alpha_K H$.   □

Lemma 5.4. *Let* $G, K \in \mathbf{G}$, $i \in \mathbb{N}$ *such that* $G\beta_i K$. *Then* $\mathrm{Comp}_K(G) \subseteq [G]_{\beta_{i+1}}$.

Proof. Consider again $H \in \mathrm{Comp}(In_{\mathcal{S}(K)}(G))$. As in the proof above, we have exactly $G\alpha_K H$. Since $G\beta_i K$, by definition of β_{i+1}, we get $G\beta_{i+1}H$.   □

From Lemma 5.2, we get in particular

Proposition 5.5. *Let* $G, K \in \mathbf{G}$, $i \in \mathbb{N}$ *such that* $G\beta K$. *Then* $\mathrm{Comp}_K(G) \subseteq [G]_\beta$.

5.2 A Generic Consensus Algorithm

The idea of the algorithm is the following. From the execution, starting from $\alpha^* = \beta_0$, we will select some β_i−classes recursively. When a β_i−class $\mathcal{C}_i$ has been selected by the algorithm, we select a β_{i+1}−class $\mathcal{C}_{i+1}$ that is included in $\mathcal{C}_i$. The selection is based on the number of occurrences of events belonging to the classes. By waiting long enough, it is possible to compute the β_i−class of *some* of the events. The Consensus Algorithm with some integer parameters is the Algorithm 1. The selection function `SelectClass` is described in Algorithm 2. We introduce the following notation. If Var is a variable of the algorithm then Var_r^u is the value of this variable at node u and round r. Given a full information history Hist value and a round r, it is possible to retrieve from Hist a set of arcs (ie messages) that where transmitted in round r. We denote $\mathsf{View}[r] = \{(v', v) \mid$ in Hist, at round r, v received a non *null* message from $v'\}$.

We can now present the selection function `SelectClass`. The integer r is the value of the current round. Some integers $k_0, \ldots, k_p$ are chosen (the correct values will be given later). By convention, the minimum of the empty set is $+\infty$. The idea is to select recursively a β_i−class by choosing always the class refined from the selected β_{i-1}−class to be the first to appear (*as computed*) k_i times.

5.3 Proof of Correctness of the Algorithm

We denote by $\mathfrak{D} = 2 \times |\mathbf{G}| \times |V|$. We now show that with $k_p = |V|$ the number of vertices and $k_i = |\mathbf{G}| \times k_{i+1} + \mathfrak{D}$, and $r_{max} = |\mathbf{G}| \times k_0 + \mathfrak{D}$, we have the Agreement Condition.

Consider a run with the sequence of events $G_1 G_2 \cdots G_{r_{max}}$. Note that View $[j]$ does not always give exactly the event that occurred at round j but we have this important lemma.

Algorithm 1. ConsensusAlgorithm($r_{max}, k_0, \ldots, k_p$)

   Input: val $\in \{0,1\}$;
   $r_{max}, k_0, \ldots, k_p \in \mathbb{N}$

1   Hist:= {val};
2   **for** $r = 1$ **to** r_{max} **do**
3   send $(*, \text{Hist})$ (* *broadcast to out-neighbours* *);
4   Hist := Hist.{**recv**()};
5   **for** $j = 1$ **to** r **do**
6   compute View $[j]$ from Hist;
7   **for** $i = 0$ **to** p **do**
8   (* *Compute β_i−classes if possible* *);
9   **if** *exists $\mathcal{C}$ a β_i−class such that* $\text{Comp}(\text{View}[j]) \subseteq \mathcal{C}$ **then**
10           $\lfloor$ Class $[i][j] := \mathcal{C}$;
11   **else**
12           $\lfloor$ Class $[i][j] := \bot$;

13 **foreach** *node* v **do**
14   compute Val $[v]$ from the initial value of v in Hist, if it is in Hist
15 **Decide:** Val $[s_{\texttt{SelectClass}(\text{Class}, k_0 \cdots k_p)}]$;

Algorithm 2. Selection function `SelectClass`

   Input: classes Class;
   $k_0, \cdots, k_p \in \mathbb{N}$;
1   Selected$[-1] = \mathbf{G}$;
2   **for** $i = 0$ **to** p **do**
3   **foreach** β_i−*class* $\mathcal{C} \subset$ Selected$[i-1]$ **do**
4         $\lfloor$ Reach$[\mathcal{C}] := \min\{j \leq r$ such that $\#\{j' \leq j \mid \text{Class}[i][j'] = \mathcal{C}\} \geq k_i\}$;
5   Selected$[i] :=$ the β_i−class $\mathcal{C}$ with minimal Reach$[\mathcal{C}]$
   Output: Selected$[p]$

Lemma 5.6. *Let $j \in \mathbb{N}$. For all $u \in V$, $\exists r_0, \forall r, j < r < r_0$, $\text{Class}_r^u[i][j] = \bot$ and $\forall r \geq r_0$ $\text{Class}_r^u[i][j] = [G_j]_{\beta_i}$.*

Proof. Using the full information protocol, we have obviously $\text{View}_r^u[j] \in G_j$ for all $j \leq r$. So $G_j \in \text{Comp}(\text{View}_r^u[j])$ and when the test at line 9 of the algorithm is satisfied, then $\text{Comp}(\text{View}_r^u[j]) \subseteq \text{Class}_r^u[i][j]$ and the lemma follows. □

In the following lemma, we prove that $\mathfrak{D}$ is a kind of delay for the computations of β_{i+1}−classes from the β_i−classes.

Lemma 5.7. *Consider a β_i−class $\mathcal{C}$. Let $r \in \mathbb{N}$, and let $k = \#\{\ell \leq r \mid G_\ell \in \mathcal{C}\}$. Consider $\tau_1 \ldots, \tau_k \in \mathbb{N}$ the subsequence of indices ($\tau_j < \tau_{j+1}$ for all $1 \leq j < k$) such that $G_{\tau_j} \in \mathcal{C}$ for all $1 \leq j \leq k$. For all $u \in V$, for all $1 \leq j \leq k - \mathfrak{D}$, $\text{Class}_r^u[i+1][\tau_j] \neq \bot$.*

Proof. We call $\mathcal{C}$-round a round where an event in $\mathcal{C}$ occurs. From the pigeonhole principle, there exists an event $K \in \mathcal{C}$ that occurs at least $2 \times |V|$ times in the last $\mathfrak{D}$ $\mathcal{C}$-rounds. Therefore, in the first $|V|$ occurrences of these occurrences of K, all information about the $k - \mathfrak{D}$ first $\mathcal{C}$-rounds was exchanged between sources of K. So $\text{Comp}(\text{View}_r^v[j]) \in \text{Comp}_K(G_j)$ for any source v of K. Therefore all these sources are able to compute locally at least the β_{i+1}-classes of these first $r - \mathfrak{D}$ $\mathcal{C}$-rounds from Lemma 5.4. Then the remaining $|V|$ occurrences of K broadcast these computed values (more exactly the Hist sets to compute these values) to every other node u. $\qquad\square$

The same technique shows that everybody can compute the α^*- classes of the first $r - \mathfrak{D}$ rounds.

Proposition 5.8. *There exists $\mathcal{C}_0$ a α^*-class such that for all u, $\text{Selected}[0] = \mathcal{C}_0$.*

Proof. From Lemma 5.3, we have that every node can compute the α^*- class of the first $k_0 \times |\mathbf{G}|$ round. Again, by pigeonhole principle $\text{Selected}^u[0]$ will be defined for all node u. As the computed α^*-classes correspond to the actual ones by Lemma 5.6, they are the same, $\text{Selected}^u[0]$ will be the same class for every node. $\qquad\square$

We will extend this result to all i.

Proposition 5.9. *For $i \leq p$, there exists $\mathcal{C}_i$ a β_i-class such that for all u, $\text{Selected}[i] = \mathcal{C}_i$.*

Proof. The proof is done by induction. The case $i = 0$ is the proposition 5.8 above.

Suppose the assertion is true for i. Now we consider the class $\mathcal{C}_i$. From the selection criterion, there is a round r where this class has occurred k_i times. Consider r' the round where $\mathcal{C}_i$ has occurred only $k_i - \mathfrak{D}$ times.

From Lemma 5.7, we have that every node can compute the $\beta_{i+1}-$class of the first $k_{i+1} \times |\mathbf{G}|$ $\mathcal{C}_i-$round. Again, by pigeonhole principle $\text{Selected}^u[i+1]$ will be defined for all node u. As the computed $\beta_{i+1}-$classes are the same, it will be the same selected class for every node. The assertion for $i+1$ has been proved. $\qquad\square$

Proposition 5.10. *Algorithm 1 with constant r_{max} and $k_1, \ldots, k_p$ as defined solves the Consensus problem on the dynamic network $\mathbf{G}$.*

Proof. Given the full information protocol is a *for* loop, Termination is obvious. From the selection criterion and Prop. 5.9, $\mathcal{C}_p = \text{Selected}^u[p]$ is well defined for all node u at time r_{max} and $\mathcal{C}_p$ appears at least $k_p = |V|$ times. Therefore node $s_{\mathcal{C}_p}$ have been able to broadcast its initial value to every node. Hence the decision value is defined and identical on every node. The Agreement Condition is satisfied. Since the decided value is always the initial value of some node, the Integrity Condition is also obviously satisfied by the algorithm. $\qquad\square$

The reader should note that it is possible to change the full information protocol from a *for* to a *while* loop by halting when `SelectClass` outputs a value. Finally, we underline that the k_i bounds are not optimal but were chosen to give a simple proof. It is an open question to get optimal bounds or more generally, to get an optimal Consensus algorithm.

References

[BRS12]   Biely, M., Robinson, P., Schmid, U.: Agreement in directed dynamic networks. In: Even, G., Halldórsson, M.M. (eds.) SIROCCO 2012. LNCS, vol. 7355, pp. 73–84. Springer, Heidelberg (2012)

[CBS09]   Charron-Bost, B., Schiper, A.: The heard-of model: computing in distributed systems with benign faults. Distributed Computing 22(1), 49–71 (2009)

[CFQS12] Casteigts, A., Flocchini, P., Quattrociocchi, W., Santoro, N.: Time-varying graphs and dynamic networks. International Journal of Parallel, Emergent and Distributed Systems 27(5), 387–408 (2012)

[FG11]   Fevat, T., Godard, E.: About minimal obstructions for the coordinated attack problem. In: Proc. of 25th IEEE International Parallel & Distributed Processing Symposium, IPDPS 2011 (2011)

[FHMV95] Fagin, R., Halpern, J.Y., Moses, Y., Vardi, M.Y.: Reasoning about Knowledge. MIT Press (1995)

[GP11]   Godard, E., Peters, J.: Consensus vs. Broadcast in communication networks with arbitrary mobile omission faults. In: Kosowski, A., Yamashita, M. (eds.) SIROCCO 2011. LNCS, vol. 6796, pp. 29–41. Springer, Heidelberg (2011), Updated version on http://arxiv.org/abs/1106.3579

[KLO10]   Kuhn, F., Lynch, N.A., Oshman, R.: Distributed computation in dynamic networks. In: STOC, pp. 513–522 (2010)

[KMO11]   Kuhn, F., Moses, Y., Oshman, R.: Coordinated consensus in dynamic networks. In: PODC 2011, pp. 1–10. ACM (2011)

[Lyn96]   Lynch, N.A.: Distributed Algorithms. Morgan Kaufmann Publishers Inc., San Francisco (1996)

[MR02]   Moses, Y., Rajsbaum, S.: A layered analysis of consensus. SIAM Journal on Computing 31(4), 989–1021 (2002)

[PSL80]   Pease, L., Shostak, R., Lamport, L.: Reaching agreement in the presence of faults. Journal of the ACM 27(2), 228–234 (1980)

[Ray02]   Raynal, M.: Consensus in synchronous systems: A concise guided tour. In: Pacific Rim Intern. Symp. on Dependable Computing, p. 221. IEEE (2002)

[SW89]   Santoro, N., Widmayer, P.: Time is not a healer. In: Cori, R., Monien, B. (eds.) STACS 1989. LNCS, vol. 349, pp. 304–313. Springer, Heidelberg (1989)

[SW07]   Santoro, N., Widmayer, P.: Agreement in synchronous networks with ubiquitous faults. Theor. Comput. Sci. 384(2-3), 232–249 (2007)

[SWK09]   Schmid, U., Weiss, B., Keidar, I.: Impossibility results and lower bounds for consensus under link failures. SIAM J. on Computing 38(5), 1912–1951 (2009)

Self-adjusting Grid Networks
to Minimize Expected Path Length

Chen Avin[1], Michael Borokhovich[1], Bernhard Haeupler[2], and Zvi Lotker[1]

[1] Ben-Gurion University of the Negev, Israel
{avin,borokhom,zvilo}@cse.bgu.ac.il
[2] Massachusetts Institute of Technology, USA
haeupler@mit.edu

Abstract. Given a network infrastructure (e.g., data-center or on-chip-network) and a distribution on the source-destination requests, the expected path (route) length is an important measure for the performance, efficiency and power consumption of the network. In this work we initiate a study on *self-adjusting networks*: networks that use local-distributed mechanisms to adjust the position of the nodes (e.g., virtual machines) in the network to best fit the route requests distribution. Finding the optimal placement of nodes is defined as the minimum expected path length (MEPL) problem. This is a generalization of the minimum linear arrangement (MLA) problem where the network infrastructure is a line and the computation is done centrally. In contrast to previous work, we study the distributed version and give efficient and simple approximation algorithms for interesting and practically relevant special cases of the problem. In particular, we consider grid networks in which the distribution of requests is a symmetric product distribution. In this setting, we show that a simple greedy policy of position switching between neighboring nodes to locally minimize an objective function, achieves good approximation ratios. We are able to prove this result using the useful notions of expected rank of the distribution and the expected distance to the center of the graph.

1 Introduction

In the last decade we have witnessed two new major and related phenomena in distributed computing. The first is the emerge of huge data centers and warehouse-scale computers. The second phenomenon is the decentralization and parallelism of workload in single multi-core computers. In both cases (but on different scale) the system is a network of computing primitives that share global computational goals. In data centers networks, as well as in modern multiprocessor computers, multiple processes run in parallel to execute common tasks so, in many cases, these processes need to communicate with each other to work on their shared tasks.

Reducing energy waste, and in particular the power consumption of computing is one of the major challenges of the 21st century. Both data centers and

T. Moscibroda and A.A. Rescigno (Eds.): SIROCCO 2013, LNCS 8179, pp. 36–54, 2013.

single computers are no exception, and constantly increasing their energy and power usage. For example, the total cost of power consumption of data centers in the USA alone is estimated to be 50 billion dollars [1]. Moreover, the energy consumed by data centers is estimated to double every five years [2]. The focus of this work is to improve upon the energy that is consumed by routing in such systems. It is estimated that in data centers the energy consumed by routing is about 20%-30% of the total energy [3]. Routing in network-on-chip (NoC) consumes even up to 50% of the total energy [4]. These numbers pose our community both an opportunity and a challenge. The opportunity is to gain significant energy savings for these systems; the challenge is to design and implement clever and simple algorithms that can improve routing efficiency.

Another common property of these systems is that they all operate in a fixed network infrastructure. This means that we cannot change the structure of the network by, for example, rewiring links. But instead, what we can do, is to move the locations of processes (e.g., virtual machines) between the different computers (or CPUs). In this paper, we formulate the problem of saving energy on a fixed infrastructure network using migration of processes. The basic idea is that the energy cost of routing in a network is proportional to the length of the routes which suggests the following: If we can make the routes lengths (or the *expected* route length) shorter, then we can save energy. We devise local and distributed algorithms that (re-)place processes in the network to reduce the expected path length. This can be achieved, for example, by Software Defined Networking (SDN) [5] – the concept, which provides, among others, much better control over the network functionality. In SDN, a software management platform may support an abstraction for moving a selected process from one physical machine to another. Recently, this approach became practical, when Google announced [6] the implementation of OpenFlow [7] in its own backbone.

The problem of minimizing the total energy consumed by routing is dependent on two major properties of the system: (i) the infrastructure (topology) of the communication network and (ii) the statistical pattern of route requests between sources and destinations. We first show that even in a very simple pattern such as every node has an activity level and the probability to send or receive a message is proportional to its level, the problem is NP-complete on general network topologies. Secondly, even when the network is "simpler" or regular, like a grid network, the problem can still be hard if the request distribution is "complex" in some sense. With this in mind we turn to analyze approximation algorithms for the setting where both the topology and the requests have nice properties. Our routing and activity distributions are partially justified from real data [8,9]. We concentrate on local and distributed algorithms, namely, processes can be exchanged (i.e., relocated) only between nearby nodes without any centralized coordination.

1.1   Overview of Our Results

First, we formulate the discussed problem as the *minimum expected path length* (MEPL) problem, that is, given a network infrastructure and a distribution of requests, minimizing route costs by finding an optimal *placement* for processes in the network. When the network is a line, MEPL is identical to the minimum linear arrangement (MLA) [10] which is known to be NP-complete. In this work we consider d-dimensional grid networks, $d \geq 1$, and requests that comes independently from a *symmetric product distribution* where the frequency of a route request (u, v) is a multiplication of the *activity* levels of both u and v. In contrast to previous works, our goal is to design simple distributed algorithms for these more realistic settings.

We first show that MEPL is NP-complete if (i): we only assume that the network is a 2-dimensional grid, and (ii): we only assume that the requests come from a *symmetric product distribution*. But, somewhat surprisingly, if both conditions hold, we are able to present a simple, local, distributed algorithm that achieves good approximation to the optimal solution for the MEPL problem. Our algorithm is self-adjustable in the sense that nodes switch processes based on the continuously observed sequence of route requests each node is involved in. This approach was inspired and bears some similarity to self-adjusting data structures like splay trees [11]. In particular we are able to show (informal):

Theorem. *For a d-dimensional grid network and a symmetric product distribution of requests there is a simple distributed algorithm, which defines a local switching policy between a process and its neighbors that achieves a constant approximation to the minimum expected path length (MEPL) problem.*

Interestingly, we prove this theorem using a measure called *expected rank* which is related to the uncertainty of a random variable in a similar manner as entropy is.

We then turn to more complex distributions of requests and discuss requests that are *clustered* into disjoint groups. While for this setting few extremely unstable bad local minima can exist we present promising simulation results. In particular we show that for the 2-dimensional grid that starting from a random and thus almost worst case initial state of processes locations in the network our local algorithms converge to an almost optimal local minimum.

Organization: In Section 2 we discuss related work and somewhat similar approaches. Section 3 introduces the formal problem and definitions. The hardness of MEPL is proved in Section 4 and then in Section 5 we prove our main result, a constant factor approximation in d-dimensional meshes with product route distributions. In Section 6 we discuss a more complex setting: clustered requests; and we end the paper with a short conclusion in Section 7.

2   Related Work

Energy saving along with green computing is an active topic of research in the recent years. In a recent paper [12] Lis et al. study memory architectures of microprocessors. The authors suggest that processes will migrate to a location that is closer to the data instead of what is common in today architectures, i.e., coping the data to be closer to the process. The logic behind this idea is that programs are much smaller than their data. We take this idea one step further by reducing the communication distance between two processes. Improving the energy efficiency of routing in networks was also considered. Batista et al. [13] used traffic engineering on grids to self-adjust to routing requests. In [3] and [14] different authors considered data centers and tried to save energy by turning off routers and links when demand in the network is low. Other self-adjusting routing schemes were considered, e.g., in scale-free networks, to overcome congestion [15,16]. In [17], authors studied virtual machines migration in tree topology with the goal of minimizing servers' load, while in [18], the problem of mapping processes to physical servers, in order to minimize congestion, was presented. In the current work, we optimize another important metric – average routing path length, which minimizes the number of lookups and transmissions performed by routers.

The most related areas of research to our study are graph *arrangement, embedding* and *labeling* problems [19,20]. The basic question there is to embed a guest graph G into a host graph H in order to minimize some objective function like the *bandwidth* or the *cutwidth*; we relate our study to this settings in the model section. In particular, some VLSI design problems were considered on a two dimensional grids.[21,22]. There are two significant differences here: first, we consider *distributions* on the route requests which restrict our guest graphs and second, and more importantly, we are interested in distributed, self-adjusting algorithms to solve the problem and not a centralized solution. In [23], the authors dealt with a problem that is similar to a special case of MEPL, where all nodes have the same activity level (uniform requests distribution), moreover, the proposed solution is centralized and not distributed.

As described in the introduction the self-adjusting nature of our solution was inspired by self-adjusting data structures like splay trees [11] which adjust their structure according to requests made to the data structure in such a way that the amortized cost matches the cost of the optimal (static) solution. Recently, Avin et al. in [24], extended the idea of splay trees to splay networks, i.e., self-adjusting trees that adapt themselves according to the pattern of routing requests to minimize the length of routing paths. Such a solution can be successfully used in overlay p2p networks.

The local greedy switch strategy we use is related to physics and natural dynamics which indirectly try to minimize energy. Using this analogy for optimization purposes has a long history. E.g., simulated annealing [25] can be seen as simulating physics while cooling the temperature, i.e., the local moves selected shift over time more and more bias from mostly random behavior to greedy energy minimization. Here we only look at greedy steps. In a networking context similar approaches were used for load balancing via diffusive paradigms [26] and for routing via gradient mechanisms [27].

Another very related research is about self-stabilizing graphs [28,29]. The goal there is also to maintain some objective using local edge exchanges, mostly in an overlay network. In a similar manner we would like to extend the current work to solve MEPL on overlay and peer-to-peer networks, using edge rewiring as well.

3 Model and Problem Definition

We model the communication network by an undirected, unweighted and connected *host* graph H. Given a graph, its vertex set is denoted $V(H)$ and its edge set by $E(H)$. We denote the number of nodes with $n = |V(H)|$ and the number of edges with $m = |E(H)|$. Let $d_H(\cdot)$ be the distance function between nodes in H, i.e., for two nodes $u, v \in H$ we define $d_H(u, v)$ to be the number of edges of a *shortest* path between u and v in H.

We assume that the network serves route requests drawn independently from an arbitrary distribution $\mathcal{P}$ and messages are routed along the shortest paths in H. Alternately, we represent the distribution $\mathcal{P}$ as a weighted directed *guest* graph G where $|V(G)| = n$. For a directed edge $(u, v) \in E(G)$ let the weight of the edge $\mathrm{p}(u, v)$ denote the probability of a route request for a message from node u to v.

Given a network infrastructure host graph H and a distribution on the route requests represented by a guest graph G, a **placement** (or *labeling* [19]) function is a bijective[1] function $\varphi : V(G) \to V(H)$ which determines the locations of nodes of G (processes) in the network H (hosts). Given G, H and a placement function φ the expected path length of route requests is defined as:

$$\mathrm{EPL}(G, H, \varphi) = \sum_{(u,v) \in E(G)} \mathrm{p}(u, v) \cdot d_H(\varphi(u), \varphi(v))$$

When H and G are clear from the context we may write just $\mathrm{EPL}(\varphi)$. Note a special case of this definition, when $\mathcal{P}$ is the uniform distribution: this gives the *average path length* in the network which is often used in the literature instead of the diameter, for example to show that a network is a *small world network* [30].

For H and $\mathcal{P}$ we would like to find an optimal placement of the nodes in the network to minimize the expected path length. Formally:

Definition 1 (*Minimum Expected Path Length problem*). *Given a host graph H and a probability distribution represented by a guest graph G, find a placement function that minimize the expected path length:*

$$\mathrm{MEPL} = \min_{\varphi} \mathrm{EPL}(G, H, \varphi)$$

As mentioned earlier, this problem is motivated by the network serving point-to-point routing requests that are independently sampled from a distribution $\mathcal{P}$. If we assume that the cost for a request u, v is $d(\varphi(u), \varphi(v))$ then the MEPL problem tries to minimize the expected cost of a route. Note that this is also equivalent to minimizing the expected number of lookups performed by routers, and

---

[1] In this work we consider the classic case where every host machine can run at most one process (e.g., one virtual machine).

minimizing the expected total number of transmissions - all important metrics in terms of energy saving and efficiency. More formally, if we denote by $\mathcal{E}(u, v)$ energy spent by the network on serving request (u, v), then $\mathcal{E}(u, v) = (\mathcal{E}_{lookup} + \mathcal{E}_{trans})\mathrm{d}(\varphi(u), \varphi(v))$, where $\mathcal{E}_{lookup}$ and $\mathcal{E}_{trans}$ is the energy required by a router for a lookup and transmission operations respectively, and $\mathrm{d}(\varphi(u), \varphi(v))$ is the number of routers on the path (u, v) in H. By assuming that $\mathcal{E}_{lookup} + \mathcal{E}_{trans} = 1$ unit of energy, we obtain: $\mathcal{E}(u, v) = \mathrm{d}(\varphi(u), \varphi(v))$ and thus, the expectation of the total spent energy $\mathcal{E}$ is:

$$\mathbb{E}[\mathcal{E}] = \sum_{(u,v) \in E(G)} \mathrm{p}(u, v) \cdot \mathcal{E}(u, v) = \sum_{(u,v) \in E(G)} \mathrm{p}(u, v) \cdot \mathrm{d}_H(\varphi(u), \varphi(v)) = \mathrm{EPL}(G, H, \varphi).$$

In this work, we mostly consider local and distributed switching rules to find a good placement: rules where a process is only allowed to switch places with processes that are in its neighborhood (i.e., close to it). The goal is that after a sequence of local switches the network will reach its minimum expected path length and solve the MEPL problem. On the one hand, our results from Section 4 will show that this is not possible (efficiently) in a general setting even with global knowledge and non-local switches. Throughout the paper we thus consider at times simpler forms of networks and requests distributions, i.e., *grid networks* and the *symmetric product distributions*:

Definition 2 (d-dimensional grid networks). *A mesh network of size $n = k^d$ with nodes embedded on all locations $[k]^d$ where $[k]$ is the set of integers $1, 2, \ldots, k$. Each node is connected to all the nodes at ℓ_1-distance one from it, i.e., each node has at most two neighbors in each of the d dimensions.*

Definition 3 (symmetric product distribution). *In symmetric product distribution, each node of G (i.e., process) has a level of activity and the more two nodes u and v are active the more likely that the route $\{u, v\}$ gets requested. More precisely, we scale the activity levels of the nodes such that they form a distribution with an activity level $\mathrm{p}(u)$ for each node u and assume that the request distribution is induced by the product of the activity levels, i.e., $\mathrm{p}(u, v) = \mathrm{p}(u) \cdot \mathrm{p}(v)$.*

In order to allow EPL optimization by means of the local switching rules, we make the following assumptions. First, in case of a *symmetric product distribution* of requests, a node (process) u can measure the activity level of any other node v by simply calculating the ratio between the frequency of the requests (u, v) and sum of the frequencies of all the other requests (u, w), where $w \in V$. Second, in order to be able to make EPL calculation locally, a node needs to know the locations of all other nodes in H. Thus, we assume the existence of some central directory service, similar to a DNS service, that keeps track of nodes' locations (addresses). Note that if the requests distribution is not *symmetric product*, activity levels cannot be measured locally, and, in this case, additional centralized directory may be used to track nodes' activity.

4   Hardness of MEPL

In this section we show that solving the general MEPL problem is hard. Indeed, we prove two results that demonstrate how the hardness of the problem can come from either an involved network topology G or the structure of the routing request distribution $\mathcal{P}$. This serves also as an additional motivation why in the rest of the paper we turn to graphs and distributions with more realistic structure.

For both our examples it suffices to use probability distributions that only have one non-zero probability value. In our first statement, we show that even if we restrict ourselves to symmetric product distributions, the MEPL problem is hard on general networks:

Lemma 1. *Given a host graph H and a symmetric product distribution of requests $\mathcal{P}$, it is NP-complete to decide whether the MEPL is smaller than a given value.*

Proof. We describe a reduction from the k-CLIQUE problem. In the k-CLIQUE problem, one is given a graph H' and has to decide whether H' contains a k-clique, that is, whether H' contains a complete graph on k nodes as a subgraph. This is one of Karp's 21 NP-complete problems [31]. For the reduction we take the graph H' as the network's host graph H. As a request distribution we use a symmetric product probability distribution that puts $1/k^2$ probability weight on each of the pairs $V' \times V'$ formed by a subset of the nodes of size k and zero probability on any other pair. If H contains a k-clique, then the unique optimal solution to MEPL with value $k(k-1)/k^2 = 1 - 1/k$ will be obtained if all k nodes are placed in this clique. If H does not contain a k-clique the there will be at least one request pair $u, v \in V' \times V'$ that is at least two far apart and the total cost will be at least $1 - 1/k + 1/k^2$. Thus, deciding whether the MEPL is smaller than $1 - 1/k + 1/k^2$ is equivalent to deciding whether H' has a k-CLIQUE and thus, NP-hard. Lastly, it is easy to see that deciding whether the MEPL is smaller than a given value problem can be easily achieved in NP by guessing and then verifying a solution with smaller value. $\square$

This lemma shows that solving the MEPL problem for general network topologies is hard. Next, we show that even if we restrict the graph to be nice, e.g., 2-dimensional grid, a lack of structure in the probability distribution can make the MEPL problem hard, too:

Lemma 2. *Given a probability distribution of requests $\mathcal{P}$, it is NP-complete to decide whether the MEPL is smaller than a given value on a 2-dimensional grid network.*

Proof. We describe a reduction from the problem of embedding a tree in a 2-dimensional grid which was shown to be NP-hard by Bhatt and Cosmadakis [21]. More precisely it is NP-hard to decide whether a given tree T (with maximum degree 4) is a subgraph of the grid. Given an instance of this problem in form of a tree T we construct a hard MEPL instance as follows: We take the two-dimensional grid $[k]^2$ as a topology where k equals the number of nodes in T. As a request distribution we take a subset of k nodes to correspond to nodes in T and put a probability mass of $1/(k-1)$ on each pair of nodes that corresponds to two neighbors in the tree T; all other $k^2 - (k-1)$ node pairs have a probability of zero. The MELP for such an instance is 1 if and only if the tree can be embedded in the grid. If this is not the case then at least one request pair will be separated by a path of length at least two increasing the average to at least $1 + 1/(k-1)$. Thus deciding whether the MEPL is smaller than $1 + 1/(k-1)$ is equivalent to deciding whether T can be embedded into the 2-dimensional grid. This proves solving the MEPL problem on the 2-dimensional grid NP-hard.   □

Contrasting these two hardness results, the next sections will show that if one assumes a grid graph *and* a symmetric product distribution, nice algorithmic results can be obtained.

5   Distributed MEPL with Symmetric Product Distributions

For general request distribution it is hard to find a good or optimal solution even when one is not restricted to local and distributed switching rules. With this in mind, we first restrict ourselves to a simpler model of requests, namely, symmetric product distributions (Definition 3). Second, we assume d-dimensional grid topologies, and in particular the line and a 2-dimensional grid. We assume that a process learns the distribution from requests it is involved in and thus, it can decide whether the switching (exchanging positions) with a neighbor will increase or decrease the objective function, the expected path length of the network. The main result of this section is that under the above settings, a good approximation to the objective function can be found using only simple (greedy) local switching rules. To prove this result, we need the following definitions.

5.1   Expected Distance to Center and Expected Rank

To find a good placement for nodes (processes) which gives a good approximation to the MEPL, we define the *expected center* and the *expected distance* to it.

Definition 4 (center and expected distance to the center). *Consider a symmetric product distribution $\mathcal{P}$ (with its corresponding guest graph G) and a host graph H. Then, for a placement φ, the expected center of H is a node c^*, s.t.:*

$$c^*(G, H, \varphi) = \operatorname*{argmin}_{x \in V(H)} \sum_{u \in V(G)} \mathrm{p}(u)\mathrm{d}(\varphi(u), x).$$

The expected distance to the center for φ, $c^*(\varphi)$ is:

$$C(G, H, \varphi) = \min_{x \in V(H)} \sum_{u \in V(G)} \mathrm{p}(u)\mathrm{d}(\varphi(u), x),$$

or equally:

$$C(G, H, \varphi) = \sum_{u \in V(G)} p(u)d(\varphi(u), c^*).$$

When H and $\mathcal{P}$ are clear from the context (recall that $\mathcal{P}$ defines the guest graph G), both C and c^* can be written simply as $C(\varphi)$ and $c^*(\varphi)$. The minimum expected distance to the center is defined then as $C_{\min} = \min_\varphi C(\varphi)$. The next lemma describes the relation between C and EPL.

Lemma 3. *Consider a symmetric product distribution $\mathcal{P}$ and a host graph H. For any given placement φ, $2C(\varphi) \geq \mathrm{EPL}(\varphi) \geq C(\varphi)$.*

Proof. To see the upper bound, we suppose that instead of routing between two nodes directly, we route every request via the center c^*. Routing a request in this way results in sampling two requests and summing up their distances to the center. In expectation, this is exactly $2C$. Formally:

$$\mathrm{EPL}(\varphi) = \sum_{(u,v) \in E(G)} p(u)p(v) \cdot d(\varphi(u), \varphi(v))$$

$$\leq \sum_{(u,v) \in E(G)} p(u)p(v)(d(\varphi(u), c^*) + d(c^*, \varphi(v)))$$

$$= \sum_{u \in V(G)} p(u)d(\varphi(u), c^*) + \sum_{v \in V(G)} p(v)d(c^*, \varphi(v)) = 2C(\varphi)$$

The fact that C is a lower bound, we show as follows:

$$\mathrm{EPL}(\varphi) = \sum_{(u,v) \in E(G)} p(u)p(v) \cdot d(\varphi(u), \varphi(v))$$

$$= \sum_{u \in V(G)} p(u) \sum_{v \in V(G)} p(v)d(\varphi(u), \varphi(v))$$

$$\geq \sum_{u \in V(G)} p(u) \sum_{v \in V(G)} p(v)d(c^*, \varphi(v))$$

$$= \sum_{v \in V(G)} p(v)d(c^*, \varphi(v)) = C(\varphi)$$

□

Corollary 1. MEPL $\geq C_{\min}$

This follows since for the optimal placement φ^*: MEPL $= \mathrm{EPL}(\varphi^*) \geq C(\varphi^*) \geq C_{\min}$.

An important ingredient in bounding the performance of our local rules will be the following measure of *expected rank*. This quantity is an interesting measure on the concentration and uncertainty of a distribution.

Definition 5 (*Rank of nodes and the Expected rank*). *The rank of a node is the position of the node in the ordered list of nodes' probabilities (breaking ties arbitrarily). The node with the highest probability has rank 0. The rank of the node $u \in V$ is denoted as $r(u)$. The expected rank of a probability distribution on the nodes of graph G is:* $\mathbb{E}[R] = \sum_{u \in V(G)} p(u)r(u)$.

We next describe the local switching rules by which our distributed algorithm works.

5.2   (Greedy) Local Switching Strategies

For two nodes $u, v \in V(G)$ and a placement φ, we say that u is a *neighbor* of v if and only if $(\varphi(u)\varphi(v)) \in E(H)$. A switching of u and v is taken to be understood as a new placement φ' where for each $w \in V(G), w \neq u, v \; \varphi'(w) = \varphi(w)$ and $\varphi'(u) = \varphi(v)$ and $\varphi'(v) = \varphi(u)$, i.e., u and v switch places on H.

We propose the following greedy strategy. A node switches with a neighbor if, according to the (observed) marginal distribution on the requests involving itself and its neighbor, switching positions improves the objective value. In this work, we consider two simple optimization rules:

1. **M-rule**: Node will switch locations with its neighbor if the switch will minimize the objective function: the expected path-length between all pairs of nodes. This criterion is exactly the MEPL objective.
2. **C-rule**: Node will switch location with its neighbors if the switch will minimize the expected path-length between the *center* node and all the other nodes. This objective does not give us a solution for the MEPL problem, but it will be used as an upper bound for it.

If nodes switch only when this decreases the expected path-length (or some other criterion), then it is clear that this, strictly monotone, potential can not drop indefinitely (or too often) and thus, a (quick) convergence is guaranteed. A placement φ is said to be ***local minimum*** (or local optimum) if and only if no node in G can switch according to the rule they operate (i.e., M-rule or C-rule). When using the C-rule, we can prove the following about the local minimum placement.

Lemma 4. *Any local minimum placement φ with respect to the C-rule, is center monotone: If there is a path of possible switchings from $\varphi(u)$ to $\varphi(v)$ that is distance-decreasing with respect to c^*, i.e., a path for which every step goes strictly closer to c^*, then $p(u) \leq p(v)$.*

Proof. Assume for sake of contradiction that φ is a local minimum, that $p(u) > p(v)$ and that there is a distance-decreasing path P of possible switchings from $\varphi(u)$ and to $\varphi(v)$. By induction there has to be two nodes u' and v' such that $\varphi(u')$ and $\varphi(v')$ are neighbors on the path P with $p(u') > p(v')$ but v' is closer to c^* than u'. By assumption it is possible to switch u' and v' and it is easy to see that this is an improvement with regards to the C-rule. This contradicts the assumption that φ is a local minimum.   □

Note that according to the C-rule, two neighbors switch locations only if the switch decreases C: the expected distance to the center. The improvement of the switch can be found locally, since the center location can be computed locally at each node via the expected position of its requests (which are identical to all nodes because of the product distribution). Therefore, the C-rule will greedily minimize, for each node, the distance to the expected position of its requests.

In order to implement our local and distributed switching strategies, there is a need for a protocol, in which at each step (synchronous or asynchronous) a node will choose a neighbor to negotiate the switching. Additionally, all the nodes should know the locations and activity levels of all the other nodes. As already discussed in Section 3, the location information can be provided using a centralized directory service, and the activity level, in case of a *symmetric product distribution*, can be obtained locally by just counting the number of requests to and from specific peer. Detailed description of such protocol is out of the scope of the current work which deals with the performance analysis of the local and distributed switching strategies compared to a global optimal nodes placement.

Throughout the rest of this paper we assume that the system converges against a local minimum and analyze the performance of such a solution in this stable state. On the other hand, we do NOT assume anything about the starting position OR the specific order of the dynamics (node switches). Thus, in many cases, an initially random starting position converges (e.g., using random improving switches) to a (near) optimal solution; we make no such assumptions and assume a worst case sequence of improvements and a worst-case initialization.

5.3   The Line - Linear Placement

First, we study a greedy local switch strategy on a 1-dimensional grid - the line. We assume that the C-rule switching strategy is sequentially applied (in arbitrary order) on an arbitrary initial state and continuously adjust the network by switching neighbors. The strategy will converge against a local optimum from which no switch of two neighboring nodes improves the objective value in expectation. We are interested in quantifying how far such a locally optimal solution can be from the global optimum. The following theorem gives an answer for this question.

Theorem 1. *Let H be the line and $\mathcal{P}$ a symmetric product distribution, then any locally optimal solution achieved by the C-rule (or M-rule) is at most a factor of four larger than the global optimum of MEPL.*

We prove this theorem for the C-rule but this could be done similarly to the M-rule. Assume H and $\mathcal{P}$ as in the theorem. We first give an upper bound on the expected path length achieved by the C-rule in terms of the expected rank of the distribution (Definition 5).

Lemma 5. *For any locally optimal solution φ achieved by the C-rule: $C(\varphi) \leq \mathbb{E}[R]$, and $\mathrm{EPL}(\varphi) \leq 2\mathbb{E}[R]$.*

Proof. Let $d(\varphi(v), c^*)$ be the distance of $\varphi(v)$ from c^*. We want to bound it in terms of $r(v)$, the rank of v. From Lemma 4 we get that all nodes between $\varphi(v)$ and c^* on the line have higher probability than v and thus $d(\varphi(v), c^*) \leq r(v)$. So: $C(\varphi) = \sum_{v \in V(G)} p(v)d(\varphi(v), c^*) \leq \sum_{v \in V(G)} p(v)r(v) = \mathbb{E}[R]$. From Lemma 3, $EPL(\varphi) \leq 2C(\varphi) \leq 2\mathbb{E}[R]$. $\qquad\square$

We now prove a lower bound for MEPL on the line and any symmetric product distribution of requests.

Lemma 6. *MEPL* $\geq C_{\min} \geq \frac{1}{2}\mathbb{E}[R]$.

Proof. Let φ^* be the placement such that $C_{\min} = C(\varphi^*)$. Note that by definition, φ^* minimizes the expected path length to the center. Given the center $c^*(\varphi^*)$ and an arbitrary node v with a distance $d(\varphi^*(v), c^*)$ from c^*, we want to find an upper bound on the rank of v by bounding how many nodes can have a higher activity level than v. Clearly, all such nodes will be at most at the distance $d(\varphi^*(v), c^*)$ from the center, since otherwise, φ^* will not be optimal. Since in a line there at most two nodes at distance i from the center $d(\varphi^*(v), c^*) \geq r(v)/2$ we obtain as desired: $C_{\min} = \sum_{v \in V} p(v)d(\varphi^*(v), c^*) \geq \sum_v p(v)r(v)/2 = \mathbb{E}[R/2]$. $\qquad\square$

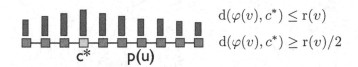

$$d(\varphi(v), c^*) \leq r(v)$$
$$d(\varphi(v), c^*) \geq r(v)/2$$

Fig. 1. Illustration of rank properties used in Lemmas 5 and 6

To conclude the proof of Theorem 1, we combine Lemmas 3, 5 and 6 to get that for a local minimum φ: $2\mathbb{E}[R] \geq EPL(\varphi) \geq MEPL \geq \frac{1}{2}\mathbb{E}[R]$. Thus, the ratio between the worst case local solution and the optimal solution is at most 4.

5.4   The d-Dimensional Grid

In this section, we extend the ideas from the line to grid networks. Our results apply readily to grids of arbitrary dimension but, for sake of simplicity, we stick to two dimensions here. We first show that using the same greedy approach as in the line, namely switching neighboring nodes using the M-rule, leads to a drastically worse ratio between local and global minima.

Lemma 7. *On the d-dimensional grid, there is a local minimum with regards to the M-rule and the C-rule that is a factor of $\Omega(n^{1/d - 1/d^2})$ worse than the global minimum.*

Note that the last lemma implies a $\Omega(n^{1/4})$ worst-case ratio for the 2-dimensional grid. Surprisingly, we can avoid this, locally stable but highly suboptimal solution, by allowing only slightly longer switches. The rule we propose is that a node can also switch with any of the neighbors in ℓ_1-distance three that differs two in one axis and one in the other (similar to the chess knight moves) and the switching is according to the C-rule. In this case we can prove the following.

Theorem 2. *Let H be the 2-dimensional grid and $\mathcal{P}$ a symmetric product distribution, then any locally optimal solution achieved by the C-rule (and allowing "chess knight move" switches) is at most a factor of 4.62 larger than the global optimum of MEPL.*

The proof of this theorem is similar in spirit to the 1-dimensional grid, where we provide bounds on C for the optimal and locally optimal placements. First, we show that we get a fat set from this strategy and also prove a stronger rank-property of any locally optimal solution; namely, nodes that are far away from the center have to have a (quadratically) high rank.

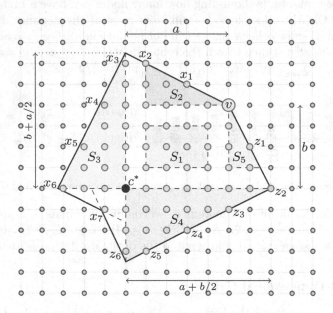

Fig. 2. A local minimum when optimizing the distance to the expected position of a routing request (black node c^*). The red nodes mark the positions that dominate the green node v, that is, that have a path consisting of possible switchings each goes strictly closer to the black node. In a local minimum these nodes have to have a larger probability than the green node making the rank of the green node at least as large as the number of red nodes.

Lemma 8. *For any local minimum of the C-rule $C(\varphi) \leq \frac{4}{\sqrt{6}}\mathbb{E}[\sqrt{R}]$, where R is the rank of a given distribution (see Definition 5).*

Proof. Consider a node v in general relative position to the center c^* (see Figure 2). We want to estimate how many nodes have a larger probability (higher rank) than the node v. To achieve this estimation, we analyze the area of the largest polygon, such that every node inside and on the edges of the polygon belongs to some distance-decreasing path from v to c^*. According to Lemma 4, we get the guarantee that for any local minimum, with respect to the C-rule, any node u in the polygon has $p(v) \leq p(u)$. Figure 2 provides an example for this: The node x_1 is closer to the center than v and switching between x_1 and v is possible. This implies that $p(v) \leq p(x_1)$. Furthermore, if we look at the distance-decreasing paths $v, x_1, x_2, \ldots, x_7$ and $v, z_1, z_2, \ldots, z_7$ we obtain from Lemma 4 that $p(v) \leq p(x_1) \leq p(x_2) \leq \ldots \leq p(x_7)$ and that $p(v) \leq p(z_1) \leq \ldots \leq p(x_7)$. These paths can furthermore be extended to any node in the polygon. All these nodes have such higher activity levels then $p(v)$. To get a bound on the rank of v, i.e., on the number of nodes that have a larger activity level, we count the number of nodes that are inside the polygon. This number generally involves many floors and ceilings. We avoid these by first calculating the number of nodes $A(x)$ bounded by a right triangle shape that starts at a point and whose (axis parallel) legs have length x and $x/2$. We denote this number by $A(x)$ and in the following give a formula and an estimate for it that holds for any positive real x:

$$A(x) := \sum_{i=0}^{\lfloor x/2 \rfloor} (\lfloor x \rfloor + 1 - 2i) \geq \begin{cases} \frac{x^2}{4} + 1 & \text{if } x \geq 1, \\ 1 & \text{if } x < 1. \end{cases}$$

Now we are ready to calculate S_{total}. We denote the number of nodes in the middle rectangle as S_1, the number of nodes in the upper triangle as S_2 and so on according to Figure 2.

$$S_1 = (a-1)(b-1) \qquad S_2 = A(a-1) \qquad S_3 = A(b+a/2)$$
$$S_4 = A(a+b/2) \qquad S_5 = A(b-1)$$

So, we obtain that $S_{total} = S_1 + S_2 + S_3 + S_4 + S_5 - 3$, where -3 is needed since the node v should not be calculated (but was counted twice) and c^* should be calculated once (but was counted twice). By adding up the expressions we obtain for $a, b \geq 2$: $S_{total} = S_1 + S_2 + S_3 + S_4 + S_5 - 3 \geq \frac{6}{16}(a+b)^2$. Easy to verify that the inequality $S_{total} \geq \frac{6}{16}(a+b)^2$ holds also in the case where a, or b, or both equal to 1. Since the rank of v is at least S_{total}, we get $r(v) \geq \frac{6}{16}(d(v,c^*))^2$, and thus: $d(v,c^*) \leq \frac{4}{\sqrt{6}}\sqrt{r(v)}$. Hence: $C = \sum_{v \in V} p(v)d(v,c^*) \leq \sum_{v \in V} p(v) \frac{4}{\sqrt{6}}\sqrt{r(v)} = \frac{4}{\sqrt{6}}\mathbb{E}[\sqrt{R}]$. □

Next we show a lower bound for the cost of the optimum placement obtaining a similar expression in terms of the (expected) rank.

Lemma 9. MEPL $\geq C_{\min} \geq \frac{1}{\sqrt{2}}\mathbb{E}[\sqrt{R}]$.

Proof. Let φ^* be the placement such that $C_{\min} = C(\varphi^*)$. Note that by definition φ^* minimize the expected path to the center. Given the center $c^*(\varphi^*)$ and an arbitrary node v with a distance $d(v,c^*)$ from c^*, we again find an upper bound on the rank of v by showing how many nodes can have a larger activity level than v. Again, all such nodes will be at most at the distance $d(v,c^*)$ from the center, since otherwise the solution will not be a global optimum. There are exactly 4 nodes at distance 1 from c^*, 8 nodes at the distance 2 and in general $4i$ nodes at distance i. This lead to the to the following bound: $r(v) \leq \sum_{i=1}^{d(v,c^*)} 4i = 2(d(v,c^*))^2$. So, we obtain $d(v,c^*) \geq \sqrt{r(v)/2}$, and thus: $C = \sum_{v \in V} p(v)d(v,c^*) \geq \sum_v p(v)\sqrt{r(v)/2} = \mathbb{E}[\sqrt{R/2}]$ $\qquad\square$

Now we are ready to prove the result of Theorem 2. From Lemmas 3 and 9 we get that the minimum expected path length is at least $\mathbb{E}[\sqrt{R/2}]$. From Lemma 8 we obtain that $C(\varphi) \leq \mathbb{E}[\frac{4}{\sqrt{6}}\sqrt{R}]$, and thus by Lemma 3 the expected path length of a local minimum can not be larger than $2\mathbb{E}[\frac{4}{\sqrt{6}}\sqrt{R}]$. Therefore the ratio between the optimal solution and any local minimum with regards to the C-rule is at most $2\frac{4\sqrt{2}}{\sqrt{6}} \approx 4.62$.

All proofs above can be extended to the d-dimensional grid. The lower bound guarantees in this context that any solution on the d-dimensional grid has a cost of at least $\Omega(E[R^{1/d}])$, where R is the rank of a sampled node. Similarly for any constant d we get a $O(E[R^{1/d}])$ upper bound from the fact that a fat body is dominated by any node. The constant factor in the upper bound decreases like 2^{-d} since using the *chess knight* improvement switches along any dimension costs a factor of 2. By using longer improvement directions of length k instead of 3 the factor of two can be brought down to $1 - \frac{1}{k-2}$. Thus, e.g., using length d improvement switches on the d-dimensional grid results in a constant factor approximation for any dimension d.

6 Clustered Requests

We have demonstrated that self-adjusting networks and their local switching rules work well on grid networks with symmetric product distributions. We now briefly discuss some interesting preliminary results on a more general type of request distributions: clustered requests. For this we consider situations where processes can be clustered into groups such that communication takes place only or predominantly between processes belonging to the same group. This locality is inspired by practice and we believe that such a structure in the requests is quite common.

Ideally, a self-adjusting network "detects" such clusters and arranges processes such that groups will reside in separate parts of the network infrastructure. Such an arrangement facilitates short routes since requests between group members get routed quickly without leaving the group. Such a well clustered placement of nodes can be a drastic improvement of the expected path length compared to a non-optimized placement. In particular, any placement that is oblivious to the clustering, e.g., a random placement, will have a bad performance – leaving plenty of room for improvement. We believe that the simple switching strategies presented in this paper perform very well in many such settings.

Our investigations and simulations on d-dimensional torus topologies have led to several interesting preliminary results in this direction: On the negative side we were able to construct local minima in many topologies that have a poor performance compared to the global optimum. This shows that it is not possible to give the same type of strong approximation guarantee, independent of the initialization. In simulations, we observed that for $d = 1$, i.e., on a ring topology, even random initializations lead to bad local minima. The reason for this is that connectivities in the ring are too restricted to allow the resolution of distributed clusters without disturbing other, already fixed, clusters. Fortunately, this changes drastically for higher dimensions. For $d > 1$ local minima become extremely unstable and sensitive to small perturbations. They can only occur if one starts in a carefully constructed worst-case initialization. This observation is supported by our simulations which produce very promising results. Figure 3 shows an example for the improvement in a 2-dimensional torus. Overall, the greedy switching algorithm successfully detects clusters and groups them together. However, there are still some clusters that are not connected. In Figure 3 (a) we can see the initial random placement of the clustered nodes. Nodes with a black bold frame are the centers of their clusters (as was defined earlier, a center is a node that has the minimal expected distance to all other nodes in the cluster). In Figure 3 (b) we see the placement of the nodes achieved by the greedy M-rule switching strategy. Although it looks that the nodes are highly grouped, we can see that the shapes of the clusters are not optimal (an optimal placement should look like a circle around the center of the cluster). Some clusters are stretched (e.g., brown cluster) and some are even not connected (e.g, orange cluster).

When every node on the torus belongs to an active cluster, we can frequently run into situation in which two nodes will not switch even if it is improvement for one of the clusters. This suboptimal local solution can be improved if we allow some nodes on a torus to be inactive. In the following two figures we see the results of such simulation where 50% on the nodes are inactive. In Figure 3 (c) we can see the initial random placement of the clustered nodes, (the inactive nodes are white

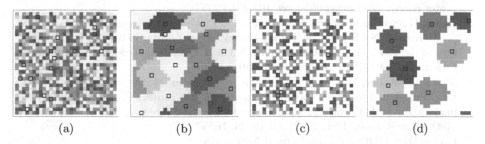

| (a) | (b) | (c) | (d) |

Fig. 3. (a) Torus with 900 nodes, 16 clusters and no nodes in the inactive cluster. Nodes are placed at random positions. (b) Final placement after applying the local greedy strategy in a round robin fashion. Clusters are grouped together, but their shapes are not optimal for a given cluster. (c) Torus with 900 nodes, 8 active clusters and half of the nodes in the inactive cluster. V_0 has $n/2$ nodes. Nodes are placed at random positions. (d) Final placement after applying the local greedy strategy. Clusters are grouped together and almost optimally shaped around their center.

colored). In Figure 3 (d) we see the placement of a local minimum of EPL achieved by the same greedy switching strategy. But now we can observe much nicer concentration of the nodes around their centers. These figures lead to many interesting future research questions about the topic. Animated version of the figures can be found here: http://www.bgu.ac.il/avin/pmwiki/pmwiki.php?n=Main.Self-AdjustingNetworks .

7 Conclusions and Future Work

In this preliminary work, we formally defined the MEPL problem which has practical significance in saving energy of fixed infrastructure network. We showed that in general cases, the problem is hard to compute, but under some realistic assumptions on network infrastructure and traffic patterns, we propose efficient local and distributed algorithms that achieve almost optimal solution. Our algorithms are based on migration of processes, which allows network optimization without changing the underlying infrastructure. This idea integrates well with an SDN concept which will probably include process migration functionality in its management platform.

In future work, we plan to extend our results to other topologies that are used in data centers networks, e.g., fat trees [32]. We also aim to investigate other types of requests distributions that are based on real data.

References

1. Poess, M., Nambiar, R.: Energy cost, the key challenge of today's data centers: a power consumption analysis of tpc-c results. Proceedings of the VLDB Endowment 1(2), 1229–1240 (2008)
2. U.S. Environmental Protection Agency: Report to congress on server and data center energy efficiency public law 109-431 (2007)
3. Heller, B., Seetharaman, S., Mahadevan, P., Yiakoumis, Y., Sharma, P., Banerjee, S., McKeown, N.: Elastictree: Saving energy in data center networks. In: Proceedings of the 7th USENIX Conference on Networked Systems Design and Implementation, p. 17. USENIX Association (2010)
4. Mirza-Aghatabar, M., Koohi, S., Hessabi, S., Pedram, M.: An empirical investigation of mesh and torus noc topologies under different routing algorithms and traffic models. In: 10th Euromicro Conference on Digital System Design Architectures, Methods and Tools, DSD 2007, pp. 19–26. IEEE (2007)
5. Greene, K.: TR10: Software-Defined Networking
6. Hoelzle, U.: Openflow @ google, Open Networking Summit (2012)
7. McKeown, N., Anderson, T., Balakrishnan, H., Parulkar, G., Peterson, L., Rexford, J., Shenker, S., Turner, J.: Openflow: enabling innovation in campus networks. SIGCOMM Comput. Commun. Rev. 38(2), 69–74 (2008)
8. Gummadi, K., Dunn, R., Saroiu, S., Gribble, S., Levy, H., Zahorjan, J.: Measurement, modeling, and analysis of a peer-to-peer file-sharing workload. In: Proceedings of the Nineteenth ACM Symposium on Operating Systems Principles, pp. 314–329. ACM (2003)

9. Klemm, A., Lindemann, C., Vernon, M., Waldhorst, O.: Characterizing the query behavior in peer-to-peer file sharing systems. In: Proceedings of the 4th ACM SIGCOMM Conference on Internet Measurement, pp. 55–67. ACM (2004)
10. Johnson, D., Garey, M.: Computers and intractability: A guide to the theory of np-completeness. Freeman&Co., San Francisco (1979)
11. Sleator, D., Tarjan, R.: Self-adjusting binary search trees. Journal of the ACM (JACM) 32(3), 652–686 (1985)
12. Lis, M., Shim, K., Cho, M., Fletcher, C., Kinsy, M., Lebedev, I., Khan, O., Devadas, S.: Brief announcement: distributed shared memory based on computation migration. In: SPAA, pp. 253–256. ACM (2011)
13. Batista, D., da Fonseca, N., Granelli, F., Kliazovich, D.: Self-adjusting grid networks. In: IEEE International Conference on Communications, ICC 2007, pp. 344–349. IEEE (2007)
14. Shang, Y., Li, D., Xu, M.: Energy-aware routing in data center network. In: Proceedings of the First ACM SIGCOMM Workshop on Green Networking 2010, pp. 1–8. ACM, New York (2010)
15. Tang, M., Liu, Z., Liang, X., Hui, P.M.: Self-adjusting routing schemes for time-varying traffic in scale-free networks. Phys. Rev. E 80(2), 026114 (2009)
16. Zhang, H., Liu, Z., Tang, M., Hui, P.: An adaptive routing strategy for packet delivery in complex networks. Physics Letters A 364(3-4), 177–182 (2007)
17. Jain, N., Menache, I., Naor, J(S.), Shepherd, F.B.: Topology-aware VM migration in bandwidth oversubscribed datacenter networks. In: Czumaj, A., Mehlhorn, K., Pitts, A., Wattenhofer, R. (eds.) ICALP 2012, Part II. LNCS, vol. 7392, pp. 586–597. Springer, Heidelberg (2012)
18. Bansal, N., Lee, K.W., Nagarajan, V., Zafer, M.: Minimum congestion mapping in a cloud. In: Proceedings of the 30th Annual ACM SIGACT-SIGOPS Symposium on Principles of Distributed Computing, PODC 2011, pp. 267–276. ACM, New York (2011)
19. Chung, F.: Labelings of graphs. Selected Topics in Graph Theory 3, 151–168 (1988)
20. Díaz, J., Petit, J., Serna, M.: A survey of graph layout problems. ACM Comput. Surv. 34, 313–356 (2002)
21. Bhatt, S., Cosmadakis, S.: The complexity of minimizing wire lengths in vlsi layouts. Information Processing Letters 25(4), 263–267 (1987)
22. Bhatt, S., Thomson Leighton, F.: A framework for solving vlsi graph layout problems. Journal of Computer and System Sciences 28(2), 300–343 (1984)
23. Demaine, E.D., Fekete, S.P., Rote, G., Schweer, N., Schymura, D., Zelke, M.: Integer point sets minimizing average pairwise l1 distance: What is the optimal shape of a town? Comput. Geom. Theory Appl. 44(2), 82–94 (2011)
24. Avin, C., Haeupler, B., Scheideler, C., Schmid, S.: Locally self-adjusting tree networks. In: 27th IEEE International Parallel and Distributed Processing Symposium, IPDPS (2013)
25. Kirkpatrick, S., Gelatt, C., Vecchi, M.: Optimization by simulated annealing. Science 220(4598), 671 (1983)
26. Rabani, Y., Sinclair, A., Wanka, R.: Local divergence of markov chains and the analysis of iterative load-balancing schemes. In: Focs, p. 694. IEEE Computer Society (1998)
27. Mukherjee, S., Gupte, N.: Gradient mechanism in a communication network. Phys. Rev. E 77(3), 036121 (2008)

28. Jacob, R., Richa, A., Scheideler, C., Schmid, S., Täubig, H.: A distributed poly-logarithmic time algorithm for self-stabilizing skip graphs. In: Proceedings of the 28th ACM Symposium on Principles of Distributed Computing, pp. 131–140. ACM (2009)
29. Jacob, R., Ritscher, S., Scheideler, C., Schmid, S.: A self-stabilizing and local de-launay graph construction. Algorithms and Computation, 771–780 (2009)
30. Watts, D., Strogatz, S.: Collective dynamics of 'small-world'networks. Nature 393(6684), 440–442 (1998)
31. Karp, R.: Reducibility among combinational problems. Complexity of Computer Computations, 85–104 (1972)
32. Leiserson, C.E.: Fat-trees: universal networks for hardware-efficient supercomputing. IEEE Trans. Comput. 34(10), 892–901 (1985)

On Advice Complexity of the k-server Problem
under Sparse Metrics

Sushmita Gupta[1], Shahin Kamali[2], and Alejandro López-Ortiz[2]

[1] University of Southern Denmark, Odense, Denmark
[2] Cheriton School of Computer Science, University of Waterloo, Ontario, Canada

Abstract. We consider the k-SERVER problem under the advice model
of computation when the underlying metric space is sparse. On one side,
we introduce $\Theta(1)$-competitive algorithms for a wide range of sparse
graphs, which require advice of (almost) linear size. Namely, we show
that for graphs of size N and treewidth α, there is an online algorithm
which receives $O(n(\log \alpha + \log \log N))^1$ bits of advice and optimally serves
a sequence of length n. With a different argument, we show that if a
graph admits a system of μ collective tree (q, r)- spanners, then there
is a $(q + r)$-competitive algorithm which receives $O(n(\log \mu + \log \log N))$
bits of advice. Among other results, this gives a 3-competitive algorithm
for planar graphs, provided with $O(n \log \log N)$ bits of advice. On the
other side, we show that an advice of size $\Omega(n)$ is required to obtain
a 1-competitive algorithm for sequences of size n even for the 2-server
problem on a path metric of size $N \geq 5$. Through another lower bound
argument, we show that at least $\frac{n}{2}(\log \alpha - 1.22)$ bits of advice is required
to obtain an optimal solution for metric spaces of treewidth α, where
$4 \leq \alpha < 2k$.

1 Introduction

Online algorithms have been extensively studied in the last few decades. In the
standard setting, the input to an online algorithm is a sequence of *requests*,
which should be *answered* sequentially. To answer each request, the algorithm
has to take an irreversible decision without looking at the incoming requests.
For minimization problems, such a decision involves a *cost* and the goal is to
minimize the total cost.

The standard method for analysis of online algorithms is the *competitive analysis*, which compares an online algorithm with an optimal offline algorithm, $\mathcal{OPT}$.
The competitive ratio of an online algorithm[2] $\mathcal{ALG}$ is defined as the maximum
ratio between the cost of $\mathcal{ALG}$ for serving a sequence and the cost of $\mathcal{OPT}$ for
serving the same sequence, within an additive constant factor.

Although the competitive analysis is accepted as the main tool for analysis
of online algorithms, its limitations have been known since its introduction: Inputs adversarially produced to draw out the worst performance of a particular

---

[1] We use $\log x$ to denote $\log_2(x)$.
[2] In this paper we only consider deterministic algorithms.

T. Moscibroda and A.A. Rescigno (Eds.): SIROCCO 2013, LNCS 8179, pp. 55–67, 2013.
© Springer International Publishing Switzerland 2013

algorithm are not commonplace in real life applications. Therefore, in essence competitive analysis mostly measures the benefit of knowing the future, and not the true difficulty of instances. From the perspective of an online algorithm, the algorithm is overcharged for its complete lack of knowledge about the future. Advice complexity quantifies this *gap in information* that gives $\mathcal{OPT}$ an unassailable advantage over any online strategy.

Under the advice model for online algorithms [14,6], the input sequence $\sigma = \langle r_1 \ldots r_n \rangle$ is accompanied by b bits of advice recorded on an advice tape B. For answering the request r_i, the algorithm takes an irreversible decision which is a function of $r_1, \ldots, r_i$ and the advice provided on B. The advice complexity of an online problem is the minimum number of bits which is required to optimally solve any instance of the problem. In the context of the communication complexity, it is desirable to provide an advice of small size, while achieving high quality solutions. For example, it is easy to see that for the ski-rental problem [19] a single bit of advice is sufficient to achieve an optimal solution.

We are interested in the advice complexity of the k-SERVER problem, as well as the relationship between the size of advice and the competitive ratio of online algorithms. To this end, we study the problem for a wide variety of sparse graphs.

1.1   Preliminaries

An instance of the k-SERVER problem includes a metric space M, k mobile servers, and a request sequence σ. The metric space can be modelled as an undirected, weighted graph of size $N > k$. (We interchangeably use terms 'metric space' and 'graph'.) Each request in the input sequence σ denotes a vertex of M, and an online algorithm should move one of the servers to the requested vertex to *serve* the request. The cost of the algorithm is defined as the total distance moved by all k servers over σ.

For any graph $G = (V, E)$, a tree decomposition of G with width α is a pair $(\{X_i \| i \in I\}, T)$ where $\{X_i \| i \in I\}$ is a family of subsets of V (bags), and T is a tree whose nodes are the subsets X_i such that

- $\bigcup_{i \in I} X_i = V$ and $\max_{i \in I} |X_i| = \alpha + 1$.
- for all edges $(v, w) \in E$, there exists an $i \in I$ with $v \in X_i$ and $w \in X_i$.
- for all $i, j, k \in I$: if X_j is on the path from X_i to X_k in T, then $X_i \cap X_k \subseteq X_j$.

The treewidth of a graph G is the minimum width among all tree decompositions of G. Informally speaking, the tree decomposition is a mapping a graph to a tree so that the vertices associated to each node (bag) of the tree are close to each other, and the treewidth measures how close the graph is to such tree.

We say that a graph $G = (V, E)$ admits a system of μ collective tree (q, r)-spanners if there is a set $\mathcal{T}(G)$ of at most μ spanning trees of G such that for any two vertices x, y of G, there exists a spanning tree $T \in \mathcal{T}(G)$ such that $d_T(x, y) \leq q \times d_G(x, y) + r$.

For the ease of notation, we assume k denotes the number of servers, N, the size of metric space (graph), n, the length of input sequence, and α, the treewidth of the metric space.

1.2   Existing Results

The advice model for the analysis of the online algorithm was first proposed in [14]. Under that model, each request is accompanied by an advice of fixed length. A slight variation of the model was proposed in [7,6], which assumes that the online algorithm has access to an advice tape. At any time step, the algorithm may refer to the tape and read any number of advice bits. The advice-on-tape model has an advantage that enables algorithms to use sublinear advice. This model has been used to analyze the advice complexity of many online problems, which includes paging [7,18,21], disjoint path allocation [7], job shop scheduling [7,21], k-server [6], knapsack [8], bipartite graph coloring [4], online coloring of paths [16], set cover [20,5], maximum clique [5], and graph exploration [11]. In this paper, we adopt this definition of the advice model.

For the k-SERVER problem on general metrics, there is an algorithm which achieves a competitive ratio of $k^{O(1/b)}$ for $b \leq k$, when provided with bn bits of advice [14]. This ratio was later improved to $2\lceil\lceil\log k\rceil/(b-2)\rceil$ in [6], and then to $\lceil\lceil\log k\rceil/(b-2)\rceil$ in [23]. Comparing these results with the lower bound k for the competitive ratio of any online algorithm [22], one can see how an advice of linear size can dramatically improve the competitive ratio.

The k-SERVER problem has been studied under specific metric spaces which include trees [10], metric spaces with $k+2$ points [1], Manhattan space [2], the Euclidean space [2], and the cross polytope space [3]. For trees, it is known that the competitive ratio of any online algorithm is at least k, while there are online algorithms which achieve this ratio [10]. Under the advice model, the k-SERVER problem has been studied when the metric space is the Euclidean plane, and an algorithm with constant competitive ratio is presented, which receives n bits of advice for sequences of length n [6]. In [23], tree metric spaces are considered and a 1-competitive algorithm is introduced which receives $2n + 2\lceil\log(p+2)\rceil n$ bits of advice, where p is the *caterpillar dimension* of the tree. There are trees for which p is as large as $\lceil\log N\rceil$. Thus, the 1-competitive algorithm of [23] needs $\Theta(n \log \log N)$ bits of advice.

1.3   Contribution

We introduce an online algorithm which optimally serves any input sequence on a metric of treewidth α, when provided $O(n(\log \alpha + \log \log N))$ bits of advice. We provide another algorithm that achieves a competitive ratio of at most $q + r$, when the metric space admits a system of μ collective tree (q, r)-spanners. The algorithm receives $O(n(\log \mu + \log \log N))$ bits of advice. This yields competitive algorithms for a large family of graphs, e.g., a 3-competitive algorithm for planar graphs which reads $O(n(\log \log N))$ bits of advice.

Our the other side, we show that a sublinear advice does not suffice to provide close-to-optimal solution, even if we restrict the problem to 2-server problem on paths of size $N \geq 5$. Precisely, we show that $\Omega(n)$ bits of advice are required to obtain a c-competitive algorithm for any value of $c \leq 5/4 - \epsilon$ (ϵ is an arbitrary small constant). Since there is a 1-competitive algorithm which receives $O(n)$

bits of advice for paths [23], we conclude that $\Theta(n)$ bits of advice are necessary and sufficient to obtain a 1-competitive algorithm for these metrics. Through another lower bound argument, we show that any online algorithm requires an advice of size at least $\frac{n}{2}(\log \alpha - 1.22)$ bits to be optimal on a metric of treewidth α, where $4 \leq \alpha < 2k$.

For graphs with constant treewidth, the advice size our algorithm (the first algorithm) is almost linear. Considering that an advice of linear size is required for 1-competitive algorithms (our first lower bound), the algorithm has an advice of nearly optimal size. For graphs with treewidth $\alpha \in \Omega(\lg N)$, the advice size is $O(n \log \alpha)$, which is asymptotically tight when $4 \leq \alpha < 2k$, because at least $\frac{n}{2}(\log \alpha - 1.22)$ bits are required to be optimal in this case (our second lower bound).

Due to space restrictions, many proofs have been removed. They will appear in the long version of the paper.

2   Upper Bounds

2.1   Graphs with Small Treewidth

We introduce an algorithm called *Graph-Path-Cover*, denoted by $\mathcal{GPC}$, to show that $O(n(\log \alpha + \log \log N))$ bits of advice are sufficient to optimally serve a sequence of length n on any metric space of treewidth α. We start with the following essential lemma.

Lemma 1. *Let T be a tree decomposition of a graph G. Also, let x and y be two nodes of G and $P = (x = p_0, p_1, \ldots p_{l-1}, y = p_l)$ be the shortest path between x and y. Let X and Y be two bags in T which respectively contain x and y. Any bag on the unique path between X and Y in T contains at least one node p_i $(0 \leq i \leq l)$ from P.*

Similar to the Path-Cover algorithm introduced for trees in [23], $\mathcal{GPC}$ moves its servers on the same trajectories as $\mathcal{OPT}$ moves its. Suppose that $\mathcal{OPT}$ uses a server s_i to serve the requests $[x_{a_{i,1}}, \ldots, x_{a_{i,n_i}}]$ ($i \leq k$ and $n_i \leq n$). So, s_i is moved on the unique path from its initial position to $x_{a_{i,1}}$, and then from $x_{a_{i,1}}$ to $x_{a_{i,2}}$, and so on. Algorithm Path-Cover tends to move s_i on the same path as $\mathcal{OPT}$.

For any node v in G, $\mathcal{GPC}$ treats one of the bags which contains v as the *representative bag* of v. Moreover, it assumes an ordering of the the nodes in each bag. Each node in G is *addressed* via its representative bag, and its index among the nodes of that bag. A server s_i, located at a vertex v of G, is addressed via a bag which contains v (not necessarily the representative bag of v) and the index of v in that bag. Note that while there may be a unique way to address a node, there might be several different ways to address a server.

Assume that for serving a request y, $\mathcal{OPT}$ moves a server s_i from a node x to y in G. Let X and Y be respectively the representative bags of x and y, and Z be the least common ancestor of X and Y in T. By Lemma 1, the shortest path between x and y passes at least one node z in Z, and that node can be indicated by $\lceil \log h \rceil + \lceil \log \alpha \rceil$ bits of advice (h denotes the height of the tree associated with the tree decomposition), with $\lceil \log h \rceil$ bits indicating Z and $\lceil \log \alpha \rceil$ bits indicating the index of the said node z in Z. After serving x, $\mathcal{GPC}$ moves s_i to z, provided that the address of z is given as part of the advice for x. For serving y, $\mathcal{GPC}$ moves s_i to y, provided that the address of s_i (address of z) is given as part of the advice for y. In what follows, we elaborate this formally.

Before serving any request, $\mathcal{GPC}$ moves each server s_i from its initial position x_0 to a node z_0 on the shortest path between x_0 and the first node $x_{a_{i,1}}$ served by s_i in $\mathcal{OPT}$'s scheme. $\mathcal{GPC}$ selects z_0 in a way that it will be among the vertices in the least common ancestor of the representative bags of x_0 and $x_{a_{i,1}}$ in the tree decomposition (by Lemma 1 such a z_0 exists). To move all servers as described, $\mathcal{GPC}$ reads $(\lceil \log h \rceil + \lceil \log \alpha \rceil) \times k$ bits of advice. After these *initial* moves, $\mathcal{GPC}$ moves servers on the same trajectories of $\mathcal{OPT}$ as argued earlier. Assume that x, y and w denote three requests which are consecutively served by s_i in $\mathcal{OPT}$'s scheme. The advice for serving x contains $\lceil \log h \rceil + \lceil \log \alpha \rceil$ bits which represents a node z_1, which lies on the shortest path between x and y and is situated inside the least common ancestor of the respective bags in T. $\mathcal{GPC}$ moves s_i to z_1 after serving x. The first part of advice for y contains $\lceil \log h \rceil + \lceil \log \alpha \rceil$ bits indicating the node z_1 from which s_i is moved to serve y. The second part of advice for y indicates a node z_2 on the shortest path between y and w in the least common ancestor of their bags in T. This way, $2(\lceil \log h \rceil + \lceil \log \alpha \rceil)$ bits of advice per request are sufficient to move servers on the same trajectories as $\mathcal{OPT}$.

The above argument implies that an advice of size $2(\lceil \log h \rceil + \lceil \log \alpha \rceil) \times n + (\lceil \log h \rceil + \lceil \log \alpha \rceil) \times k$ is sufficient to achieve an optimal algorithm. The value of h (the height of the tree decomposition) can be as large as N, however we can apply the following lemma to obtain height-restricted tree decompositions.

Lemma 2. *[9,15] Given a tree decomposition with treewidth α for a graph G with N vertices, one can obtain a tree decomposition of G with height $O(\log N)$ and width at most $3\alpha + 2$.*

If we apply $\mathcal{GPC}$ on a height-restricted tree decomposition, we get the following theorem.

Theorem 1. *For any metric space of size N and treewidth α, there is an online algorithm which optimally serves any input sequence of size n, provided with $O(n(\log \alpha + \log \log N))$ bits of advice.*

2.2 Graphs with Small Number of Collective Tree Spanners

In this section we introduce an algorithm which receives an advice of almost linear size and achieves constant competitive ratio for a large family of graphs.

Theorem 2. *If a metric space of size N admits a system of μ collective tree (q, r)-spanners, then there is a deterministic online algorithm which on receiving $O(n(\log \mu + \log \log N))$ bits of advice, achieves a competitive ratio of at most $q + r$ on any sequence of length n.*

Proof. When there is only one tree T in the collection (i.e., $\mu = 1$), we can apply the PathCover algorithm of [23] on T to obtain the desired result. To be precise, for the optimal algorithm $\mathcal{OPT}$, we denote the path taken by it to serve a sequence of requests with the server s_i to be $P_G = \left[x_{a_{i,1}}, x_{a_{i,2}}, \ldots, x_{a_{i,n_i}} \right]$. PathCover algorithm moves s_i on the path $P_T = \left[x_{a_{i,1}}, \ldots, x_{a_{i,2}}, \ldots, x_{a_{i,n_i}} \right]$ in T. Since T is a spanner of G, the total length of P_T does not exceed that of P_G by more than a factor of $q + r$ for each edge in P_G, and consequently the cost of the algorithm is at most $q + r$ times that of $\mathcal{OPT}$'s. Thus, the algorithm is $(q + r)$-competitive. After serving a request x with server s_i, PathCover can move s_i to the least common ancestor of x and y, where y is the next request at which $\mathcal{OPT}$ uses s_i. This requires $\lceil \log h \rceil$ bits of advice per request (h being the height of the tree). Instead, the algorithm can use the caterpillar decomposition of T and move servers on the same set of paths while using only $O(\log \log N)$ bits of advice. The main idea is the same, whether we use a rooted tree or the caterpillar decomposition. Here for the ease of explanation, we will only argue for the rooted tree, but the statement of the theorem holds when the caterpillar decomposition is used.

We introduce an algorithm that mimics $\mathcal{OPT}$'s moves for each server by picking suitable trees from the collection to move the server through. The advice for each request indicates which tree from the collection would best approximate the edges traversed by the server in $\mathcal{OPT}$'s scheme to reach the next node at which it is used. To this end, we look at the tree spanners as rooted trees. If $\mathcal{OPT}$ moves a server s_i on the path $P_G = \left[x_{a_{i,1}}, x_{a_{i,2}}, \ldots, x_{i,n_i} \right]$, then for each edge $(x_{a_{i,j}}, x_{a_{i,j+1}})$ on this path, our algorithm moves s_i on the shortest path of (one of) the tree spanners which best approximates the distance between the vertices $x_{a_{i,j}}$ and $x_{a_{i,j+1}}$. As explained below, the selection of suitable spanners at every step can be ensured by providing $2 \lceil \log \mu \rceil$ bits of advice with each request.

Let us denote the initial location of the k servers by $z_1, \ldots, z_k$, and let $z_1', \ldots, z_k'$ respectively denote the first requested nodes served by them. Before serving any request, for any server s_i, the algorithm reads $\lceil \log \mu \rceil + \lceil \log h \rceil$ bits of advice to detect the tree $T_p(1 \le p \le \mu)$ that preserves the distance between z_i and z_i' in G, and moves s_i to the least common ancestor of z_i and z_i'. Moreover, the algorithm labels s_i with index p. These labels are used to move the correct servers on the trees in order to cover the same paths as $\mathcal{OPT}$. Let w and y be two vertices which are served respectively before and after x with the same server in $\mathcal{OPT}$'s scheme. To serve the request to x the algorithm works as follows:

- Find the spanner T_p which best approximates the length of the shortest path between w and x in G. This can be done if provided with $\lceil \log \mu \rceil$ bits of advice with x.
- Read $\lceil \log h \rceil$ bits of advice to locate a server s labeled as p on the path between node x and the root of T_p. Move s to serve x. In case of caterpillar decomposition, the algorithm reads roughly $\log \log N$ bits.

- After serving x, find the spanner T_q which best approximates the length of the shortest path between x and y in G. This can be done if provided with $\lceil \log \mu \rceil$ bits of advice with x.
- Find the least common ancestor of x and y in T_q. This can be done by adding $\lceil \log h \rceil$ bits of advice for x, where h is the height of T_q. In case of caterpillar decomposition, this would require roughly $\log \log N$ bits.
- Move s to the least common ancestor of x and y and label it as q.

Thus, since $\mathcal{OPT}$ moves the server s_i on the path $P_G = [x_{a_{i,1}}, x_{a_{i,2}}, \ldots, x_{a_{i,n_i}}]$, our algorithm moves s_i from $x_{a_{i,j}}$ to $x_{a_{i,j+1}}$ for each j $(1 \le j \le n_i - 1)$, on the path in the tree which approximates the distance between these two vertices within a multiplicative factor of $q + r$. The labels on the servers ensure that the algorithm moves the 'correct' servers on the trees. i.e, the ones which were intended to be used. Consequently, the cost of an algorithm for each server is increased by a multiplicative factor, at most $q + r$. Therefore, the total cost of the algorithm is at most $(q + r) \times \mathcal{OPT}$. The size of advice for each request is $2\lceil \log \mu \rceil + O(\log \log N)$, assuming that the caterpillar decomposition is used. Adding to that an additional $k(\log \mu + O(\log \log N))$ bits for the initial movement of servers completes the proof. □

In recent years, there has been wide interest in providing collective tree spanners for various families of graphs. The algorithms which create these spanners run in polynomial time. It is known that any planar graph of size N has a system of $O(\log N)$ collective (3,0)-spanners [17]; every AT-free graph (including interval, permutation, trapezoid, and co-comparability graphs) admits a system of two (1,2)-spanners [12]; every chordal graph admits a system of at most $\log N$ collective (1,2)-spanners [13]; and every Unit Disk Graphs admits a system of at most $2 \log_{1.5} N + 2$ collective (3,12)-spanners [24].

Corollary 1. *For metric spaces of size N and sequences of length n, $O(n \log \log N)$ bits of advice are sufficient to obtain I) a 3-competitive algorithm for planar graphs II) a 3-competitive algorithm for AT-free graph (including interval, permutation, trapezoid, and co-comparability graphs) III) a 3-competitive algorithm for chordal graphs IV) a 15-competitive algorithm for Unit Disk Graphs.*

3   Lower Bounds

3.1   2-server Problem on Path Metric Spaces

In this section, we show that an advice of sublinear size does not suffice to achieve close-to-optimal solutions, even for the 2-server problem on a path metric space of size $N \ge 5$. Without loss of generality, we only consider online algorithms which are *lazy* in the sense that they move only one server at the time of serving a request. It is not hard to see that any online algorithm can be converted to a lazy algorithm without an increase in its cost. Hence, a lower bound for the performance of lazy algorithms applies to all online algorithms. In the reminder of this section, the term *online algorithm* means a lazy algorithm.

Consider a path of size $N \geq 5$ which is horizontally aligned and the vertices are indexed from 1 to N. Assume that the servers are initially positioned at vertices 2 and 4. We build a set of instances of the problem, so that each instance is formed by $m = n/7$ rounds of requests. Each round is defined by requests to vertices $(3, 1|5, 3, 2, 4, 2, 4)$, where the second request of a round is either to vertex 1 or vertex 5. Each round ends with consecutive requests to vertices 2 and 4. So, it is reasonable to move servers to these vertices for serving the last requests of each round. This intuition is formalized in the following lemma.

Lemma 3. *Consider an algorithm A that serves an instance of the problem as defined above. There is another algorithm A' with a cost which is not more than that of A, for which the servers are positioned at vertices 2 and 4 before starting to serve each round.*

According to the above lemma, to provide a lower bound on the performance of online algorithms, we can consider only those algorithms which keep servers at positions 2 and 4 before each round. For any input sequence, we say a round R_t has type 0 if the round is formed by requests to vertices $(3, 1, 3, 2, 4, 2, 4)$ and has type 1 otherwise, i.e., when it is formed by requests to vertices $(3, 5, 3, 2, 4, 2, 4)$. The first request of a round is to vertex 3. Assume the second request is to vertex 5, i.e., the round has type 1. An algorithm can move the left vertex s_l positioned at 2 to serve the first request (to vertex 3) and the right server s_r positioned at 4 to serve the second request (to vertex 5). For serving other requests of the round, the algorithm can move the servers to their initial positions, and pay a total cost of 4 for the round (see Figure 1(a)). Note that this is the minimum cost that an algorithm can pay for a round. This is because there are four requests to distinct vertices and the last two are request to the initial positions of the servers (i.e., vertices 2 and 4). Next, assume that the algorithm moves the right vertex s_r to serve the first request (to vertex 3). The algorithm has to serve the second request (to vertex 5) also with s_r. The third request (to vertex 3) can be served by any of the servers. Regardless, the cost of the algorithm will not be less than 6 for the round (see Figure 1(b)). With a symmetric argument, in case the second request is to vertex 1 (i.e., the round has type 0), if an algorithm moves the right server to serve the first request it can pay a total cost of 4, and if it moves the left server for the first request, it pays a cost of at least 6 for the round.

In other words, an algorithm should 'guess' the type of a round at the time of serving the first request of the round. In case it makes a right guess, it can pay a total cost of 4 for that round, and if it makes a wrong guess, it pays a cost of at least 6. This relates the problem to the *Binary String Guessing Problem*.

Definition 1 ([14,5]). *The* Binary String Guessing Problem with known history (2-SGKH) *is the following online problem. The input is a bitstring of size n, and the bits are revealed one by one. For each bit b_t, the online algorithm $\mathbb{A}$ must guess if it is a 0 or a 1. After the algorithm has made a guess, the value of b_t is revealed to the algorithm.*

Lemma 4 ([5]). *On an input of length m, any deterministic algorithm for 2-SGKH that is guaranteed to guess correctly on more than αm bits, for $1/2 \leq \alpha < 1$, needs to read at least $(1 + (1 - \alpha) \log(1 - \alpha) + \alpha \log \alpha)m$ bits of advice.*

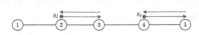

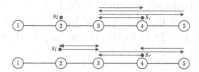

(a) In case of a right guess for the type of a round, the algorithm can pay a cost of 4.

(b) In case of a wrong guess for the type of a round, the algorithm pays a cost of at least 6.

Fig. 1. The cost of an algorithm for a round of type 1, i.e., request to vertices $(3, 5, 3, 2, 4, 2, 4)$. The servers are initially located at 2 and 4.

We reduce the 2-SGKH problem to the 2-server problem on paths.

Lemma 5. *If there is a 2-server algorithm with cost at most γn ($\gamma \geq 4/7$) for an instance of length n (as defined earlier), then there is a 2-SKGH algorithm which guesses at least $\frac{6-7\gamma}{2}m$ bits correctly for any input bit string of size $m = n/7$.*

Proof. Let B denote a bit string of length $m = n/7$, which is the input for the 2-SKGH problem. Consider the instance of the 2-server problem in which the types of rounds are defined by B. Precisely, the tth round has type 0 if the tth bit of B is 0, and has type 1 otherwise. We run the 2-server algorithm on such an instance. At the time of serving the first request of the tth round, the 2-server algorithm guesses the type of round t by moving the left or right server. In particular, it guesses the type of the round to be 0 if it moves the right server for the first request, and 1, otherwise. Define a 2-SGKH algorithm which performs according to the 2-server algorithm, i.e., it guesses the tth bit of B as being 0 (resp. 1) if the 2-server algorithm guesses the tth round as having type 0 (resp. 1). As mentioned earlier, the 2-server algorithm pays a cost of 4 for the round for each right guess, and pays cost of at least 6 for each wrong guess. So, the cost of the algorithm is at least $4\beta m + 6(1 - \beta)m = (6 - 2\beta)m$, in which βm is the number of correct guesses ($\beta \leq 1$). Consequently, if an algorithm has cost at most equal to this value, it correctly guesses the types of at least βm rounds, i.e., it correctly guesses at least βm bits of a bit string of length m. Defining γ as $(6 - 2\beta)/7$ completes the proof.   □

Lemmas 4 and 5 give the following theorem.

Theorem 3. *On input of length n, any deterministic algorithm for the 2-server problem which has a competitive ratio smaller than τ ($1 < \tau < 5/4$), needs to read at least $(1 + (2\tau - 2)\log(2\tau - 2) + (3 - 2\tau)\log(3 - 2\tau))n/7$ bits of advice, even if the metric space is restricted to being a path of size $N \geq 5$.*

Proof. There is an offline 2-server algorithm which pays a cost of 4 for each round and consequently pays a total cost of $4m = 4n/7$. Hence, in order to have a competitive ratio of τ, the cost of an algorithm should be at most $4\tau n/7$. According to Lemma 5, this requires the existence of a 2-SKGH algorithm which correctly guesses at least $(3 - 2\tau)m$ bits of a bit string of length m. By Lemma

4, this needs reading at least $(1 + (1 - (3 - 2\tau)) \log(1 - (3 - 2\tau)) + (3 - 2\tau) \log(3 - 2\tau))m = (1 + (2\tau - 2) \log(2\tau - 2) + (3 - 2\tau) \log(3 - 2\tau))n/7$ bits of advice. Note that $3 - 2\tau$ is in the range required by the lemma when $1 < \tau < 5/4$.   □

3.2   Metrics with Small Treewidth

We show that there are instances of the k-SERVER problem in a metric space with treewidth α, for which any online algorithm requires at least $\frac{n}{2}(\log \alpha - 1.22)$ bits of advice to perform optimally. Our construction is based on the one described in [6], where a lower bound for a general metric space is provided.

We introduce *units graphs* and *module graphs* as follows. A γ-unit graph is a bipartite graph $G = (U \cup W, E)$ where $U = \{u_1, \ldots, u_\gamma\}$ contains γ vertices, and W contains $2^\gamma - 1$ vertices each representing a proper subset of U. There is an edge between two vertices $u \in U$ and $w \in W$ iff $u \notin Set(w)$, where $Set(w)$ denotes the set associated with a vertex $w \in W$. Let $B_i \subseteq W$ denote the set of vertices of W whose associate sets have size i. i.e., for $w \in B_i$ we have $|Set(w)| = i$. A *valid request sequence* is defined as $\langle x_0, x_1, \ldots, x_{\gamma-1} \rangle$ so that for each i, $x_i \in B_i$ and $Set(x_i) \subseteq Set(x_{i+1})$. In other words, a valid sequence starts with a request to the vertex associated with the empty set, and with each step one element is added to get a larger set defining the next request. With this definition, one can associate every input sequence I with a unique permutation π of set $\{1, 2, \ldots, \gamma\}$.

A γ-module graph G includes two γ-unit graphs $G_1 = (U_1 \cup W_1, E_1)$ and $G_2 = (U_2 \cup W_2, E_2)$. In such a graph, those vertices in W_1 which represent sets of size i are connected to the $(i+1)$th vertex of U_2; the vertices of W_2 and U_1 are connected in the same manner. Consider an instance of the k-SERVER problem defined on a k-module graph, where initially all servers are located at the vertices of U_1. A valid sequence for the module graph is defined by repetition of *rounds of requests*. Each round starts with a valid sequence for G_1 denoted by π_1, followed by k requests to distinct vertices of U_2, a valid sequence for G_2, and k requests to distinct vertices of U_1. It can be verified that there is a unique optimal solution for serving any valid sequence on G, and consequently a separate advice string is required for each sequence [6]. Since there are $(k!)^{(n/(2k))}$ valid sequences of length n, at least $(n/(2k)) \log(k!) \geq n(\log k - \log e)/2$ bits of advice are required to separate all valid sequences.

Lemma 6. *Any γ-module graph has a tree decomposition of width 2γ.*

In what follows, we consider metric spaces with treewidth α such that $4 \leq \alpha \leq 2k$. Assume that α is an even integer and we have $k = m\alpha/2$ for some positive integer m. Consider a metric space G_b defined by a set of γ-modules where $\gamma = \alpha/2$. There are $k/\gamma = m$ such modules in G_b. Let $M^1, \ldots, M^m$ denote these modules, and let $G_1^i = (U_1^i \cup W_1^i, E_1), G_2^i = (U_2^i \cup W_2^i, E_2)$ denote the unit graphs involved in the construction of M^i ($i \leq m$). For each module M^i, select exactly one vertex from U_1^i, and connect all of the selected vertices to a common *source*. This makes G_b a connected graph.

Since there are m modules and in the ith module U_1^i contains γ vertices, there are $m \times \gamma = k$ vertices in all of the U_1^is. Assume that the k servers are initially placed at separate nodes in the U_1^is. A valid sequence for G_b is defined by a sequence of rounds of requests in which each round has the following structure:

$$f(\pi_1^1, \ldots, \pi_1^m), (b_1^1, \ldots, b_1^m), \ldots, (b_\gamma^1, \ldots, b_\gamma^m), f(\pi_2^1, \ldots, \pi_2^m), (a_1^1, \ldots, a_1^m), \ldots, (a_\gamma^1, \ldots, a_\gamma^m)$$

Here, f is a function that combines the requests from m permutations. Let $(\pi^1, \ldots, \pi^m)$ denote m permutations such that π^i contains γ requests $\langle r_1^i, \ldots, r_\gamma^i \rangle$ which defines a permutation in the module M^i. Thus, f gives a sequence of length $m \times \gamma$ starting with m requests to r_1^is, followed by m requests to r_2^js, and so on. For each j, $(1 \leq j \leq \gamma)$ we have fixed orderings on the vertices such that $(a_j^1, \ldots, a_j^m) \in (U_1^1, \ldots, U_1^m)$ and $(b_j^1, \ldots, b_j^m) \in (U_2^1, \ldots, U_1^m)$. With this definition, when a valid sequence of G_b is projected to the requests arising in a module M, the resulting subsequence is a valid sequence for M.

Lemma 7. *There is a unique optimal solution to serve a valid sequence on the metric space G_b. Also, each valid sequence requires a distinct advice string in order to be served optimally.*

To find a lower bound for the length of the advice string, we count the number of distinct valid sequences for the metric space G_b. In each round, there are $(\gamma!)^2$ valid sequences for each γ-module. Since there are m such modules, there are $(\gamma!)^{2m}$ possibilities for each round. A valid sequence of length n involves $n/(4\gamma m)$ rounds; hence there are $(\gamma!)^{n/(2\gamma)}$ valid sequences of length n. Each of these sequences need a distinct advice string. Hence, at least $\log((\gamma!)^{n/(2\gamma)}) \geq (n/2)\log(\gamma/e) = (n/2)\log(\alpha/(2e))$ bits of advice are required to serve a valid sequence optimally. This proves the following theorem.

Theorem 4. *Consider the k-SERVER problem on a metric space of treewidth α, such that $4 \leq \alpha < 2k$. At least $\frac{n}{2}(\log\alpha - 1.22)$ bits of advice are required to optimally serve an input sequence of length n.*

Concluding Remarks

For path metric spaces, we showed any 1-competitive algorithm requires an advice of size $\Omega(n)$. This bound is tight as there is an optimal algorithm which receives $O(n)$ bits of advice [23]. The same lower bound applies for trees, however, the best algorithm for tree receives an advice of $O(n \lg \lg N)$. We conjecture that the lower bound argument can be improved for trees to match it with the upper bound.

References

1. Bartal, Y., Koutsoupias, E.: On the competitive ratio of the work function algorithm for the k-server problem. Theoret. Comput. Sci. 324, 337–345 (2004)
2. Bein, W.W., Chrobak, M., Larmore, L.L.: The 3-server problem in the plane. Theoret. Comput. Sci. 289(1), 335–354 (2002)
3. Bein, W.W., Iwama, K., Kawahara, J., Larmore, L.L., Oravec, J.A.: A randomized algorithm for two servers in cross polytope spaces. Theoret. Comput. Sci. 412(7), 563–572 (2011)
4. Bianchi, M.P., Böckenhauer, H.-J., Hromkovič, J., Keller, L.: Online coloring of bipartite graphs with and without advice. In: Gudmundsson, J., Mestre, J., Viglas, T. (eds.) COCOON 2012. LNCS, vol. 7434, pp. 519–530. Springer, Heidelberg (2012)
5. Böckenhauer, H.-J., Hromkovič, J., Komm, D., Krug, S., Smula, J., Sprock, A.: The string guessing problem as a method to prove lower bounds on the advice complexity. In: Du, D.-Z., Zhang, G. (eds.) COCOON 2013. LNCS, vol. 7936, pp. 493–505. Springer, Heidelberg (2013)
6. Böckenhauer, H.J., Komm, D., Královič, R., Královič, R.: On the advice complexity of the k-server problem. In: Aceto, L., Henzinger, M., Sgall, J. (eds.) ICALP 2011, Part I. LNCS, vol. 6755, pp. 207–218. Springer, Heidelberg (2011)
7. Böckenhauer, H.-J., Komm, D., Královič, R., Královič, R., Mömke, T.: On the advice complexity of online problems. In: Dong, Y., Du, D.-Z., Ibarra, O. (eds.) ISAAC 2009. LNCS, vol. 5878, pp. 331–340. Springer, Heidelberg (2009)
8. Böckenhauer, H.-J., Komm, D., Královič, R., Rossmanith, P.: On the advice complexity of the knapsack problem. In: Fernández-Baca, D. (ed.) LATIN 2012. LNCS, vol. 7256, pp. 61–72. Springer, Heidelberg (2012)
9. Bodlaender, H.L.: A tourist guide through treewidth. Acta Cybernet. 11, 1–23 (1993)
10. Chrobak, M., Larmore, L.L.: An optimal online algorithm for k-servers on trees. SIAM J. Comput. 20, 144–148 (1991)
11. Dobrev, S., Královič, R., Markou, E.: Online graph exploration with advice. In: Even, G., Halldórsson, M.M. (eds.) SIROCCO 2012. LNCS, vol. 7355, pp. 267–278. Springer, Heidelberg (2012)
12. Dragan, F.F., Yan, C., Corneil, D.G.: Collective tree spanners and routing in at-free related graphs. J. Graph Algorithms Appl. 10(2), 97–122 (2006)
13. Dragan, F.F., Yan, C., Lomonosov, I.: Collective tree spanners of graphs. SIAM J. Discrete Math. 20(1), 241–260 (2006)
14. Emek, Y., Fraigniaud, P., Korman, A., Rosén, A.: Online computation with advice. Theor. Comput. Sci. 412(24), 2642–2656 (2011)
15. Farzan, A., Kamali, S.: Compact navigation and distance oracles for graphs with small treewidth. In: Aceto, L., Henzinger, M., Sgall, J. (eds.) ICALP 2011, Part I. LNCS, vol. 6755, pp. 268–280. Springer, Heidelberg (2011)
16. Forišek, M., Keller, L., Steinová, M.: Advice complexity of online coloring for paths. In: Dediu, A.-H., Martín-Vide, C. (eds.) LATA 2012. LNCS, vol. 7183, pp. 228–239. Springer, Heidelberg (2012)
17. Gupta, A., Kumar, A., Rastogi, R.: Traveling with a pez dispenser (or, routing issues in mpls). SIAM J. Comput. 34(2), 453–474 (2004)
18. Hromkovič, J., Královič, R., Královič, R.: Information complexity of online problems. In: Hliněný, P., Kučera, A. (eds.) MFCS 2010. LNCS, vol. 6281, pp. 24–36. Springer, Heidelberg (2010)

19. Karlin, A.R., Manasse, M.S., McGeoch, L.A., Owicki, S.: Competitive randomized algorithms for non-uniform problems. In: Johnson, D.S. (ed.) SODA 1990, pp. 301–309 (1990)
20. Komm, D., Královič, R., Mömke, T.: On the advice complexity of the set cover problem. In: Hirsch, E.A., Karhumäki, J., Lepistö, A., Prilutskii, M. (eds.) CSR 2012. LNCS, vol. 7353, pp. 241–252. Springer, Heidelberg (2012)
21. Komm, D., Královič, R.: Advice complexity and barely random algorithms. In: Černá, I., Gyimóthy, T., Hromkovič, J., Jefferey, K., Králović, R., Vukolić, M., Wolf, S. (eds.) SOFSEM 2011. LNCS, vol. 6543, pp. 332–343. Springer, Heidelberg (2011)
22. Manasse, M., McGeoch, L.A., Sleator, D.: Competitive algorithms for online problems. In: Simon, J. (ed.) STOC 1988, pp. 322–333 (1988)
23. Renault, M.P., Rosén, A.: On online algorithms with advice for the k-server problem. In: Solis-Oba, R., Persiano, G. (eds.) WAOA 2011. LNCS, vol. 7164, pp. 198–210. Springer, Heidelberg (2012)
24. Yan, C., Xiang, Y., Dragan, F.F.: Compact and low delay routing labeling scheme for unit disk graphs. Comput. Geom. 45(7), 305–325 (2012)

Connected Surveillance Game*

Frédéric Giroire[1], Dorian Mazauric[2], Nicolas Nisse[1],
Stéphane Pérennes[1], and Ronan Soares[1,3]

[1] COATI, INRIA, I3S(CNRS/UNS), Sophia Antipolis, France
[2] ACRO, Laboratoire d'Informatique Fondamentale de Marseille, France
[3] ParGO Research Group, UFC, Fortaleza, Brazil

Abstract. The *surveillance game* [Fomin *et al.*, 2012] models the problem of web-page prefetching as a pursuit evasion game played on a graph. This two-player game is played turn-by-turn. The first player, called the *observer*, can mark a fixed amount of vertices at each turn. The second one controls a *surfer* that stands at vertices of the graph and can slide along edges. The surfer starts at some initially marked vertex of the graph, her objective is to reach an unmarked node The *surveillance number* $sn(G)$ of a graph G is the minimum amount of nodes that the observer has to mark at each turn ensuring it wins against any surfer in G. Fomin *et al.* also defined the *connected surveillance game* where the marked nodes must always induce a connected subgraph. They ask if there is a constant $c > 0$ such that $\frac{csn(G)}{sn(G)} \leq c$ for any graph G. It has been shown that there are graphs G for which $csn(G) = sn(G) + 1$. In this paper, we investigate this question.

We present a family of graphs G such that $csn(G) > sn(G) + 1$. Moreover, we prove that $csn(G) \leq sn(G)\sqrt{n}$ for any n-node graph G. While the gap between these bounds remains huge, it seems difficult to reduce it. We then define the *online surveillance game* where the observer has no *a priori* knowledge of the graph topology and discovers it little-by-little. Unfortunately, we show that no algorithm for solving the online surveillance game has competitive ratio better than $\Omega(\Delta)$.

Keywords: Surveillance game, Cops and robber games, Cost of connectivity, Online strategy, Competitive ratio, Prefetching.

1 Introduction

In this paper, we study two variants of the *surveillance game* introduced in [1]. This two-player game involves one Player moving a mobile agent, called *surfer*, along the edges of a graph, while a second Player, called *observer*, marks the vertices of the graph. The surfer wins if it manages to reach an unmarked vertex. The observer wins otherwise.

Surveillance Game. More formally, let $G = (V, E)$ be an undirected simple n-node graph, $v_0 \in V$, and $k \in \mathbb{N}^*$. Initially, the surfer stands at v_0 which is

---

* This work has been partially supported by European Project FP7 EULER, ANR CEDRE, ANR AGAPE, Associated Team AlDyNet, and project ECOS-Sud Chile.

T. Moscibroda and A.A. Rescigno (Eds.): SIROCCO 2013, LNCS 8179, pp. 68–79, 2013.
© Springer International Publishing Switzerland 2013

marked and all other nodes are not marked. Then, turn-by-turn, the observer first marks k unmarked vertices and then the surfer may move to a neighbor of her current position. Once a node has been marked, it remains marked. The surfer wins if, at some step, she reaches an unmarked vertex; and the observer wins otherwise. Note that the game lasts at most $\lceil \frac{n}{k} \rceil$ turns. When the game is played on a directed graph, the surfer has to follow arcs when it moves [1]. A *k-strategy for the observer from* v_0, or simply a *k-strategy from* v_0, is a function $\sigma : V \times 2^V \to 2^V$ that assigns the set $\sigma(v, M) \subseteq V$ of vertices, $|\sigma(v, M)| \leq k$, that the observer should mark in the *configuration* (v, M), where $M \subseteq V$, $v_0 \in M$, is the set of already marked vertices and $v \in M$ is the current position of the surfer. We emphasis that σ depends implicitly on the graph G, i.e., it is based on the full knowledge of G. A k-strategy from v_0 is *winning* if it allows the observer to win whatever be the sequence of moves of the surfer starting in v_0. The *surveillance number* of a graph G with initial node v_0, denoted by $\mathrm{sn}(G, v_0)$, is the smallest k such that there exists a winning k-strategy starting from v_0.

Let us define some notations used in the paper. Let Δ be the maximum degree of the nodes in G and, for any $v \in V$, let $N(v)$ be the set of neighbors of v. More generally, the neighborhood $N(F)$ of a set $F \subseteq V$ is the subset of vertices of V which have a neighbor in F. Moreover, we define the closed neighborhood of a set F as $N[F] = N(F) \cup F$.

As an example, let us consider the following *basic strategy*: let σ_B be the strategy defined by $\sigma_B(v, M) = N(v) \setminus M$ for any $M \subseteq V$, $v_0 \in M$, and $v \in M$. Intuitively, the basic strategy σ_B asks the observer to mark all unmarked neighbors of the current position of the surfer. It is straightforward, and it was already shown in [1], that σ_B is a winning strategy for any $v_0 \in V$ and it easily implies that $\mathrm{sn}(G, v_0) \leq \max\{|N(v_0)|, \Delta - 1\}$.

Web-Page Prefetching, Connected and Online Variants. The surveillance game has been introduced because it models the web-page prefetching problem. This problem can be stated as follows. A web-surfer is following the hyperlinks in the digraph of the web. The web-browser aims at downloading the web-pages before the web-surfer accesses it. The number of web-pages that the browser may download before the web-surfer accesses another web-page is limited due to bandwidth constraints. Therefore, designing efficient strategies for the surveillance game would allow to preserve bandwidth while, at the same time, avoiding the waiting time for the download of the web-page the web-surfer wants to access.

By nature of the web-page prefetching problem, in particular because of the huge size of the web digraph, it is not realistic to assume that a strategy may mark any node of the network, even nodes that are "far" from the current position of the surfer. For this reason, [1] defines the *connected* variant of the surveillance game. A strategy σ is said *connected* if $\sigma(v, M) \cup M$ induces a connected subgraph of G for any M, $v_0 \in M \subseteq V(G)$. Note that the basic strategy σ_B is connected. The *connected surveillance number* of a graph G with initial node v_0, denoted by $\mathrm{csn}(G, v_0)$, is the smallest k such that there exists a winning connected k-strategy starting from v_0. By definition, $\mathrm{csn}(G, v_0) \geq \mathrm{sn}(G, v_0)$ for any graph G and $v_0 \in V(G)$. In [1], it is shown that there are graphs G and $v_0 \in V(G)$ such that $\mathrm{csn}(G, v_0) = \mathrm{sn}(G, v_0) + 1$. Only the trivial upper bound $\mathrm{csn}(G, v_0) \leq \Delta \, \mathrm{sn}(G, v_0)$ is known and a natural question is how big the gap between $\mathrm{csn}(G, v_0)$ and $\mathrm{sn}(G, v_0)$ may be [1]. This paper provides a partial answer to this question.

Still the connected surveillance game seems unrealistic since the web-browser cannot be asked to have the full knowledge of the web digraph. For this reason, we define the *online surveillance game*. In this game, the observer discovers the considered graph while marking its nodes. That is, initially, the observer only knows the starting node v_0 and its neighbors. After the observer has marked the subset M of nodes, it knows M and the vertices that have a neighbor in M and the next set of vertices to be marked depends only on this knowledge, i.e., the nodes at distance at least two from M are unknown. In other words, an *online strategy* is based on the current position of the surfer, the set of already marked nodes and knowing only the subgraph H of the marked nodes and their neighbors (a more formal definition is postponed to Section 3). By definition, the next nodes marked by such a strategy must be known, i.e., adjacent to an already marked vertex. Therefore, an online strategy is connected. We are interested in the competitive ratio of winning online strategies. The competitive ratio $\rho(\mathcal{S})$ of a winning online strategy $\mathcal{S}$ is defined as $\rho(\mathcal{S}) = \max_{G, v_0 \in V(G)} \frac{S(G, v_0)}{\mathrm{sn}(G, v_0)}$, where $S(G, v_0)$ denotes the maximum number of vertices marked by $\mathcal{S}$ in G at each turn, when the surfer starts in v_0. Note that, because any online winning strategy $\mathcal{S}$ is connected, $\mathrm{csn}(G, v_0) \leq \rho(\mathcal{S}) \, \mathrm{sn}(G, v_0)$ for any graph G and $v_0 \in V(G)$.

1.1   Related Work

The surveillance game has mainly been studied in the computational complexity point of view. It is shown that the problem of computing the surveillance number is NP-hard in split graphs [1]. Moreover, deciding whether the surveillance number is at most 2 is NP-hard in chordal graphs and deciding whether the surveillance number is at most 4 is PSPACE-complete. Polynomial-time algorithms that compute the surveillance number in trees and interval graphs are designed in [1]. All previous results also hold for the connected surveillance number. Finally, it is shown that, for any graph G and $v_0 \in V(G)$, $\max \lceil \frac{|N[S]|-1}{|S|} \rceil \leq \mathrm{sn}(G, v_0) \leq \mathrm{csn}(G, v_0)$ where the maximum is taken over every subset $S \subseteq V(G)$ inducing a connected subgraph with $v_0 \in S$. Moreover, both previous inequalities turn into an equality in case of trees. [1] asks for an example where the inequalities are strict.

In the literature, there are mainly three types of prefetching: server based hints prefetching [2–4], local prefetching [5] and proxy based prefetching [6]. In local prefetching, the client has no aid from the server when deciding which documents to prefetch. In the server based hints prefetching, the server can aid the client to decide which pages to prefetch. Lastly, in the proxy based prefetching, a proxy that connects its clients with the server decides which pages to prefetch. Moreover, some studies consider that the prefetching mechanism has perfect knowledge of the web-surfer's behaviour [7, 8]. In these studies, the objective is to minimize the waiting time of the web-surfer with a given bandwidth, by designing good prediction strategies for which pages to prefetch. In the context of prefetching web-pages, the surveillance game is a model to study a local prefetching scheme to guarantee that a websurfer never has to wait a web-page to be downloaded, whilst minimizing the bandwith necessary to achieve this.

1.2 Our Results

In this paper, we study both the connected and online variants of the surveillance game. First, we try to evaluate the gap between non-connected and connected surveillance number of graphs. We give a new upper bound, independent from the maximum degree, for the ratio csn / sn. More precisely, we show that, for any n-node graph G and any $v_0 \in V(G)$, $\mathrm{csn}(G, v_0) \leq \mathrm{sn}(G, v_0)\sqrt{n}$. Then, we describe a family of graphs G such that $\mathrm{csn}(G, v_0) = \mathrm{sn}(G, v_0) + 2$. Note that, contrary to the simple example that shows that connected and not connected surveillance number may differ by one, a larger difference seems much more difficult to obtain.

As mentioned above, the online variant of the surveillance game is a more constraint version of the connected game. We prove that any online strategy has competitive ratio at least $\Omega(\Delta)$. More formally, we describe a familily of trees with constant surveillance number such that, for any online winning strategy, there is a step when the strategy has to mark at least $\frac{\Delta}{4}$ vertices. Unfortunately, this shows that the best (up to constant ratio) online strategy is the basic one.

2   Cost of Connectedness

In this section, we investigate the cost of the connectivity constraint. We first prove the first non-trivial upper bound for the ratio csn / sn. More precisely, we show that for any n-node graph G, $\mathrm{csn}(G, v_0) \leq \mathrm{sn}(G, v_0)\sqrt{n}$. Then, we improve the lower bound of [1]. That is, we show a family of graphs where $\mathrm{csn}(G, v_0) > \mathrm{sn}(G, v_0) + 1$. Finally, we disprove a conjecture in [1].

2.1   Upper Bound

In this section, we give the first non-trivial upper bound (independent from the degree) of the cost of the connectivity in the surveillance game.

Theorem 1. *Let G be any connected n-node graph and $v_0 \in V(G)$, then*

$$\mathrm{csn}(G, v_0) \leq \mathrm{sn}(G, v_0)\sqrt{n}.$$

Proof. $\mathrm{sn}(G, v_0) = 1$ if and only if G is a path with v_0 as an end. In this case, $\mathrm{csn}(G, v_0) = \mathrm{sn}(G, v_0)$ and the result holds.

Let us assume that $k = \mathrm{sn}(G, v_0) > 1$. We describe a connected strategy σ marking at most $k\sqrt{n}$ nodes per turn. Let $M^0 = \{v_0\}$ and let M^t be the set of vertices marked after $t \geq 1$ turns. Assume moreover that M^t induces a connected graph of G containing v_0. Finally, let v_t be the vertex occupied by the surfer after turn t. The set $\sigma(v_t, M^t)$ of nodes marked by the observer at step $t+1$ is defined as follows. If $|V(G) \setminus M^t| \leq k\sqrt{n}$, then let $\sigma(v_t, M^t) = V(G) \setminus M^t$. Otherwise, let $H \subseteq V(G) \setminus M^t$ such that $|H| = k\sqrt{n}$, $H \cup M^t$ induces a connected subgraph and $|H \cap N(v_t)|$ is maximum. Then, $\sigma(v_t, M^t) = H$, i.e., the strategy marks $k\sqrt{n}$ new nodes in a connected way and, moreover, mark as many unmarked nodes as possible among the neighbors of v_t. In particular, if $|N(v_t) \setminus M^t| \leq k\sqrt{n}$, then all neighbors of v_t are marked after turn $t+1$.

By definition, σ is connected and marks at most $k\sqrt{n}$ nodes per turn. We need to show σ is winning.

For purpose of contradiction, let us assume that the surfer wins against σ by following the path $P = (v_0, \ldots, v_t, v_{t+1})$. At its $t+1^{th}$ turn, the surfer moves from a marked vertex v_t to an unmarked vertex v_{t+1}.

Therefore, $n > tk\sqrt{n}$, otherwise the observer marking $k\sqrt{n}$ nodes at each turn would have already marked every vertex on the graph by the end of turn t. Moreover, by definition of sigma, $|N(v_t) \setminus M^t| > k\sqrt{n}$

Since, $\text{sn}(G, v_0) = k$, let $\mathcal{S}$ be any k-winning (non necessarily connected) strategy for the observer. Assume that the observer follows $\mathcal{S}$ against the surfer following $P \setminus \{v_{t+1}\}$. Since, $\mathcal{S}$ is winning, all vertices of $N(v_t)$ must be marked after turn t, otherwise the surfer would win by moving to an unmarked neighbor of v_t. Therefore, since $\mathcal{S}$ can mark at most k vertices each turn, $|N(v_t)| \leq kt$.

Taking both inequalities, we have that $k\sqrt[3]{n} < |N(v_t)| \leq kt$. Hence, $\sqrt[3]{n} < t$. Therefore, $n > tk\sqrt[3]{n} > nk$, a contradiction. $\square$

2.2   Lower Bound

This section is devoted to proving the following theorem.

Theorem 2. *There exists a family of graphs G and $v_0 \in V(G)$ such that*

$$\text{csn}(G, v_0) > \text{sn}(G, v_0) + 1.$$

We use the following result proved in [1]. For any graph $G = (V, E)$ and any vertex $v_0 \in V$, a k-strategy for G with initial vertex v_0 is winning if and only if it is winning against a surfer that is constrained to follow induced paths on G. In other words, the walk of the surfer is contrained to be an induced path.

In the following theorem, by *adding a path* $P = (v_1, \cdots, v_r)$ *between two vertices u and v of G*, we mean that the induced path P is added as an induced subgraph of G and the edges $\{u, v_1\}$ and $\{v_r, v\}$ are added.

Let x, α, β and γ be four strictly positive integers satisfying the following:

(1) $\max\{\beta, \dfrac{\beta}{2} + \gamma + 1\} < \alpha < \min\{\beta + \gamma + 1, 2\gamma + 2\}$ (2) $\beta < 2\gamma + 2$

(3) $3x \geq \alpha + \beta + 2\gamma + 12$ (4) $x > \dfrac{4}{5}(\alpha + \beta + \gamma) + 10$ (5) $2\alpha \geq 73 + \beta + 2\gamma$.

For instance, $x = 250$, $\alpha = 146$, $\beta = \gamma = 73$ satisfy all the above inequalities.

For proving the main theorem in this section we mainly rely in the family of graphs built in the following the procedure described below.

Let $\mathcal{G} = (V, E)$ be a graph with 10 isolated vertices $\{v_0, w_0, w_1, w_2, w_0', w_1', w_2', s_0, s_1, s_2\}$. Then, for all $i \in \{0, 1, 2\}$ do the following:

1. $4x - 9$ vertices of degree one are added and made adjacent to s_i;
2. $3x - 2$ vertices of degree one are added and made adjacent to w_i, respectively $3x - 2$ neighbors of degree one are added to w_i';
3. two disjoint paths $A^i = (a_1^i, \cdots, a_\alpha^i)$ and $A'^i = (a_1'^i, \cdots, a_\alpha'^i)$ are added between v_0 and s_i;
4. a path $B^i = (b_1^i, \cdots, b_\beta^i)$ is added between v_0 and w_i, and a path $B'^i = (b_1'^i, \cdots, b_\beta'^i)$ is added between v_0 and w_i';
5. for any $j \in \{i, i+1 \mod 3\}$ a path $C^{i,j} = (c_1^{i,j}, \cdots, c_\gamma^{i,j})$ is added between s_j and w_i, and a path $C'^{i,j} = (c_1'^{i,j}, \cdots, c_\gamma'^{i,j})$ is added between s_j and w_i';
6. for any $1 \le j \le \alpha$, $3x - 1$ vertices of degree one are added and made adjacent to a_j^i, respectively $3x - 1$ neighbors of degree one are added to $a_j'^i$;
7. for any $1 \le j \le \beta$, $3x - 1$ vertices of degree one are added and made adjacent to b_j^i, respectively $3x - 1$ neighbors of degree one are added to $b_j'^i$;
8. for any $1 \le j \le \gamma$, $\ell \in \{i, i+1 \mod 3\}$, $3x - 1$ vertices of degree one are added and made adjacent to $c_j^{i,\ell}$, respectively $3x - 1$ neighbors of degree one are added to $c_j'^{i,\ell}$.

The shape of $\mathcal{G}$ is depicted in Figure 1. $\mathcal{G}$ has $(30 + 18(\alpha + \beta) + 36\gamma)x - 29$ vertices. For any $i \in \{0, 1, 2\}$, the node s_i has $4x - 3$ neighbors, v_0 has 12 neighbors, and any other non-leaf node has degree $3x + 1$.

Claim. [9] If $\max\{\beta, \frac{\beta}{2} + \gamma + 1\} < \alpha < \min\{\beta + \gamma + 1, 2\gamma + 2\}$ and $\beta < 2\gamma + 2$, the unique (up to symmetries) minimum Steiner-tree for $S = N[v_0] \cup \{s_0, s_1, s_2\}$ in $\mathcal{G}$ has $15 + \alpha + \beta + 2\gamma$ vertices and consists of the vertices of the paths $A^0, B^1, C^{1,1}, C^{1,2}$ and the vertices in $S \cup \{w_1\}$.

In Fig. 1, the scheme of a minimum Steiner-tree for $S = N[v_0] \cup \{s_0, s_1, s_2\}$ is depicted with dashed lines.

For any $i \in \{0, 1, 2\}$, let $\mathcal{A}_i = N[v_0] \cup N[A^i] \cup N[s_i]$ (resp., $\mathcal{A}_i' = N[v_0] \cup N[A'^i] \cup N[s_i]$). Note that $|\mathcal{A}_i| = |\mathcal{A}_i'| = (3\alpha + 4)x + 9$ and that the $\mathcal{A}_i$ and $\mathcal{A}_j$, $i \ne j$, pairwise intersect only in $N[v_0]$.

For any $i \in \{0, 1, 2\}$, let $\mathcal{B}_i = N[v_0] \cup N[B^i] \cup N[w_i] \cup N[C^{i,i}] \cup N[C^{i,i+1 \mod 3}] \cup N[s_i] \cup N[s_{i+1 \mod 3}]$ and $\mathcal{B}_i'$ is defined similarly. $|\mathcal{B}_i| = |\mathcal{B}_i'| = (3\beta + 6\gamma + 11)x + 5$. Finally, for any $i \in \{0, 1, 2\}$ and $j \in \{i, i+1 \mod 3\}$, let $\mathcal{B}_{i,j} = N[v_0] \cup N[B^i] \cup N[w_i] \cup N[C^{i,j}] \cup N[s_j]$ and $\mathcal{B}_{i,j}' = N[v_0] \cup N[B'^i] \cup N[w_i'] \cup N[C'^{i,j}] \cup N[s_j]$.

Lemma 1. *For any $i \in \{0, 1, 2\}$ and $j \in \{i, i+1 \mod 3\}$, during its first step, any winning $(3x + y)$-strategy for $\mathcal{G}$ must mark at least*

- $x + 8 - y(\alpha + 1)$ *nodes in $\mathcal{A}_i$ (resp., in $\mathcal{A}_i'$), and*
- $x + 8 - y(\beta + \gamma + 2)$ *nodes in $\mathcal{B}_{i,j}$ (resp., in $\mathcal{B}_{i,j}'$), and*
- $2x + 4 - y(\beta + 2\gamma + 3)$ *nodes in $\mathcal{B}_i$ (resp., in $\mathcal{B}_i'$).*

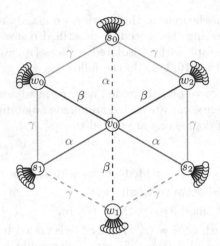

Fig. 1. Graph Family Scheme. Here we show only one "layer" of the graph.

The proof can be found in [9].

Lemma 2. $\mathrm{sn}(\mathcal{G}, v_0) \leq 3x$.

Proof. To show that $\mathrm{sn}(\mathcal{G}, v_0) \leq 3x$, consider the following strategy for the observer. For any $i \in \{0, 1, 2\}$, in the first step, it marks $x - 4$ one-degree neighbots of s_i and the 12 neighbors of v_0. Then, at subsequent step, marks all unmarked neighbors of the current position of the surfer. It is easy to see (see details in [9]), by induction on the number of steps that, each time that the surfer arrives at a new node, this node is marked and has at most $3x$ unmarked neighbors. □

Lemma 3. $\mathrm{csn}(\mathcal{G}, v_0) > 3x + 1$.

Proof. For purpose of contradiction, let us assume that there is a winning connected $3x + 1$-strategy. Let F be the set of vertices marked by this strategy during the first step. Clearly, $N(v_0) \subseteq F$ and $|F| \leq 3x + 1$.

For any $0 \leq i \leq 2$, let $f_i = |F \cap N[s_i]|$ and let $f_{min} = \min_i f_i$. Without loss of generality, $f_{min} = f_0$. We first show that $f_{min} > 3$.

By Lemma 1, for any $i \in \{0, 1, 2\}$, $|F \cap (\mathcal{A}_i \setminus N[v_0])| \geq x - 5 - \alpha$ and, for any $i \in \{0, 2\}$, $|F \cap (\mathcal{B}_{i,0} \setminus N[v_0])| \geq x - 6 - (\beta + \gamma)$ and $|F \cap (\mathcal{B}'_{i,0} \setminus N[v_0])| \geq x - 6 - (\beta + \gamma)$. Therefore,

$$3x + 1 \geq |F \cap (\mathcal{A}_0 \cup \mathcal{A}'_0 \cup \mathcal{A}_1 \cup \mathcal{A}_2 \cup \mathcal{B}_{0,0} \cup \mathcal{B}_{2,0} \cup \mathcal{B}'_{0,0} \cup \mathcal{B}'_{2,0})|$$
$$\geq 12 + 4(x - 5 - \alpha) + 4(x - 6 - (\beta + \gamma)) - 5|F \cap N[s_0]|$$
$$\geq 8x - 4(\alpha + \beta + \gamma) - 32 - 5f_{min}$$

Hence, $5f_{min} \geq 5x - 4(\alpha + \beta + \gamma) - 33$, and $f_{min} \geq x - \frac{4}{5}(\alpha + \beta + \gamma) - 7 > 3$ by the above inequality.

Therefore, by definition of f_{min}, $|F \cap N[s_i]| \geq 4$ for any $i \in \{0, 1, 2\}$. By connectivity of the strategy, $s_i \in F \cap N[s_i]$ for any $i \in \{0, 1, 2\}$. Hence, F must contain a subset of vertices inducing a subtree spanning $S = N[v_0] \cup \{s_0, s_1, s_2\}$. Let T be an inclusion-minimal subset of F that induces a subtree spanning S. By Claim 2.2, $|T| \geq \alpha + \beta + 2\gamma + 15$. Let $T' = T \setminus (N[v_0] \cup \bigcup_{0 \leq i \leq 2} N[s_i])$. Then, $|T'| \geq \alpha + \beta + 2\gamma - 4$. Moreover, because of the symmetries, we may assume w.l.o.g., that $T' \subseteq \bigcup_{0 \leq i \leq 2}(\mathcal{A}_i \cup \mathcal{B}_i)$.

By Lemma 1 and because $N(v_0) \subseteq F$, for any $0 \leq i \leq 2$, $|F \cap (\mathcal{A}'_i \cup \mathcal{B}'_{i+1 \bmod 3})| \geq x + 8 - (\alpha + 1) + 2x + 4 - (\beta + 2\gamma + 3) - 12 = 3x - (\alpha + \beta + 2\gamma) - 4$. Hence, $|T'| + |F \cap (\mathcal{A}'_i \cup \mathcal{B}'_{i+1 \bmod 3})| \geq 3x - 8$. Let $W_i = F \setminus (\mathcal{A}'_i \cup \mathcal{B}'_{i+1 \bmod 3} \cup T')$. Since $|F| \leq 3x + 1$, it follows that $|W_i| \leq 9$.

Let $f_{max} = \max_i f_i$ and assume w.l.o.g. that $f_{max} = f_2$. Since $\sum_{0 \leq i \leq 2} f_i \leq |F \setminus T'|$, we get that $f_0 + f_1 \leq \lfloor \frac{2}{3}(5 + 3x - (\alpha + \beta + 2\gamma)) \rfloor$.

To conclude, $|F \cap \mathcal{B}'_0| = |N(v_0)| + f_0 + f_1 + |W_0| \leq 21 + \lfloor \frac{2}{3}(5 + 3x - (\alpha + \beta + 2\gamma)) \rfloor$. On the other hand, Lemma 1 implies that $|F \cap \mathcal{B}'_0| \geq 2x + 1 - (\beta + 2\gamma)$. Therefore, $22 + \frac{2}{3}(5 + 3x - (\alpha + \beta + 2\gamma)) > 2x + 1 - (\beta + 2\gamma)$ and it follows $73 > 2\alpha - \beta - 2\gamma$. This contradicts the inequalities. □

Lemmas 2 and 3 are sufficient to prove Theorem 2. More precisely, it shows that there exists a family of graphs G and $v_0 \in V(G)$ such that $\text{csn}(G, v_0) \geq \text{sn}(G, v_0) + 2$. However, the family of graphs we described does not allow to increase further the cost of connectivity. Indeed, $\text{csn}(\mathcal{G}, v_0) \leq 3x + 2$ [9].

To conclude this section, we answer negatively a question in [1]. We show that there is a graph $\mathcal{G}$ such that $\text{sn}(\mathcal{G}, s) = k$ and $\max_{S \subseteq V(\mathcal{G})} \lceil \{ \frac{|N[S]| - 1}{|S|} \} \rceil < k$ [9].

3   Online Surveillance Number

In this section, we study the online variant of the surveillance game motivated by the web-page prefetching problem where the observer (the web-browser) discovers new nodes through hyperlinks in already marked nodes. In this variant, the observer does not know a priori the graph in which it is playing. That is, initially, the observer only knows v_0 and the identifiers of its neighbors. Then, when a new node is marked, the observer discovers all its neighbors that are not yet marked. Note that the degree of a node is not known before it is marked.

Another property of an online strategy that must be defined concerns the moment when the observer discovers the unmarked neighbors of a node that it has decided to mark. There are two natural models. Assume that the set M of nodes have been marked and this is the turn of the observer, and let $N(M)$ be the set of nodes with a neighbor in M. Either it first chooses the k nodes that will be marked among the set $N(M) \setminus M$ of the unmarked neighbors of the nodes that were already marked and then the observer marks each of these k nodes and discover their unknown neighbors simultaneously. Or, the observer first chooses one node x in $N(M) \setminus M$, marks it and discovers its unmarked neighbors, then it chooses a new node to be marked in $N(M \cup \{x\}) \setminus (M \cup \{x\})$ and so on until the observer finishes its turn after marking k nodes. We choose to consider the second model because it is less restricted, i.e., the observer has more power, and, even in this case, our result is pessimistic since we show that the basic strategy is the best one with respect to the competitive ratio.

Formal Definition of Online Strategy. Now we are ready to formally define an online strategy. Let $k \geq 1$, let $G = (V, E)$ be a graph, $v_0 \in V$, and let $\mathcal{G}$ be the set of subgraphs of G.

Given $M \subseteq V$ be a subset of nodes inducing a connected subgraph containing v_0 in G. Let $G_M \in \mathcal{G}$ be the subgraph of G known by the observer when M is the set of marked nodes. That is, $G_M = (M \cup N(M), E_M)$ where $E_M = \{(u, v) \in E \mid u \in M\}$. For any $u, v \in N(M) \setminus M$, let us set $u \sim_M v$ if and only if $N(u) \cap M = N(v) \cap M$. Let χ_M be the set of equivalent classes, called *modules*, of $N(M) \setminus M$ with respect to $\sim_M$. The intuition is that two nodes in the same module of χ_M are known by the observer but cannot be distinguished. For instance, $\chi_{\{v_0\}} = \{N(v_0)\}$.

A *k-online strategy for the observer starting from* v_0 is a function $\sigma : \mathcal{G} \times V \times 2^V \times \{1, \cdots, k\} \to 2^V$ such that, for any subset $M \subseteq V$ of nodes inducing a connected subgraph containing v_0 in G, for any $v \in M$, and for any $1 \leq i \leq k$, then $\sigma(G_M, v, M, i) \in \chi_M$. This means that, if M is the set of nodes already marked and thus the observer only knows the subgraph G_M, if v is the position of the surfer and it remains $k - i + 1$ nodes to be marked by the observer before the surfer moves, then the observer will mark one node in $\sigma(G_M, v, M, i)$.

More precisely, we say that the observer *follows* the k-online strategy σ if the game proceeds as follows. Let $M = M^0$ be the set of marked nodes just after the surfer has moved to $v \in M$. Initially, $M^0 = \{v_0\}$ and $v = v_0$. Then, the strategy proceeds sequentially in k steps for $i = 1$ to k. First, the observer marks an arbitrary node $x_1 \in \sigma(G_{M^0}, v, M^0, 1)$. Let $M^1 = M^0 \cup \{x_1\}$. Sequentially, after having marked $1 < i < k$ nodes at this turn, the observer marks one arbitrary node $x_{i+1} \in \sigma(G_{M^i}, v, M^i, i+1)$ and $M^{i+1} = M^i \cup \{x_{i+1}\}$. When the observer has marked k nodes, that is after choosing $x_k \in \sigma(G_{M^{k-1}}, v, M^{k-1}, k)$, it is the turn of the surfer, when it may move to a node adjacent to its current position and then a new turn for the observer starts. Note that because we are interested in the worst case for the observer, each marked node $x_i \in \sigma(G_{M^{i-1}}, v, M^{i-1}, i)$ is chosen by an adversary.

The *online surveillance number* of a graph G with initial node v_0, denoted by $\mathrm{on}(G, v_0)$, is the smallest k such that there exists a winning k-online strategy starting from v_0. In other words, there is a winning k-online strategy σ starting from v_0 such that an observer following σ wins whatever be the trajectory of the surfer and the choices done by the adversary at each step. Note that, since we consider the worst scenario for the observer, we may assume that the surfer has full knowledge of G.

Theorem 3. *There exists an infinite family of rooted trees such that, for any T with root $v_0 \in V(T)$ in this family, $\mathrm{sn}(T, v_0) = 2$ and $\mathrm{on}(T, v_0) = \Omega(\Delta)$ where Δ is the maximum degree of T.*

Proof. We first define the family $(T_k)_{k \geq 1}$ of rooted trees as follows.

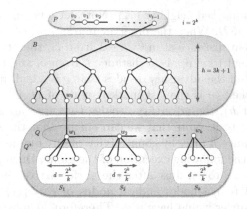

Fig. 2. Tree T_k described in the proof of Theorem 3

Let $k \geq 4$ be a power of two and let $i = 2^k$ and $d = \frac{2^k}{k}$.

Let us consider a path $P = (v_0, v_1, \ldots, v_{i-1})$ with i nodes Let B be a complete binary tree of height $h = 3k+1$ and rooted at some vertex v_i, i.e., B has $2^{h+1} - 1$ vertices. Let w_0 be any leaf of B. Finally, let $Q = (w_1, \ldots, w_k)$ be a path on k nodes. Note that, P, B and Q depend on k.

The tree T_k is obtained from P, B and Q by adding an edge between v_{i-1} and v_i, an edge between w_0 and w_1. Finally, for any $1 \leq j \leq k$, let us add an independent set, S_j, with d vertices and an edge between each vertex of S_j and w_j (i.e., each node in S_j is a leaf). T_k is then rooted in v_0.

Let Q^+ denote the union of vertices of Q and $\bigcup_{j=1}^{k} S_j$. The maximum degree Δ of T_k is reached by any node w_j, $1 \leq j < k$, and $\Delta = d + 2 = \frac{2^k}{k} + 2$.

Clearly, $\text{sn}(T_k, v_0) > 1$. We show that $\text{sn}(T_k, v_0) = 2$.

Consider the following (offline) strategy for the observer. At each turn $j \leq i$, the surfer marks the vertex v_j and one unmarked vertex of Q^+ that is closest to the surfer. For each turn $j > i$ and while the surfer does not occupy a node in $Q^+ \cup \{w_0\}$, the observer marks the neighbors of the current position of the surfer if they are not already marked. Finally, if the surfer occupies a node in $Q^+ \cup \{w_0\}$, the observer marks two unmarked nodes of Q^+ that are closest to the surfer. It is easy to see, by induction on the number of steps that, each time that the surfer arrives at a new node, this node is marked and has at most 2 unmarked neighbors. Hence, $\text{sn}(T_k, v_0) = 2$.

Now it remains to show that $\text{on}(T_k, v_0) = \Omega(\Delta)$. Let γ be any online strategy for T_k and marking at most $\frac{d}{4} = \frac{2^{k-2}}{k}$ nodes per turn. We show that γ fails.

For this purpose, we model the fact that the observer does not know the graph by "building" the tree during the game. More precisely, each time the observer marks a node v, then the adversary may add new nodes adjacent to v or decide that v is a leaf. Of course, the adversary must satisfy the constraint that eventually the graph is T_k. Initially, the observer only knows v_0 that has one neighbor v_1. Now, for any $1 \leq j < i$, when the observer marks the node v_j of P, then the adversary "adds" a new node v_{j+1} adjacent to v_j, i.e., the observer discovers its single unmarked neighbor v_{j+1}. Now, let v be any node of B. Recall that h is the height of B. When the observer marks v, there are three cases to be considered: if v is at distance at most $h-1$ from v_i, then the adversary adds two new nodes adjacent to v; if v is at distance h from v_i and not all nodes of B have been marked then the adversary decides that v is a leaf; finally, if all nodes of B have been marked (v is the last marked node of B, i.e., B is a complete binary tree of height h), the adversary decides that $v = w_0$ and add one new neighbor w_1 adjacent to it. Note that we can ensure that the last node of B to be marked is at distance h of v_i by connectivity of any online strategy.

Now, let consider the following execution of the game. During the first i steps, the surfer goes from v_0 to v_i. Just after the surfer arrives in v_i, the observer has marked at most $(di)/4$ nodes and all nodes of $P \cup \{v_i\}$ must be marked since otherwise the surfer would have won. Therefore, at most $i(d/4-1)+1 = 2^{2k-2}/k - 2^k + 1$ nodes of B are marked when it is the turn of the surfer at v_i. Since B has $2^{h+1}-1 = 2^{3k+2}-1$ nodes, at least one node of B is not marked.

From v_i, the surfer always goes toward w_0. Note that the observer may guess this strategy but it does not know where is w_0 while all nodes of B have not been marked.

Then let $0 \le t \le h$ and let $v'_t \in V(B)$ be the position of the surfer at step $i+t$ and B^t the subtree of B rooted at v'_t. Note that, at step i, $v'_0 = v_i$ and $B^0 = B$. Let B^t_l and B^t_r be the subtrees of B rooted at the children of v'_t. W.l.o.g., let us assume that the number of marked nodes in B^t_l is at most the number of marked nodes in B^t_r, when it is the turn of the surfer standing at v'_t. Then, the surfer moves to the root of B^t_l. That is, v'_{t+1} is the child of v_t whose subtree contains the minimum number of marked nodes.

Let m_t be the number of marks in the subtree of B rooted at v'_t when it is the turn of the surfer at v'_t. Since, at beginning of step i there are at most $2^{2k-2}/k - 2^k + 1$ nodes of B that are marked and $k \ge 4$, $m_0 \le 2^{2k-2}/k - 2^k + 1 \le 2^{2k-2}/k$. Note that, for any $t > 0$, $m_t \le (m_{t-1}-1+\frac{d}{4})/2 \le (m_{t-1}+\frac{d}{4})/2$. Simply expanding this expression we get that, for any $t > 0$,

$$m_t \le \frac{m_0}{2^t} + \frac{2^k}{k}\sum_{j=3}^{t+2}2^{-j} \le \frac{2^{2k-(t+2)}}{k} + \frac{2^k}{k}\sum_{j=3}^{t+2}2^{-j}.$$

Therefore, for any $t \ge 2k$:

$$m_t \le \frac{1}{k} + \frac{2^k}{k}\sum_{j=3}^{t+2}2^{-j} \le \frac{2^k+1}{k}.$$

In particular, at step $i+2k$ (when it is the turn of the surfer), the surfer is at v'_{2k} which is at distance $k+1$ from w_0. Hence, $|B^{2k}| \ge 2^{k+1}-1$ and at most $\frac{2^k+1}{k} < 2^{k+1}-1$ of its nodes are marked. Hence, w_0 neither no nodes in Q^+ are marked.

From this step, the surfer directly goes to w_k unless she meets an unmarked node, in which case she goes to it and wins. When the surfer is at w_k and it is her turn, the observer may have marked at most $(2k+2)\frac{d}{4} \le \frac{kd}{2} + \frac{d}{2} \le 2^{k-1} + \frac{2^{k-1}}{k}$ nodes in Q^+. Since $|Q^+| = (d+1)k = 2^k + k$ and $k \ge 4$, at least one neighbor of w_k is not marked yet and the surfer wins. $\square$

Theorem 3 implies that, for any online strategy $\mathcal{S}$, $\rho(\mathcal{S}) = \Omega(\Delta)$. Recall that the basic strategy $\mathcal{B}$, that marks all unmarked neighbors of the surfer at each step, is an online strategy. $\mathcal{B}$ has trivially competitive ratio $\rho(\mathcal{B}) = O(\Delta)$. Hence, no online winning strategy has better competitive ratio than the basic strategy up to a constant factor. In other words:

Corollary 1. *The best competitive ratio of online winning strategies is $\Theta(\Delta)$, with Δ the maximum degree.*

As mentioned in the introduction, any online strategy is connected and therefore, for any graph G and $v_0 \in V(G)$, $\mathrm{csn}(G, v_0) \le \mathrm{on}(G, v_0)$. Moreover, we recall that, for any tree T and for any $v_0 \in V(T)$, $\mathrm{csn}(T, v_0) = \mathrm{sn}(T, v_0)$ [1]. Hence, there might be an arbitrary gap between $\mathrm{csn}(G, v_0)$ and $\mathrm{on}(G, v_0)$.

4   Conclusion

Despite our results, the main question remains open. Can the difference or the ratio between the connected surveillance number of a graph and its surveillance number be bounded by some constant?

References

1. Fomin, F.V., Giroire, F., Jean-Marie, A., Mazauric, D., Nisse, N.: To satisfy impatient web surfers is hard. In: FUN, pp. 166–176 (2012)
2. Aumann, Y., Etzioni, O., Feldman, R., Perkowitz, M.: Predicting event sequences: Data mining for prefetching web-pages (1998)
3. Albrecht, D., Zukerman, I., Nicholson, A.: Pre-sending documents on the www: a comparative study. In: 16th Int. Joint Conf. on Artificial Intelligence, pp. 1274–1279 (1999)
4. Mogul, J.C.: Hinted caching in the web. In: 7th Workshop on ACM SIGOPS European Workshop: Systems Support for World Wide Applications, pp. 103–108 (1996)
5. Wang, Z., Lin, F.X., Zhong, L., Chishtie, M.: How far can client-only solutions go for mobile browser speed? In: 21st Int. Conf. on World Wide Web, pp. 31–40 (2012)
6. Fan, L., Cao, P., Lin, W., Jacobson, Q.: Web prefetching between low-bandwidth clients and proxies: potential and performance. In: ACM SIGMETRICS Int. Conf. on Measurement and Modeling of Computer Systems, pp. 178–187 (1999)
7. Padmanabhan, V.N., Mogul, J.C.: Using predictive prefetching to improve world wide web latency. SIGCOMM Comput. Commun. Rev. 26(3), 22–36 (1996)
8. Kroeger, T.M., Long, D.D.E., Mogul, J.C.: Exploring the bounds of web latency reduction from caching and prefetching. In: USENIX Symposium on Internet Technologies and Systems, p. 2 (1997)
9. Giroire, F., Mazauric, D., Nisse, N., Pérennes, S., Pardo Soares, R.: Connected Surveillance Game. Rapport de recherche RR-8297. INRIA (2013)

Non-Additive Two-Option Ski Rental

Amir Levi and Boaz Patt-Shamir

Dept. of Electrical Engineering, Tel Aviv University
Tel Aviv 69978, Israel
{amirlevi,boazps}@tau.ac.il
http://www.eng.tau.ac.il/algs-lab

Abstract. We consider a generalization of the classical problem of ski rental. There is a game that ends at an unknown time, and the algorithm needs to decide how to pay for the time until the game ends. There are two "payment plans" or "options," such that the cost of t time units under option i (for $i = 1, 2$) is given by $a_i t + b_i$, where b_i is a one-time cost to start using option i, and a_i is the ongoing cost per time unit. We assume w.l.o.g. that $a_1 > a_2$ and $b_1 < b_2$. (The classical version is $b_1 = 0$ and $a_2 = 0$, so that option 1 is "pure rent" and option 2 is "pure buy.") We give deterministic and randomized algorithms for the general setting and prove matching lower bounds on the competitive ratios for the problem. This is the first non-trivial result for the non-additive model of ski rental, which models non-refundable one-time costs.

Keywords: Online computation, competitive analysis.

1 Introduction and Summary

To commit or not to commit? This classical dilemma is formalized quantitatively by the classical "ski rental" problem [6]: one alternative is better for the long term, another for the short term, and only the future will tell us which is the right choice. Ski rental is an elementary on-line problem, as it allows us to better understand how to minimize the cost of predicting time duration. This abstraction is useful in many computer-related scenarios, e.g., snoopy caching and TCP acknowledgment batching (see [7]), but obviously it applies also to many real-life situations, e.g., payment plans [4].

The basic setting is as follows. We are given two ways to pay for some resource we need. In the "buy" option there is a one-time fee and that's it, and in the "rent" option, we pay proportionally to the actual usage. (These options are sometimes called "slopes".) The algorithm needs only decide how to pay for the usage, which boils down to decide if and when to switch from the rent to the buy option, and the challenge is that the duration of the time we need to pay for is determined on-line, i.e., it is unknown in advance. From the competitive analysis point of view [3], one would like to bound the competitive ratio, namely the worst-case ratio, over all possible instances, of the cost paid by the algorithm to the optimal cost (which can be known only in hindsight). It is straightforward

T. Moscibroda and A.A. Rescigno (Eds.): SIROCCO 2013, LNCS 8179, pp. 80–91, 2013.

to see that the deterministic competitive ratio is 2 for this setting. A deeper result shows that the randomized competitive ratio is $e/(e-1) \approx 1.58$ [8]. Intuitively, it turns out that it is a good idea to guess when will the game end; the key is the distribution of the guesses.

Recently there was some renewed interest in the problem, motivated by power-saving models: Augustin et al. [1] mapped the "buy" and "rent" options to different operational modes of a system where cost models energy consumption. Inspired by this correspondence, they generalized the problem to include options whose associated cost is a general linear function of time, i.e., any one-time fee and any on-going payment rate. Formally, using an (a, b)-option for t time units in this model has cost $at + b$, where $a, b \geq 0$ are given constants.

To formalize generalized ski rental, one has to state precisely what happens when the algorithm switches from one option to another. Augustin et al. [1] distinguish between two variants of this issue: in the *additive* model, the immediate cost of switching from an (a, b) option to an (a', b') option is $b' - b$, namely the one-time cost paid in the past is deducted from future one-time cost. In the *non-additive* model, by contrast, it is assumed that there is an arbitrary transition cost for switching between any two options.

In this work we present the first complete study of a non-trivial variant of non-additive ski rental. Our model may be the simplest ski rental problem which is not additive. Specifically, the problem we study, called NA2SSR, is defined next.

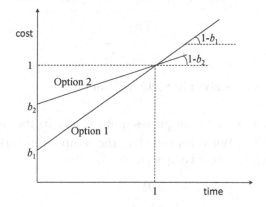

Fig. 1. A graphic representation of the input. Option i intersects the y axis at b_i and its slope is $1 - b_i$.

The Non-Additive Two-Slope Ski Rental Problem (NA2SSR). We are given two options: the payment for t time units under option i is $a_i t + b_i$ for $i = 1, 2$. The algorithm may either start with Option 1 and possibly at some time switch Option 2, or immediately start with Option 2. We assume that $a_1 > a_2$ and $b_1 < b_2$ (so that Option 1 is more "rent" and Option 2 is closer to "buy"). The cost to the algorithm for using option i for t_i time is $a_i t_i + b_i$, where t_i is the length of time Option i was in use. To avoid multiple redundant parameters, and without loss of generality, we shall assume that $0 \leq b_1 < b_2 \leq 1$, and that the $a_i = 1 - b_i$ for $i = 1, 2$ (see Figure 1).[1] Thus, NA2SSR is fully specified by the two parameters, b_1 and b_2.

Intuitively, our model formalizes simple situations in which changing one's mind, even instantly, has a cost. In the additive model, the cost of starting with Option 1 and switching to Option 2 almost immediately is almost the same as starting with Option 2. However, in the non-additive model this is not the case: If the algorithm starts with Option 1 and switches to Option 2 almost immediately, the resulting cost is at least b_1 units more than the cost of starting with Option 2 upfront. Intuitively, the one-time cost b_1 is "non-refundable."

Our results. In this paper we determine the deterministic and randomized competitive ratios of (b_1, b_2)-NA2SSR, for $0 \leq b_1 < b_2 \leq 1$. Specifically, we prove the following results, each stating both an upper and a lower bound.

Theorem 1. *The deterministic competitive ratio of (b_1, b_2)-NA2SSR is*

$$\min \left\{ b_2 + 1, \frac{b_2}{b_1}, \frac{1 - b_1}{1 - b_2} \right\}.$$

The proof of this result is given in Section 2.

Theorem 2. *The randomized competitive ratio of (b_1, b_2)-NA2SSR is*

$$\frac{e^{1 - \frac{b_1}{b_2}}}{e^{1 - \frac{b_1}{b_2}} - b_2 + b_1}.$$

The proof of this result is given in Sections 3 and 4.

Notation. The following notation proves quite handy in the remainder of the paper: $B \stackrel{\text{def}}{=} e^{(b_2 - b_1)/b_2}$. With this notation, the competitive ratio of Theorem 2 can be rewritten slightly more compactly as

$$\frac{B}{B - b_2 + b_1}.$$

It may be pleasing to observe that in the classical case, where $b_1 = 0$ and $b_2 = 1$, we have that $B = e$ and the competitive ratio attains its maximum, the well-known value of $e/(e - 1)$. Figure 2 depicts the optimal competitive ratio as a function of b_1 and b_2. We note here that all calculations in this paper (and the plots in Figure 2) were carried out mechanically using Mathematica software [12].

---

[1] Note that this can always be attained by change of units in time and cost, under the assumption that $a_1 > a_2$ and $b_1 < b_2$. If this assumption does not hold then the instance is trivial, i.e., one option is always better than the other.

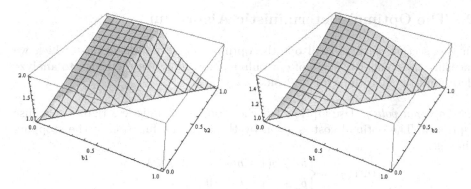

Fig. 2. *The competitive ratio of* NA2SSR *as a function of b_1 and b_2. Left: Deterministic algorithms. Right: Randomized algorithms. Note that the competitive ratio is 1 when $b_1 = b_2$, and that the worst-case value for randomized algorithms is $\frac{e}{e-1} \approx 1.58$, when $b_1 = 0$ and $b_2 = 1$.*

Related work. For general non-additive ski rental instance, the best published results we know of are as follows. Augustine et al. [1] give an optimal deterministic algorithm (derived by dynamic programming), whose exact competitive ratio is unknown. Fujiwara et al. [5] give bounds on deterministic competitive ratio of the worst and best instances with a given number of slopes, assuming that there are always pure-rent and pure-buy options. They argue that the "easiest" instances are additive, and that the "hardest" instances are the ones where the one-time cost of one option does not deduct the one-time cost of any other option. It has been observed that any non-additive ski rental problem can be solved using the generic "repeated doubling" technique, which yields a 4-competitive deterministic algorithm [2] and an e-competitive randomized algorithm [11]. These algorithms are in some sense oblivious to the particular structure of the given instance, and are therefore non-optimal in many cases.

The additive model appears more amenable to analysis: [11] presents an algorithm which computes the optimal randomized ski-rental algorithm for any given additive instance.[2] The value of the resulting optimal competitive ratio is not know in general (it's never worse than $\frac{e}{e-1}$). One exception is the special case of two options: one is pure rent, and the other is general (one time fee and on-going payment). In our notation, this is an instance of NA2SSR with $b_1 = 0$ and $b_2 \in [0, 1]$. For this case, [10] shows that the competitive ratio is exactly (upper- and lower-bounded) $\frac{e}{e-b_2}$, coinciding with our more general bounds.

Organization. The remainder of this paper is organized as follows. In Section 2 we discuss the optimal (offline) strategy and derive an optimal deterministic algorithm. In Section 3 we prove the upper bound of Theorem 2, and in Section 4 we prove the lower bound.

---

[2] A randomized algorithm for ski rental is a sequence of probability distributions of when to switch from each slope to another. The algorithm in [11] computes these optimal distributions.

2   The Optimal Deterministic Algorithm

In this section we first spell out the optimal offline policy, against which we measure the competitiveness of our online algorithms. We then turn to analyze deterministic algorithms for NA2SSR.

The optimal policy: Use Option 1 if the stopping time is less than 1, else use Option 2. The optimal cost is given by the following function of the stopping time y:

$$\text{OPT}(y) = \begin{cases} b_1 + y(1 - b_1) & \text{if } 0 \leq y \leq 1, \text{ and} \\ b_2 + y(1 - b_2) & \text{if } y > 1. \end{cases}$$

Deterministic algorithms. We now prove Theorem 1. To do that, we analyze the competitive ratios of all deterministic online algorithms, and show that the competitive ratio of all of them is at least $\min\left\{b_2 + 1, \frac{b_2}{b_1}, \frac{1-b_1}{1-b_2}\right\}$.

A deterministic algorithm is characterized by its time of switching from Option 1 to Option 2. We denote the algorithm that switches at time z by A_z. We proceed by case analysis.

Consider first A_0, the algorithm that immediately uses Option 2 and never uses Option 1. This algorithm is optimal if the stopping time t satisfies $t \geq 1$. For $0 \leq t \leq 1$, the competitive ratio is $\frac{b_2 + (1-b_2)t}{b_1 + (1-b_1)t}$ which is decreasing in t, so the worst-case competitive ratio of A_0 is attained at $t = 0$, with the value b_2/b_1.

Next, consider an algorithm that switches at time $0 < z \leq 1$. A_z is optimal if the stopping time satisfies $t < z$, and for $t \geq z$ the competitive ratio of A_z is maximized at $t = z$ for the value of

$$\frac{b_1 + (1 - b_1)z + b_2}{b_1 + (1 - b_1)z} = 1 + \frac{b_2}{b_1 + (1 - b_1)z} \geq 1 + b_2$$

because $b_1 + (1 - b_1)z \leq 1$ for $z \leq 1$.

Now consider an algorithm A_z that switches at time $z > 1$. A_z is optimal if the stopping time satisfies $t \leq 1$, and for $t \geq 1$ the competitive ratio of A_z is maximized at $t = z$ for the value of

$$\frac{b_1 + (1 - b_1)z + b_2}{b_2 + (1 - b_2)z} = 1 + \frac{b_1 + (b_2 - b_1)z}{b_2 + (1 - b_2)z} . \tag{1}$$

Differentiating (1), we find that the competitive ratio, as a function of z, is increasing if $b_2^2 > b_1$ and decreasing if $b_2^2 < b_1$. Suppose first that $b_2^2 > b_1$. In this case we have that the competitive ratio of A_z is lower-bounded at the lower end of the range of z, namely at $z = 1$, where the ratio is $1 + b_2$. Finally, if $b_2^2 < b_1$, the competitive ratio is lower bounded by the limit as $z \to \infty$ (i.e., by the competitive ratio of A_∞, the algorithm that never switches to Option 2). Therefore, the competitive ratio in this case is at least

$$\lim_{z \to \infty} \left(\frac{b_1 + (1 - b_1)z + b_2}{b_2 + (1 - b_2)z} \right) = \frac{1 - b_1}{1 - b_2},$$

assuming that $b_2 < 1$. Note that if $b_2^2 < b_1$, then $\frac{1-b_1}{1-b_2} \leq \frac{1-b_2^2}{1-b_2} = 1 + b_2$. Note also that $\frac{1-b_1}{1-b_2} \leq \frac{b_2}{b_1}$ iff $b_2 < 1 - b_1$. We can therefore conclude that the optimal deterministic algorithm is as follows.

(1) If $b_1 \geq b_2^2$ and $b_1 \geq 1 - b_2$, then A_∞ is optimal.
(2) If $b_1 \leq b_2^2$ and $b_1 \leq \frac{b_2}{1+b_2}$, then A_1 is optimal.
(3) If $b_1 \geq \frac{b_2}{1+b_2}$ and $b_1 \leq 1 - b_2$, then A_0 is optimal.

In case of ties, we can choose arbitrarily (the algorithms guarantee the same competitive ratio in these cases). This concludes the proof of Theorem 1.

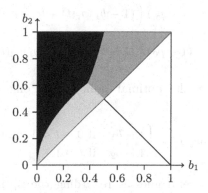

Fig. 3. *Partition of the unit square to regions according to the conditions in Theorem 1. The lines are $b_1 = b_2, b_1 = 1 - b_2, b_1 = b_2^2$ and $b_1 = \frac{b_2}{b_2+1}$. A_0 is optimal in the blue region (top right), A_1 in the gray region (top left), and A_∞ in the pink region (left bottom).*

Figure 3 shows the partition of the (b_1, b_2) plane into regions according to the optimality of the deterministic algorithm (this is a top projection of the left chart in Figure 2). We note that when $b_2 = \frac{-1+\sqrt{5}}{2} \approx 0.618$ and $b_1 = 1 - b_2 \approx 0.382$ all conditions are satisfied and all three algorithms are optimal. (In this case the competitive ratio is the golden ratio $\phi \approx 1.618$.)

Finally, we note that unlike additive ski rental, in NA2SSR the optimal deterministic algorithm depends on the parameters of the problem. In the additive version of 2 slopes, we have that $b_1 = 0$ and $b_2 > 0$, and hence A_0 and A_∞ are never optimal.

3   The Optimal Randomized Algorithm

In this section we derive an optimal randomized algorithm step-by-step, thus proving the upper bound of Theorem 2. We model the optimal algorithm as follows. First, there is some probability that the algorithm starts with Option 2. Let p_0 denote that probability. Next, let $p_1(t)$ be the probability of being at Option 1 by time $t \geq 0$, *conditioned on the event that the algorithm did not start with Option 2*. Thus $p_1(0) = 1$. We shall focus on $p_2(t) \stackrel{\text{def}}{=} 1 - p_1(t)$, namely

$p_2(t)$ is the probability that the algorithm is at Option 2 at time t, given that the algorithm did not start with Option 2.

To find explicit expressions for p_0 and p_2, we consider the *cost-spending rate* of the algorithm at time t. This cost has two components: the expected spending on one-time cost toward switching to Option 2, and the ongoing cost, apportioned to Options 1 and 2. The expected ongoing cost rate at time t is precisely $p_0(1 - b_2) + (1 - p_0)(p_1(t)(1 - b_1) + p_2(t)(1 - b_2))$. The instantaneous rate of spending due to the one-time cost at time t is precisely $(1 - p_0)b_2 \frac{d}{dt} p_2(t)$. Therefore, the expected rate of cost to the algorithm for $t \geq 0$ is:

$$\frac{d}{dt} \text{ALG}(t) = p_0(1 - b_2) + (1 - p_0)\left((1 - b_1)p_1(t) + b_2\frac{d}{dt}p_2(t) + (1 - b_2)p_2(t)\right)$$

$$= p_0(1 - b_2) + (1 - p_0)\left((1 - b_1) - (b_2 - b_1)p_2(t) + b_2\frac{d}{dt}p_2(t)\right). \quad (2)$$

On the other hand, for the optimal policy, only ongoing costs affect the derivative:

$$\frac{d}{dt}\text{OPT}(t) = \begin{cases} 1 - b_1 & \text{if } 0 < t < 1, \text{ and} \\ 1 - b_2 & \text{if } t > 1. \end{cases} \quad (3)$$

To proceed, we make several guesses regarding the optimal solution. These guesses are justified by arriving at a solution and verifying its properties. First, we guess that in the optimal solution, for all $t \geq 0$ we have a constant competitive ratio $c \stackrel{\text{def}}{=} \text{ALG}(t)/\text{OPT}(t)$. Rearranging and differentiating, this guess takes the following form.

$$\frac{d}{dt}\text{ALG}(t) = c \cdot \frac{d}{dt}\text{OPT}(t). \quad (4)$$

We now proceed by cases, to solve for p_0, p_1, p_2 and c.

Case 1: $0 \leq t \leq 1$. Plugging the expressions of (2) and (3) for this case in (4), we obtain the differential equation

$$\frac{d}{dt}p_2(t) - \frac{b_2 - b_1}{b_2} \cdot p_2(t) = \frac{p_0(b_2 - b_1) + (1 - b_1)(c - 1)}{b_2(1 - p_0)}$$

which we can easily solve for $p_2(t)$:

$$p_2(t) = \frac{p_0(b_2 - b_1) + (1 - b_1)(c - 1)}{(b_2 - b_1)(1 - p_0)} \cdot (B^t - 1), \quad (5)$$

where the invisible 1-factor of B^t was determined by the boundary condition $p_2(0) = 0$.

Case 2: $t \geq 1$. We now make another guess: that the probability of switching to Option 2 after time 1 vanishes, namely $p_2'(t) = 0$ for all $t > 1$. Plugging the expressions from (2) and (3) in (4) we obtain another expression for $p_2(t)$, for $t \geq 1$:

$$p_2(t) = \frac{(1 - b_1) - (1 - b_2)c - p_0(b_2 - b_1)}{(1 - p_0)(b_2 - b_1)} . \qquad (6)$$

(Note that $p_2(1)$ is the final probability that the algorithm is in Option 2, given that it didn't start with Option 2.) Assigning $t \leftarrow 1$ in (5) and in (6), we obtain an expression for c in terms of b_1, b_2 and p_0:

$$c = \frac{B\left((1 - b_1) - p_0(b_2 - b_1)\right)}{B(1 - b_1) - (b_2 - b_1)} . \qquad (7)$$

Special case: $t = 0$. For $t = 0$ we have directly from definitions that

$$\text{ALG}(0) = p_0 b_2 + (1 - p_0)b_1 \quad \text{and}$$
$$\text{OPT}(0) = b_1 ,$$

which means that $c = (p_0 b_2 + (1 - p_0)b_1)/b_1$. Using also (7) we can solve for p_0, and then obtain a simplified expression for the competitive ratio c. This results in the following solution.

$$p_0 = \frac{b_1}{B - (b_2 - b_1)} , \text{ and} \qquad (8)$$

$$c = \frac{B}{B - (b_2 - b_1)} . \qquad (9)$$

We summarize with a complete description of the algorithm.
Recall that $B = \exp(1 - \frac{b_1}{b_2})$.

Algorithm A^*: a randomized solution to NA2SSR.

1. With probability p_0 (cf. (8)), start with Option 2.
2. Else (with probability $1 - p_0$):
   (a) Choose a number $t \in [0, 1]$ using the cumulative probability function p_2, described by (5), which simplifies by (8) and (9) to

   $$p_2(t) = \frac{B^t - 1}{B - b_2} . \qquad (10)$$

   With probability $p_1(1) = 1 - p_2(1) = \frac{1 - b_2}{B - b_2}$, no number is chosen. In this case we set $t \leftarrow \infty$.
   (b) Start with Option 1, and switch to Option 2 at the chosen time t if the game is not over yet. If $t = \infty$, the algorithm never switches to Option 2.

Verification. To justify our guesses, we now verify the competitive ratio of our algorithm. This means we need to show that the expected cost to the algorithm, for any stopping time z of the game, is at most $\frac{B}{B+b_1-b_2}$ times the optimal cost. (We note again that the calculations were done mechanically [12].)

Consider first the special case where the game stops at time 0. In this case the expected cost to the algorithm is

$$A^*(0) = p_0 b_2 + (1-p_0)b_1 = \frac{b_1 B}{B + b_1 - b_2},$$

and since $\text{OPT}(0) = b_1$, the competitive ratio in this case is $\frac{B}{B+b_1-b_2}$. Next, we consider stopping time $z \in (0,1)$. Letting $p_2'(t)$ denote the derivative of $p_2(t)$ w.r.t. t, the expected cost to the algorithm is

$$A^*(z) = p_0(b_2 + (1-b_2)z) + (1-p_0)\left[\int_0^z (b_1 + (1-b_1)t + b_2 + (1-b_2)(z-t))p_2'(t)dt \right.$$

$$\left. + p_1(z)(b_1 + (1-b_1)z) \right]$$

$$= \frac{B(b_1 + (1-b_1)z)}{B + b_1 - b_2},$$

which is what we expect, since $\text{OPT}(z) = b_1 + (1 - b_1)z$ in this case.

Finally, we consider stopping time $z' \geq 1$. In this case the expected cost to the algorithm is

$$A^*(z') = p_0(b_2 + (1-b_2)z') + (1-p_0)\left[\int_0^1 (b_1 + (1-b_1)t + b_2 + (1-b_2)(z'-t))p_2'(t)dt \right.$$

$$\left. + p_1(1)(b_1 + (1-b_1)z') \right]$$

$$= \frac{B(b_2 + (1-b_2)z')}{B + b_1 - b_2}.$$

(The differences from the expression for $z \in (0,1)$ above is the upper limit of the integral and that $p_1(1)$ replaces $p_1(z)$.) This result is what we expect, because $\text{OPT}(z') = b_2 + (1 - b_2)z'$. This concludes the proof of the upper bound claimed in Theorem 2.

Remark: We note that the results above coincide with two special cases: in [10], the case of $b_1 = 0$ is considered, and the single parameter used is $a \stackrel{\text{def}}{=} 1 - b_2$. In this case we have $B = e$. The competitive ratio stated in [10] is $\frac{e}{e-(1-a)} = \frac{e}{e-b_1}$, which coincides with our (9). Another special case is considered in [9], where $b_2 = 1$. In this case $B = e^{1-b_1}$. The probability p_0 of starting at Option 2, as stated in [9] is $\frac{b_1}{e^{1-b_1}-(1-b_1)}$, and the competitive ratio is $\frac{e^{1-b_1}}{e^{1-b_1}-(1-b_1)}$, agreeing with the more general (8) and (9).

4   Lower Bound on the Randomized Competitive Ratio

We now prove that the algorithm presented in Section 3 is optimal. To this end, we employ Yao's Lemma [13], which says in our context that to prove a lower bound of ρ on randomized algorithms, it is sufficient to present a probability distribution $\mathcal{D}$ over the instances such that for any deterministic algorithm A_z, we have that $\mathbb{E}_{I \in \mathcal{D}}[A_z(I)/\text{OPT}(I)] \geq \rho$.

Now an instance is specified by the stopping time. To find the worst-case probability distribution over stopping times, we make a few guesses. First, we guess that the desired distribution has the following support: some probability density function (pdf) on the interval $[0, 1]$, and some positive probability mass at points 0 and T, for a point $T > 1$ to be determined. Second, we guess that with respect to that distribution, the expected competitive ratio of all algorithms that stop at some time $t \in [0, 1] \cup \{T\}$, is the same, and equals $\frac{B}{B - b_2 + b_1}$.

Based on these guesses, and using methods similar to the ones we used for deriving the algorithm in Section 3, we arrive at the following distribution.

- Let
$$q_0 \stackrel{\text{def}}{=} \frac{b_1^2 B}{b_2(B + b_1 - b_2)} .$$

  With probability q_0, the stopping time is 0. In this case the optimal policy is to take Option 1, for a cost of $\text{OPT}(0) = b_1$.
- Define
$$T \stackrel{\text{def}}{=} 1 + \frac{b_2}{b_2 - b_1} , \text{ and}$$
$$q_T \stackrel{\text{def}}{=} \frac{2 - b_2 - \frac{b_1}{b_2}}{B + b_1 - b_2} .$$

  With probability q_T, the stopping time is T. In this case the optimal policy is to take Option 2, for a cost of $\text{OPT}(T) = b_2 + T(1 - b_2)$.
- With probability $1 - q_0 - q_T$, the stopping time is chosen from the interval $(0, 1)$. This is done using the following pdf for $u \in (0, 1)$:
$$q(u) \stackrel{\text{def}}{=} K(b_1 + (1 - b_1)u)B^{-u} ,$$

  Where K is the appropriate constant so that $\int_0^1 q(u)du = 1 - q_0 - q_T$:
$$K \stackrel{\text{def}}{=} \left(\frac{b_2 - b_1}{b_2}\right)^2 \frac{B}{B + b_1 - b_2}$$

  In this case the optimal policy is Option 1.
- All other stopping times are assigned zero probability.

We now bound the expected competitive ratio of any deterministic algorithm when the instance is drawn randomly according to the above distribution.

Consider first the algorithm A_0 that starts with Option 2 immediately. Its expected competitive ratio is

$$\rho(0) = q_0 \frac{b_2}{b_1} + \int_0^1 \frac{b_2 + (1 - b_2)u}{b_1 + (1 - b_1)u} q(u)du + q_T \cdot 1$$
$$= \frac{B}{B + b_1 - b_2} .$$

Next, consider an algorithm A_z that switches from Option 1 to Option 2 at time $0 < z \leq 1$. The expected competitive ratio of A_z is

$$\rho(z) = q_0 \cdot 1 + \int_0^z q(u)du + \int_z^1 \frac{b_1 + (1-b_1)z + b_2 + (1-b_2)(u-z)}{b_1 + (1-b_1)u} q(u)du$$
$$+ q_T \frac{b_1 + (1-b_1)z + b_2 + (1-b_2)(T-z)}{b_2 + T(1-b_2)}$$
$$= \frac{B}{B + b_1 - b_2} .$$

Next, we consider $A_{z'}$ with $1 < z' \leq T$. In this case the expected competitive ratio is

$$\rho(z') = q_0 + \int_0^1 q(u)du + q_T \frac{b_1 + (1-b_1)z' + b_2 + (1-b_2)(T-z')}{b_2 + T(1-b_2)}$$
$$= \frac{B + (z'-1)(b_2 - b_1)^2/b_2}{B + b_1 - b_2}$$
$$\geq \frac{B}{B + b_1 - b_2} .$$

And finally, we consider $A_{z''}$ with $z'' > T$. The expected competitive ratio in this case is

$$\rho(z'') = q_0 + \int_0^1 q(u)du + q_T \frac{b_1 + T(1-b_1)}{b_2 + T(1-b_2)}$$
$$= \frac{B}{B + b_1 - b_2} .$$

This concludes the proof of the lower bound stated in Theorem 2, and the proof of Theorem 2 is complete.

5 Conclusion

In this paper we solved what may be the simplest non-trivial variant of the non-additive ski rental problem. The results reveal that the problem, in this simple variant, exhibits new features that are not present in the additive model. This is particularly evident in the structure of the optimal deterministic algorithm, in which starting with the "buy" option immediately is optimal for a non-empty range of the parameters. We hope that our results will be a first step in a more complete understanding of general non-additive ski-rental.

Acknowledgments. We thank Dror Rawitz for useful discussions.

References

1. Augustine, J., Irani, S., Swamy, C.: Optimal power-down strategies. SIAM J. Comput. 37(5), 1499–1516 (2008)
2. Bejerano, Y., Cidon, I., Naor, J.S.: Dynamic session management for static and mobile users: a competitive on-line algorithmic approach. In: 4th International Workshop on Discrete Algorithms and Methods for Mobile Computing and Communications, pp. 65–74. ACM (2000)
3. Borodin, A., El-Yaniv, R.: Online Computation and Competitive Analysis. Cambridge University Press (1998)
4. Fleischer, R.: On the bahncard problem. Theoretical Comput. Sci. 268(1), 161–174 (2001)
5. Fujiwara, H., Kitano, T., Fujito, T.: On the best possible competitive ratio for multislope ski rental. In: Asano, T., Nakano, S.-i., Okamoto, Y., Watanabe, O. (eds.) ISAAC 2011. LNCS, vol. 7074, pp. 544–553. Springer, Heidelberg (2011)
6. Karlin, A., Manasse, M., Rudolph, L., Sleator, D.: Competitive snoopy caching. Algorithmica 3(1), 79–119 (1988)
7. Karlin, A.R., Kenyon, C., Randall, D.: Dynamic TCP acknowledgment and other stories about $e/(e-1)$. Algorithmica 36(3), 209–224 (2003)
8. Karlin, A.R., Manasse, M.S., McGeoch, L.A., Owicki, S.: Competitive randomized algorithms for non-uniform problems. In: Proc. First Ann. ACM-SIAM Symp. on Discrete Algorithms, SODA 1990, pp. 301–309. SIAM, Philadelphia (1990)
9. Levi, A., Patt-Shamir, B.: Simplest non-additive ski rental (Unpublished manuscript)
10. Lotker, Z., Patt-Shamir, B., Rawitz, D.: Ski rental with two general options. Inf. Process. Lett. 108(6), 365–368 (2008)
11. Lotker, Z., Patt-Shamir, B., Rawitz, D.: Rent, lease, or buy: Randomized algorithms for multislope ski rental. SIAM J. Discrete Math. 26(2), 718–736 (2012)
12. Wolfram Research, Inc. Mathematica version 7.0 (2008)
13. Yao, A.C.: Probabilistic computations: toward a unified measure of complexity. In: Proceedings of the 18th Annual ACM Symposium on Theory of Computing, pp. 222–227 (1977)

Competitive FIB Aggregation for Independent Prefixes: Online Ski Rental on the Trie

Marcin Bienkowski[1,*] and Stefan Schmid[2]

[1] Institute of Computer Science, University of Wrocław, Poland
[2] Telekom Innovation Laboratories & TU Berlin, Germany

Abstract. This paper presents an asymptotically optimal online algorithm for compressing the *Forwarding Information Base* (FIB) of a router under a stream of updates (namely insert rule, delete rule, and change port of prefix). The objective of the algorithm is to dynamically aggregate forwarding rules into a smaller but equivalent set of rules while taking into account FIB update costs. The problem can be regarded as a new variant of ski rental on the FIB trie, and we prove that our deterministic algorithm is 3.603-competitive. Moreover, a lower bound of 1.636 is derived for any online algorithm.

1 Introduction

An Internet router typically stores a large number of *forwarding rules*: Given a packet's IP address, the router uses the so-called *Forwarding Information Base* (FIB) to determine the forwarding port (or next-hop) of the packet. These very time critical FIB lookups require a fast and expensive memory on the line card, which constitutes a major cost factor of today's routers. It is expected that the virtualization trend in the Internet will further increase the memory requirements [2,9], and also IPv6 does not mitigate the problem [3].

A simple and local solution to reduce the FIB size is the *aggregation (compression) of the FIB*, i.e., the replacement of the existing set of rules by an *equivalent but smaller* set. This solution does not affect neighboring routers and it can be done by a simple software update [18]. However, aggregation may come at the cost of a higher FIB update churn (e.g., see [5]): upon certain BGP updates, aggregated FIB entries may have to be disaggregated again. Frequent FIB updates are problematic as upon each update, internal FIB structures have to be rebuilt to ensure routing consistency. In particular, update costs are also critical in the context of *Software-Defined Networks (SDN)* (e.g., based on *OpenFlow* [10]), as the network controller is remote from the switch and FIB updates may have to be transmitted over a bandwidth-limited network [15].

While this problem is currently discussed intensively in the networking community [7,17], only heuristics and static algorithms have been proposed so far. We, in this paper, assume the perspective of competitive and worst-case analysis, and present a solution which *jointly optimizes* the FIB compression ratio and the number of FIB updates.

---

* Supported by MNiSW grant number N N206 368839, 2010-2013.

T. Moscibroda and A.A. Rescigno (Eds.): SIROCCO 2013, LNCS 8179, pp. 92–103, 2013.
© Springer International Publishing Switzerland 2013

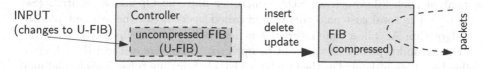

Fig. 1. Controller and FIB: the controller updates the rules in the FIB. This paper focuses on online algorithms for the controller.

1.1   The Model

An (IP) *address* is a binary string of length w (e.g., $w = 32$ for IPv4 and $w = 128$ for IPv6) or equivalently an integer from $[0, 2^w - 1]$. An (IP) *prefix* is a binary string of length at most w; we denote the empty prefix by ε. A prefix *matches* all addresses that start with it, i.e., it corresponds to a *range* of addresses of the form $[k \cdot 2^i, (k + 1) \cdot 2^i - 1]$.

Forwarding Rules. We consider a packet forwarding router with a set of *ports* (or *next-hops*). A *Forwarding Information Base (FIB)* is a set of *forwarding rules* used by the router; each rule is a *prefix-port pair* (p, c). For the presentation, we will refer to the ports by *colors*, i.e., assume a unique color for each port. For any packet processed by the router, a decision is made on the basis of its destination IP address x using the *longest prefix match* policy [11]: among the FIB rules $\{(p_i, c_i)\}_i$, the router chooses the longest p_i being a prefix of x, and forwards the packet to port c_i. (We assume that there are no two rules with the same prefixes and different ports.) If no rule matches, the packet is dropped.

The router contains two parts: the *controller* (either implemented on the route processor, or an SDN controller) and the *(compressed) FIB* (stored in a fast and expensive memory), cf. Fig. 1. The controller keeps a copy of the *uncompressed FIB (U-FIB)* and receives a stream of updates to this structure (e.g., due to various events from the *Border Gateway Protocol, BGP*). More precisely, we assume continuous time; at any time t, (1) a new forwarding rule may be *inserted*, (2) an existing rule *deleted*, or (3) a prefix may change its forwarding port (color *update*). A sequence of such *events* constitutes the *input* to our problem.

Right after a change occurs, the controller must ensure that the U-FIB and the FIB are equivalent, i.e., their forwarding and dropping behavior is the same. In this paper, we will make the simplifying assumption that the FIB prefixes are *independent*: the FIB does not contain any prefixes which overlap in their address range. To this end, the controller may insert, delete or update (change color) individual rules in the FIB. The controller may also issue these commands at any point of time (e.g., for a delayed compression of the FIB).

Costs. We associate a *fixed* cost α with any such change of a single rule in the FIB. Note that we represent the update cost as a constant to keep the model general: α is not specific for any particular FIB data structure (e.g., trie, cache, or Multibit Burrows-Wheeler [12]), but may also model the cost of transmitting a control packet between an SDN controller and the OpenFlow switch. (See also [7].) The total cost paid this way is called *update cost*; the amount paid by an algorithm ALG in a time interval I is denoted by U-COST$_I$(ALG).

The second type of cost we want to optimize is the size of the FIB, which — following [4] — is defined as the number of FIB forwarding rules. This modeling is justified by state-of-the-art approaches (see, e.g., [11, chapter 15]), where the size of such a structure is usually proportional to the number of entries in the FIB. For an algorithm ALG and time t, we denote the number of FIB rules at time t by SIZE$_t$(ALG). The total memory cost in a time interval I is then defined as M-COST$_I$(ALG) $= \int_I$ SIZE$_t$(ALG) dt.

In both objective functions (U-COST and M-COST), we drop time interval subscripts when referring to the total cost during the runtime of an algorithm. This paper focuses on minimizing the sum of these two costs, i.e., COST(ALG) $=$ U-COST(ALG) $+$ M-COST(ALG). Note that the parameter α can be used to put more emphasis on either of the two costs.

Competitive Analysis. We assume a conservative standpoint and study algorithms that do not have any knowledge of future prefix changes, and need to decide *online* on where and when to aggregate. Not relying on predictions seems to be a reasonable assumption considering the chaotic behavior of the route updates in the modern Internet [6]. We use the standard yard-stick of online analysis [1], i.e., we compare the cost of the online algorithm to the cost of an optimal offline algorithm OPT that knows the whole input sequence in advance. We call an online algorithm ALG ρ-*competitive* if there exists a constant γ, such that for any input sequence it holds that COST(ALG) $\leq \rho \cdot$ COST(OPT) $+ \gamma$. The competitive ratio of an algorithm is the *infimum* over all possible ρ such that the algorithm is ρ-competitive.

Empirical Motivation. The motivation for our simplifying assumption of independent prefixes is twofold. First, an algorithm to solve the independent case can be applied to the independent subtrees. Moreover, empirical data shows that while Internet routers typically define a default route (an empty prefix), the prefix hierarchy is typically very flat [15]: prefixes hardly overlap with more than one other prefix. As of February 2013, the Internet-wide BGP routing table contains more than 440k prefixes. In table dumps obtained from *RouteViews* [13], we observe that around half of all prefixes do not have any less specifics, and on average, a prefix has 0.64 less specifics.

Furthermore, in our modeling, we neglect the impact a FIB compression may have on IP lookup times, because they are affected only to a very limited extent. The state-of-the-art data structures used for IP lookup (see, [11, chapter 15]) use a large variety of tree-like constructs augmented with additional information. This allows for lookup times of order $O(\log w)$, with practical implementations using 2-3 memory lookups on the average). Additionally, little is known about proprietary data structures actually used in the routers of different vendors.

1.2   Related Work

There are known fast algorithms for optimal FIB aggregation of table snapshots, for example the Optimal Routing Table Constructor (ORTC) [4] and others [16]. However, as these algorithms do not support efficient handling of incremental updates, a re-computation of the optimally aggregated FIB on each forwarding rule change is needed. This is computationally expensive and can lead to high churn. There are several papers that deal with this problem by proposing heuristics that simultaneously try to limit the number of updates to the FIB while maintaining a good compression rate, including SMALTA [17] and others [7,8,18]. Moreover, some authors even proposed to only store a *subset* of rules in the FIB, leveraging Zipf's law [15]. However, none of these works give a formal bound on the achievable performance over time neither with respect to the number of updates to the aggregated FIB, nor to the aggregation gain. They also do not consider to use churn locality for their benefit.

The closest paper to ours is [14] by Sarrar et al. The authors first study the temporal and spatial locality of churn in the trie empirically, and then present an $O(w^2)$-competitive online algorithm for tries with *dependent* prefixes. It is worth noting that in the dependent case, for a large class of online algorithms, there exists a $\Omega(w)$ lower bound. This indicates that the independent prefix variant might be inherently simpler.

1.3   Our Contribution

The main contribution of this paper is a deterministic online algorithm for the FIB aggregation problem (with independent prefixes) that jointly optimizes the FIB size (by rule aggregation) as well as the number of updates to the FIB (by timed waiting). We prove that our algorithm is 3.603-competitive under a worst-case sequence of rule updates (events: *insert*, *delete*, port *change*). Furthermore, we show that there provably does not exist any online algorithm with a competitive ratio smaller than 1.636.

Technically, the problem can be regarded as a variant of online ski rental on a trie: The presented algorithm BLOCK seeks to aggregate prefixes slowly over time, amortizing aggregation costs with the memory benefits.

2   Basic Properties

Trie Representation. Throughout this paper, we will represent both the U-FIB and the FIB as one-bit tries containing all the prefixes from the forwarding rules. This affects merely the presentation: we do not assume anything about the actual implementation of the U-FIB/FIB structures. We assume that each non-leaf node has exactly two children. Each node of the tree (corresponding to some prefix p) has an associated *color* c if there is a forwarding rule (p, c); a node without any associated color is called *blank*. We assume minimal tries, that is, tries without blank sibling leaves (they may contain blank leaves, though).

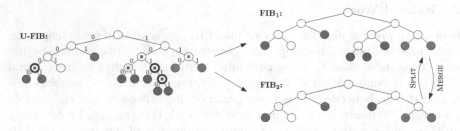

Fig. 2. One-bit tries representing a U-FIB and two possible compressions to the FIBs. Mergeable nodes are marked with small squares in the center. Nodes consolidated for the creation of FIB$_1$ are drawn with bold lines.

For any node v, we denote the subtree rooted at v by $T(v)$. A non-root node we call *left (right)* if it is a left (respectively right) son of its parent. Sometimes it is convenient to identify the nodes with the address ranges they represent.

Mergeable Nodes. Let's first assume that the state of the U-FIB is static, i.e., we are not processing an input, and study the structural properties of the possible FIB aggregations. An internal (blank) node v from U-FIB is called *c-mergeable* if all leaves of $T(v)$ are of color c. Sometimes, we will simply say that a node is *mergeable* instead of *c*-mergeable. Clearly, if a node v is *c*-mergeable, then all internal nodes from $T(v)$ are *c*-mergeable.

Mergeable nodes are the key to all compression patterns possible. Namely, any possible FIB aggregation is defined by choosing any set A of mergeable, pairwise non-overlapping nodes. For any *c*-mergeable node $v \in A$, we remove all descendants of v (recall that each of them is either an internal *c*-mergeable node or a leaf of color c) and color v with c. In the U-FIB, we call v and all the internal (mergeable) nodes of $T(v)$ *consolidated*, cf. Fig. 2.

Hence, all an algorithm may do is to choose which mergeable nodes to consolidate and when. At runtime, an algorithm may change the FIB incrementally, by consolidating or unconsolidating nodes. The only restriction is that all the (mergeable) descendants of consolidated nodes have to be consolidated as well. Therefore, there are two possible operations an algorithm may perform.

– A *merge operation* (at a mergeable, unconsolidated node v) changes the state of all, say k, mergeable, unconsolidated nodes from $T(v)$ to consolidated. These k nodes were internal ones, and hence the operation involves replacing $k+1$ leaves in the FIB by a node v, i.e., induces the update cost of $(k+2) \cdot \alpha$.
– A *split operation* has the reverse effect and is described by a tree (rooted at v) of k' consolidated nodes whose state is changed to unconsolidated. It is possible that this tree does not contain all the consolidated nodes of $T(v)$. In the FIB, this involves replacing node v by $k' + 1$ nodes, and hence the associated update cost is $(k' + 2) \cdot \alpha$.

The size of the FIB is tightly related to the number of consolidated nodes. Precisely speaking, we denote the size of the U-FIB by SIZE(U-FIB), the number of mergeable nodes at time t by $M(t)$, and nodes consolidated by an algorithm ALG by $S^{\text{ALG}}(t)$. Then, $S^{\text{ALG}}(t) \leq M(t)$ and $\text{SIZE}_t(\text{ALG}) = \text{SIZE}(\text{U-FIB}) - S^{\text{ALG}}(t)$.

So far, we only studied the compression of a static U-FIB. When at some time t a prefix (v, c) changes color to c', the following changes to the U-FIB occur. If the sibling of v has color c', then the parent of v and possibly some of its ancestors may become c'-mergeable. If the sibling of v has color c, then all the c-mergeable ancestors of v (if any) become non-mergeable. If some of these nodes were consolidated, a split operation involving these nodes is forced.

Event Costs. We fix any algorithm ALG and take a closer look at its update cost, U-COST(ALG). Whenever ALG processes a single event at time t, it has to update the FIB paying α. However, this cost can be avoided if ALG performs a related merge or split operation (as defined in Sect. 2) immediately at time t. For example, if a color update of node v changes the state of the parent of v to mergeable, ALG may merge the parent of v at the same time and pay only 3α for the cost of merging without paying the event cost of α. Another example is a forced split.

Lemma 1. *Assume that an online algorithm ALG is R-competitive if we neglect event costs. Then, it is $(R+1/3)$-competitive if we take these costs into account.*

Proof. We partition U-COST(ALG) into the sum of O-COST(ALG), the cost of all (merge or split) operations performed by ALG, and E-COST(ALG), the cost of events on nodes not accompanied by immediate operations on the same nodes. In these terms, COST(ALG) = M-COST(ALG)+O-COST(ALG)+E-COST(ALG).

Fix any input sequence. By the assumption of the lemma, M-COST(ALG) + O-COST(ALG) $\leq R \cdot$ (M-COST(OPT) + O-COST(OPT)) + γ for some constant γ. Let k be the number of events in the input sequence. Clearly, E-COST(ALG) $\leq k \cdot \alpha$. On the other hand, for any event, an optimal offline algorithm OPT either performs a merge or a split, which increases O-COST(OPT) by at least 3α, or does not perform any spatial operations on the trie, which increases E-COST(OPT) by α (color change only). Thus, $\frac{1}{3} \cdot$ O-COST(OPT) + E-COST(OPT) $\geq k \cdot \alpha$, and hence,

$$\text{COST}(\text{ALG}) \leq R \cdot (\text{M-COST}(\text{OPT}) + \text{O-COST}(\text{OPT})) + \gamma + k \cdot \alpha$$
$$\leq R \cdot \text{M-COST}(\text{OPT}) + (R + 1/3) \cdot \text{O-COST}(\text{OPT})$$
$$+ \text{E-COST}(\text{OPT}) + \gamma$$
$$\leq (R + 1/3) \cdot \text{COST}(\text{OPT}) + \gamma \ ,$$

which completes the proof. $\square$

3   The Algorithm BLOCK

Our algorithm BLOCK is very simple: With any node v, we associate a counter C_v which is a function of time. If v is c-mergeable at time t, then $C_v(t)$ measures how long (uninterruptedly) v is in this state; otherwise $C_v(t) = 0$. The algorithm

BLOCK is parameterized with two constants, $A \geq B$; we derive an optimal choice for these constants later.

As soon as there is a non-consolidated node v whose counter is $A \cdot \alpha$, BLOCK merges the tree $T(u)$ rooted at the ancestor u of v which is closest to the trie root and whose counter is at least $B \cdot \alpha$. (It is possible that $u = v$.) Furthermore, BLOCK splits only when forced, i.e., when a consolidated node changes state to non-mergeable.

Lemma 2. *Fix any values of A and B. Neglecting event costs, BLOCK(A, B) is* $\max\{(A+6)/A, (B+4)/B, (A+6)/(A+2-B), (B+4)/2, (A+4)/2\}$-*competitive.*

Proof. The proof pursues an accounting approach: We charge the non-event costs of any algorithm to particular nodes and relate the cost of BLOCK and OPT on chosen subsets of nodes. Recall that whenever an algorithm merges or splits a subtree, it pays $(k+2) \cdot \alpha$, where k is the number of leaves of this subtree. We assign the cost of 3α to the root of this tree and α to all its remaining internal nodes. Thus, whenever an algorithm consolidates a mergeable node (or changes the state back from consolidated) it pays 3α or α. Furthermore, we assume that only nodes that are mergeable but not consolidated are counted towards the size of FIB. This way, at time t, we underestimate the actual memory cost by the number of non-mergeable nodes, SIZE(U-FIB) $- M(t)$. This amount is however the same for any algorithm. Therefore, if the algorithm is R-competitive using charged costs, it is also R-competitive in the actual cost model.

We first take a mergeable period (of length τ) of node v during which it is not consolidated by BLOCK; we compare the costs of OPT and BLOCK for v in this period. In this case, BLOCK pays just τ for the memory cost. As BLOCK does not consolidate v, $\tau < A\alpha$. If OPT decides to merge v, it pays at least α for merging v and at least α for splitting it. Otherwise, it pays τ for the memory cost. In total, the ratio of the BLOCK and OPT costs is at most

$$R_1 = \tau / \min\{\tau, 2\alpha\} \leq \max\{1, \tau/2\alpha\} \leq \max\{1, A/2\} .$$

Now, we compare the costs on nodes that are consolidated (by a single merging operation) by BLOCK. More precisely, we consider any time t at which BLOCK merges a tree T rooted at node u. We analyze the total cost of BLOCK and OPT for all mergeable periods of all nodes from T, such that these periods contain time t. Let $k+1$ be the number of all consolidated nodes from T (i.e., $k \geq 0$). We denote the value of counters of these nodes at time t by C_i. For convenience, we assume these values are sorted, i.e., $C_i \leq C_{i+1}$. By the definition of the algorithm, $B\alpha \leq C_1 \leq C_2 \ldots C_k \leq C_{k+1} = A\alpha$. On the considered mergeable periods of nodes of T, BLOCK pays $\sum_{i=1}^{k+1} C_i$ (memory cost) plus $(k+3) \cdot \alpha$ (merging cost), plus $(k+1) \cdot 3\alpha$ (splitting cost as in the worst case the nodes are split individually). Thus, in total,

$$\text{COST}_T(\text{BLOCK}) \leq 2\alpha + \sum_{i=1}^{k+1}(C_i + 4\alpha) \leq (A+6)\alpha + \sum_{i=1}^{k}(C_i + 4\alpha) .$$

Now, we compute the cost of OPT on the same mergeable periods. Let $\mathcal{K}$ be the smallest time interval containing all these periods. By the algorithm definition, $\mathcal{K}$ starts at time $t - A\alpha$. We consider three cases depending on OPT actions within $\mathcal{K}$. In each bound, after computing the BLOCK-to-OPT ratio, we immediately use the relation $(a + b)/(c + d) \leq \max\{a/c, b/d\}$.

Case 1. Within $\mathcal{K}$, OPT does not merge any subtree rooted at a node from T nor at any of ancestors of u. In this case, no node of T becomes consolidated by OPT, and hence OPT just pays memory costs, i.e., $\text{COST}_T(\text{OPT}) = \sum_{i=1}^{k+1} C_i = A\alpha + \sum_{i=1}^{k} C_i$. Therefore, the BLOCK-to-OPT cost ratio is at most

$$R_2 = \frac{(A + 6)\alpha + \sum_{i=1}^{k}(C_i + 4\alpha)}{A\alpha + \sum_{i=1}^{k} C_i} \leq \max\left\{\frac{A + 6}{A}, \frac{B + 4}{B}\right\}.$$

Case 2. Within $\mathcal{K}$, OPT does not merge any subtree rooted at a node in T, but does merge a subtree rooted at an ancestor of u, say u'. By the algorithm definition, $C_{u'}(t) < B \cdot \alpha$ (otherwise BLOCK would choose u' as the root for the merging operation). This means that OPT merges not earlier than at time $t - B\cdot\alpha$, so the corresponding counters of the nodes from T are at least $C_i - B\alpha$. Hence, the cost of OPT associated with node i is at least $C_i - B\alpha$ (memory cost) plus 2α (merging and splitting). Therefore, $\text{COST}_T(\text{OPT}) \geq \sum_{i=1}^{k+1}(C_i - B\alpha + 2\alpha) = (A + 2 - B)\alpha + \sum_{i=1}^{k}(C_i + (2 - B)\alpha)$, and the ratio in this case is

$$R_3 = \frac{(A + 6)\alpha + \sum_{i=1}^{k}(C_i + 4\alpha)}{(A + 2 - B)\alpha + \sum_{i=1}^{k}(C_i + (2 - B)\alpha)} \leq \max\left\{\frac{(A + 6)}{A + 2 - B}, \frac{B + 4}{2}\right\}.$$

Case 3. Within $\mathcal{K}$, OPT merges some subtree rooted at a node from T. In this case, we split the indices of nodes of T into two sets: a set S of nodes that become consolidated sometime within $\mathcal{K}$ and a set N of nodes that remain unconsolidated for the whole period $\mathcal{K}$. Clearly, $|S| + |N| = k + 1$. For the node from S being the root of the merged subtree, OPT pays at least $3\alpha + \alpha$ (merging and splitting cost) and for the remaining nodes from S at least $\alpha + \alpha$. For any node from N, OPT pays at least C_i (memory cost). Hence, in total, $\text{COST}(\text{OPT}) = 2\alpha + \sum_{i \in S} 2\alpha + \sum_{i \in N} C_i$, and the cost ratio is

$$R_4 = \frac{2\alpha + \sum_{i \in S}(C_i + 4\alpha) + \sum_{i \in N}(C_i + 4\alpha)}{2\alpha + \sum_{i \in S} 2\alpha + \sum_{i \in N} C_i} \leq \max\left\{1, \frac{A + 4}{2}, \frac{B + 4}{B}\right\}.$$

As we all possible cases are considered above, the competitive ratio is at most $\max_{1 \leq i \leq 4} R_i$. The lemma follows by substituting the actual values of R_i.   □

Tedious case analysis and elementary algebra shows that the choice of parameters minimizing the guarantee of Lemma 2 is $A = \sqrt{13} - 1 \approx 2.606$ and $B = \frac{2}{3}A \approx 1.737$. Taking event costs into account (cf. Lemma 1), we obtain the following result.

Theorem 1. *The competitive ratio of* BLOCK$(\sqrt{13}-1, 2\sqrt{13}/3-2/3)$ *is at most* $(\sqrt{13} + 3)/2 + 1/3 \approx 3.603$.

Note that there is a simpler version of the BLOCK algorithm that consolidates only one node at a time, i.e., uses $A = B$. In such case, the optimal choice of the parameters is $A = B = 2$, and thus such algorithm is $(4 + 1/3)$-competitive.

4   Handling Insertions and Deletions

So far, we show how to handle color updates to the U-FIB. In this section, we show that it is possible to handle also insertions of new rules to the U-FIB and deletions of old rules from the U-FIB. Recall that the algorithm BLOCK simply watches the changes in the mergeability of U-FIB nodes. For completing the definition of BLOCK, it is therefore sufficient to show how insertions and deletions affect that aspect.

- A prefix (v, c) is inserted to the U-FIB. If node v already existed in the tree as a blank node, v must be a leaf. In this case, zero or more ancestors of v become c-mergeable. If, however, node v did not exist in the tree, then v is inserted as a leaf with a blank sibling. The set of mergeable nodes does not change in this case.
- A prefix (v, c) is deleted from the U-FIB. Node v becomes blank and all ancestors of v (including the root) become non-mergeable. As we require the tree to be minimal, we may have to perform an optional pruning of blank nodes. However, this does not change the mergeability of any node.

The only detail that has to be changed in the analysis of BLOCK is that now the size of the U-FIB is not constant but is a function of time, denoted $\text{SIZE}_t(\text{U-FIB})$ at time t. Then in the proof of Lemma 2, by using charged costs, we underestimate the actual memory size by $\text{SIZE}_t(\text{U-FIB}) - M(t)$, where $M(t)$ is the number of nodes mergeable at t. However, as in the original proof, this amount is the same for BLOCK and OPT, and thus R-competitiveness using charged costs implies the R-competitiveness in the actual cost model. Thus, we obtain the following result.

Theorem 2. *The competitive ratio of* $\text{BLOCK}(\sqrt{13}-1, 2\sqrt{13}/3-2/3)$ *is at most* $(\sqrt{13}+3)/2 + 1/3 \approx 3.603$ *also when insertions and deletions may occur in the input sequence.*

5   Lower Bound

The algorithm BLOCK was designed with two objectives in mind: (i) to balance the memory cost and the update cost, (ii) to exploit the possibility of merging multiple tree nodes simultaneously at a lower price. An online algorithm is bound to choose sub-optimally in both of these aspects: we will show a lower bound of 1.636 on the competitive ratio of any online algorithm.

The analysis of BLOCK suggests a straightforward lower bound: We keep a tree of two prefixes $\{0, 1\}$. By changing the color of one of them, the adversary changes the state of root from non-mergeable to mergeable, and back. When the root becomes mergeable, the algorithm may consolidate it at some time, but right after that happens, the adversary turns the root non-mergeable, enforcing a split. An analogous approach can be found in many online problems, most notably in the ski-rental problem [1]. However, unlike in the ski-rental problem, we cannot obtain the lower bound of 2 by this adversarial strategy. The first obstacle is that the memory cost (the equivalent of renting skis) is always at least 1 (even for OPT). The second obstacle are event costs that are sometimes paid also by OPT. An exact analysis would yield a lower bound of 1.5. We may improve this bound by making the tree slightly larger.

Theorem 3. *Any online algorithm* ALG *has a competitive ratio of at least* $18/11 \approx 1.636$.

Proof. The set of prefixes in the U-FIB will be constant and equal to $\{00, 01, 1\}$, initially with the colors red, green and green, respectively. A strategy of the adversary consists of phases. At the end of each phase, the state of the U-FIB will be the same as the initial one, and the ALG-to-OPT cost ratio on any phase will be at least $18/11$. The adversary may generate a sequence consisting of an arbitrary number of phases, thus ensuring that the cost ALG cannot be hidden in the additive constant γ in the definition of the competitive ratio.

In a single phase starting at time t, the adversary changes the color of prefix 00 to green, making both internal nodes of the tree mergeable. Note that in such a situation there are three possible states of ALG: a low state (no node consolidated), a middle state (the lower mergeable node consolidated) and a top state (both mergeable nodes consolidated). Two cases are possible:

- ALG changes state for the first time (either to the top or to the middle one) at time $t' \le t + \alpha$.
- ALG does not change state in the interval $[t, t+\alpha]$. In this case, let $t' > t + \alpha$ be the first time when it changes its state to the top one (it may change state to the middle one before t').

Note that if none of the two described events occurs, then ALG never changes its state to the top one. In this case, its FIB size is at least 2 whereas the optimal possible is 1. This would immediately imply a lower bound of 2 on the competitive ratio.

At time $t' + \epsilon$, the adversary changes the color of prefix 00 back to red, forcing ALG to change the state back to the low one, and ending the phase. As the adversary may choose ϵ to be arbitrarily small, in the analysis we assume $\epsilon = 0$.

To analyze the performance of ALG in a single phase, we set $\ell = t' - t$. The cost of OPT is upper-bounded by the minimum of costs of two possible strategies: (i) do nothing, (ii) change the state to top at time t and then back to low at time $t' + \epsilon$. The cost for the former strategy is 3ℓ (memory cost) plus 2α (event cost), while the cost for the latter is ℓ (memory cost) plus 4α (merging cost) + 4α (splitting cost). Altogether, $\text{COST}(\text{OPT}) \le \min\{2\alpha + 3\ell, 8\alpha + \ell\}$. We consider two cases.

1. The first event occurs, i.e, $\ell \le \alpha$. Then ALG pays at least 3α (merging cost) and another 3α (splitting cost). Additionally, the memory cost is 3ℓ as ALG is in the low state till it merges anything. Furthermore, if ALG merges after time t, then it has to pay event cost α at time t. Thus, we consider two subcases:

(a) Algorithm merges already at time t, i.e., $\ell = 0$. Then, $\text{COST}(\text{OPT}) \leq 2\alpha$, $\text{COST}(\text{ALG}) = 6\alpha$, and hence the ratio is $R_1 = 3$.

(b) Algorithm merges after time t. Then, $\text{COST}(\text{OPT}) \leq 2\alpha + 3\ell$ while $\text{COST}(\text{ALG}) = 7\alpha + 3\ell$. The ratio is then at least

$$R_2 = \frac{7\alpha + 3\ell}{2\alpha + 3\ell} \geq \frac{7\alpha + 3\alpha}{2\alpha + 3\alpha} = 2 \ .$$

2. The second event occurs, i.e., $\ell > \alpha$. We consider two subcases.

(a) ALG changes state only once, at time ℓ. Then, it pays α (event cost) plus 3ℓ (memory cost) plus 4α (merging cost) plus 4α (splitting cost). The ratio is then

$$R_3 = \frac{9\alpha + 3\ell}{\min\{2\alpha + 3\ell, 8\alpha + \ell\}} \geq \frac{18}{11} \approx 1.636 \ .$$

(b) ALG changes state more than once. Then ALG is in low state at least till time α and in state low or middle at times between α and t'. Therefore, it pays at least α (event cost) plus $3 \cdot \alpha + 2 \cdot (\ell - \alpha)$ (memory cost) plus $3\alpha + 3\alpha$ (merging cost) $+ 4\alpha$ (splitting cost). In this case, the ratio is

$$R_4 = \frac{12\alpha + 2\ell}{\min\{2\alpha + 3\ell, 8\alpha + \ell\}} \geq \frac{18}{11} \approx 1.636 \ .$$

Altogether, in either case, the competitive ratio is at least $18/11$.   □

6   Conclusions

This paper studied a novel online aggregation problem arising in the context of (classical or SDN) router optimization. The described online algorithm that provably achieves a low, constant competitive ratio. Since the derived lower bound is not tight, the main open technical question regards closing the gap between the upper and lower bound.

Acknowledgments. The authors would like to thank Magnús M. Halldórsson, Nadi Sarrar and Steve Uhlig for many interesting discussions.

References

1. Borodin, A., El-Yaniv, R.: Online Computation and Competitive Analysis. Cambridge University Press (1998)
2. Bu, T., Gao, L., Towsley, D.: On characterizing BGP routing table growth. Comput. Netw. 45, 45–54 (2004)
3. Cittadini, L., Muhlbauer, W., Uhlig, S., Bushy, R., Francois, P., Maennel, O.: Evolution of internet address space deaggregation: myths and reality. IEEE J. Sel. A. Commun. 28, 1238–1249 (2010)
4. Draves, R.P., King, C., Venkatachary, S., Zill, B.D.: Constructing optimal IP routing tables. In: Proc. of the 18th IEEE Int. Conference on Computer Communications (INFOCOM), pp. 88–97 (1999)

5. Elmokashfi, A., Kvalbein, A., Dovrolis, C.: BGP churn evolution: a perspective from the core. IEEE/ACM Transactions on Networking 20(2), 571–584 (2012)
6. Li, J., Guidero, M., Wu, Z., Purpus, E., Ehrenkranz, T.: BGP routing dynamics revisited. ACM SIGCOMM Computer Communication Review 37, 5–16 (2007)
7. Liu, Y., Zhang, B., Wang, L.: Fast incremental FIB aggregation. In: Proc. of the 32nd IEEE Int. Conference on Computer Communications, INFOCOM (2013)
8. Liu, Y., Zhao, X., Nam, K., Wang, L., Zhang, B.: Incremental forwarding table aggregation. In: Proc. of the Global Communications Conference (GLOBECOM), pp. 1–6 (2010)
9. Luo, L., Xie, G., Uhlig, S., Mathy, L., Salamatian, K., Xie, Y.: Towards TCAM-based scalable virtual routers. In: Proc. of the 8th Int. Conf. on Emerging Networking Experiments and Technologies (CoNEXT), pp. 73–84 (2012)
10. McKeown, N., Anderson, T., Balakrishnan, H., Parulkar, G., Peterson, L., Rexford, J., Shenker, S., Turner, J.: OpenFlow: enabling innovation in campus networks. ACM SIGCOMM Computer Communication Review 38, 69–74 (2008)
11. Medhi, D., Ramasamy, K.: Network Routing: Algorithms, Protocols, and Architectures. Morgan Kaufmann Publishers Inc. (2007)
12. Rétvári, G., Csernátony, Z., Körösi, A., Tapolcai, J., Császár, A., Enyedi, G., Pongrácz, G.: Compressing IP forwarding tables for fun and profit. In: Proc. of the 11th ACM Workshop on Hot Topics in Networks (HotNets), pp. 1–6 (2012)
13. RouteViews Project (2013), http://www.routeviews.org/
14. Sarrar, N., Bienkowski, M., Schmid, S., Uhlig, S., Wuttke, R.: Exploiting locality of churn for FIB aggregation. Technical Report 2012/12, Technische Universität Berlin (2012)
15. Sarrar, N., Uhlig, S., Feldmann, A., Sherwood, R., Huang, X.: Leveraging Zipf's law for traffic offloading. ACM SIGCOMM Computer Communication Review 42(1), 16–22 (2012)
16. Suri, S., Sandholm, T., Warkhede, P.R.: Compressing two-dimensional routing tables. Algorithmica 35(4), 287–300 (2003)
17. Uzmi, Z.A., Nebel, M., Tariq, A., Jawad, S., Chen, R., Shaikh, A., Wang, J., Francis, P.: SMALTA: Practical and near-optimal FIB aggregation. In: Proc. of the 7th Int. Conf. on Emerging Networking Experiments and Technologies (CoNEXT), pp. 29:1–29:12 (2011)
18. Zhao, X., Liu, Y., Wang, L., Zhang, B.: On the aggregatability of router forwarding tables. In: Proc. of the 29th IEEE Int. Conference on Computer Communications (INFOCOM), pp. 848–856 (2010)

A Nonmonotone Analysis with the Primal-Dual Approach: Online Routing of Virtual Circuits with Unknown Durations*

Guy Even and Moti Medina**

School of Electrical Engineering,Tel-Aviv Univ., Tel-Aviv 69978, Israel
{guy,medinamo}@eng.tau.ac.il

Abstract. We address the question of whether the primal-dual approach for the design and analysis of online algorithms can be applied to nonmonotone problems. We provide a positive answer by presenting a primal-dual analysis to the online algorithm of Awerbuch et al. [1] for routing virtual circuits with unknown durations.

1 Introduction

The analysis of most online algorithms is based on a potential function (see, for example, [1–4] in the context of online routing). Buchbinder and Naor [5] presented a primal-dual approach for analyzing online algorithms. This approach replaces the need to find the appropriate potential function by the task of finding an appropriate linear programming formulation.

The primal-dual approach presented by Buchbinder and Naor has a monotone nature. Monotonicity means that: (1) Variables and constraints arrive in an online fashion. Once a variable or constraint appears, it is never deleted. (2) Values of variables, if updated, are only increased. We address the question of whether the primal-dual approach can be extended to analyze nonmonotone algorithms[1].

An elegant example of nonmonotone behavior occurs in the problem of online routing of virtual circuits with unknown durations. In the problem of routing virtual circuits, we are given a graph with edge capacities. Each request r_i consists of a source-destination pair (s_i, t_i). A request r_i is served by allocating to it a path from s_i to r_i. The goal is to serve the requests while respecting the edge capacities as much as possible. In the online setting, requests arrive one-by-one. Upon arrival of a request r_i, the online algorithm must serve r_i. In the special case of unknown durations, at each time step, the adversary may introduce a new request or it may terminate an existing request. When a request terminates,

---

* The full version of this paper can be found in http://arxiv.org/abs/1304.7687.
** M.M was partially funded by the Israeli ministry of Science and Technology.
[1] The only instance we are aware of in which the primal-dual approach is applied to nonmonotone variables appears in [6]. In this instance, the change in the dual profit, in each round, is at least a constant times the change in the primal profit. In general, this property does not hold in a nonmonotone setting.

T. Moscibroda and A.A. Rescigno (Eds.): SIROCCO 2013, LNCS 8179, pp. 104–115, 2013.
© Springer International Publishing Switzerland 2013

it frees the path that was allocated to it, thus reducing the congestion along the edges in the path. The online algorithm has no knowledge of the future; namely, no information about future requests and no information about when existing requests will end. Nonmonotonicity is expressed in this online problem in two ways: (1) Requests terminate thus deleting the demand to serve them. (2) The congestion of edges varies in a nonmonotone fashion; an addition of a path increases congestion, and a deletion of a path decreases congestion.

Awerbuch et al. [1] presented an online algorithm for online routing of virtual circuits when the requests have unknown durations. In fact, their algorithm resorts to rerouting to obtain a logarithmic competitive ratio for the load. Rerouting means that the path allocated to a request is not fixed and the algorithm may change this path from time to time. Hence, allowing rerouting increases the nonmonotone characteristics of the problem.

We present an analysis of the online algorithm of Awerbuch et al. [1] for online routing of virtual circuits with unknown durations. Our analysis uses the primal-dual approach, and hence we show that the primal-dual approach can be applied in nonmonotone settings.

2 Problem Definition

2.1 Online Routing of Virtual Circuits with Unknown Durations

Let $G = (V, E)$ denote a directed or undirected graph. Each edge e in E has a capacity $c_e \geq 1$. A routing request r_k is a 4-tuple $r_k = (s_k, d_k, a_k, b_k)$, where (i) $s_k, d_k \in V$ are the source and the destination of the kth routing request, respectively, (ii) $a_k \in \mathbb{N}$ is both the arrival time and the start time of the request, and (iii) $b_k \in \mathbb{N}$ is the departure time or end time of the request. Let Γ_k denote the set of paths in G from s_k to d_k. A request r_k is served if it is allocated a path in Γ_k.

Let $[N]$ denote the set $\{0, \ldots, N\}$. The input consists of a sequence of events $\sigma = \{\sigma_t\}_{t \in [N]}$. We assume that time is discrete, and event σ_t occurs at time t. There are two types of events: (i) An *arrival* of a request. When a request r_k arrives, we are given the source s_k and the destination d_k. Note that the arrival time a_k simply equals the current time t. (ii) A *departure* of a request. When a request r_k departs there is no need to serve it anymore (namely, the departure time b_k simply equals the current time t).

The set of active requests at time t is denoted by $Alive_t$ and is defined by $Alive_t \triangleq \{r_k \mid a_k \lessgtr t \leq b_k\}$.

An *allocation* is a sequence $A = \{p_k\}_k$ of paths such that p_k is a path from the source s_k to the destination d_k of request r_k. Let $paths_t(e, A)$ denote the number of requests that are routed along edge e by allocation A at time t, formally: $paths_t(e, A) \triangleq |\{p_k : e \in p_k \text{ and } r_k \in Alive_t\}|$. The *load* of an edge e at time t is defined by $load_t(e, A) \triangleq \frac{paths_t(e, A)}{c_e}$. The *load* of an allocation A at time t is defined by $load_t(A) \triangleq \max_{e \in E} load_t(e, A)$. The *load* of an allocation A is defined by $load(A) \triangleq \max_t load_t(A)$.

An algorithm computes an allocation of paths to the requests, and therefore we abuse notation and identify the algorithm with the allocation that is computed by it. Namely, ALG(σ) denotes the allocation computed by algorithm ALG for an input sequence σ.

In the online setting, the events arrive one-by-one, and no information is known about an event before its arrival. Moreover, (1) the length N of the sequence of events is unknown; the input simply stops at some point, (2) the departure time b_k is *unknown* (and may even be determined later by the adversary), and (3) the online algorithm must allocate a path to the request as soon as the request arrives.

The *competitive ratio* of an online algorithm ALG with respect to $N \in \mathbb{N}$, and a sequence $\sigma = \{\sigma_t\}_{t \in [N]}$ is defined by $\rho(\text{ALG}(\sigma)) \triangleq \frac{load(\text{ALG}(\sigma))}{load(\text{OPT}(\sigma))}$, where OPT($\sigma$) is an allocation with minimum load. The *competitive ratio* of an online algorithm ALG is defined by $\rho(\text{ALG}) \triangleq \sup_{N \in \mathbb{N}} \max_\sigma \rho(\text{ALG}(\sigma))$.

Note that since every request has a unit demand, we may assume that $c_e \geq 1$ for every edge $e \in E$.

2.2 Rerouting

In the classical setting, a request r_k is served by a fixed single path p_k throughout the duration of the request. The term *rerouting* means that we allow the allocation to change the path p_k that serves r_k. Thus, there are two extreme cases: (i) no rerouting at all is permitted (classical setting), and (ii) total flexibility in which, a new allocation can be computed in each time step.

Following the paper by Awerbuch et al. [1], we allow the online algorithm to reroute each request at most $O(\log |V|)$ times. In the analysis of the competitive ratio, we compare the load of the online algorithm with the load of an optimal (splittable) allocation with total rerouting flexibility. Namely, the optimal solution recomputes a minimum load allocation at each time step, and, in addition may serve a request by a convex combination of paths.

3 The Online Algorithm ALG

In this section we present the online algorithm ALG that is listed in Algorithm 1. Thus algorithm is equivalent to the algorithm presented in [1].

The algorithm maintains the following variables. (1) For every edge e a variable x_e. The value of x_e is exponential in the load of edge e. (2) For every request r_k a variable z_k. The value of z_k is the complement of the "weight" of the path p_k allocated to r_k at the time the path was allocated. (3) For every routing request r_k, and for every path $p \in \Gamma_k$ a variable $f_k(p)$. The value of $f_k(p)$ indicates whether p is allocated to r_k. That is, the value of $f_k(p)$ equals 1 if path p is allocated for request r_k, and 0 otherwise.

The algorithm ALG consists of the following 5 procedures: (1) MAIN, (2) ROUTE, (3) DEPART, (4) UNROUTE, and (5) MAKEFEASIBLE.

The MAIN procedure begins with initialization. For every $e \in E$, x_e is initialized to $\frac{1}{4m}$, where $m = |E|$. For every $k \in [N]$, z_k is initialized to zero. For every $k \in [N]$, and for every path p, $f_k(p)$ is initialized to zero. Since the number of z_k and $f_k(p)$ variables is unbounded, their initialization is done in a "lazy" fashion; that is, upon arrival of the kth request the corresponding variables are set to zero.

The main procedure MAIN proceeds as follows. For every time step $t \in [N]$, if the event σ_t is an arrival of a request, then the ROUTE procedure is invoked. Otherwise, if the event σ_t is a departure of a request, then the DEPART procedure is invoked.

The ROUTE procedure serves request r_k by allocating a "lightest" path p_k in the set Γ_k (recall that Γ_k denotes the set of paths from the source s_k to the destination d_k). The allocation is done by two actions. First, the allocation of p_k to request r_k is indicated by setting $f_k(p_k) \leftarrow 1$. Second, the loads of the edges along p_k are updated by increasing the variables x_e for $e \in p_k$. The variable z_k equals the "complement" weight of the allocated path p_k. Note that this complement is with respect to half the weight of the path before its update.

The DEPART procedure "frees" the path that is allocated for p_k, by calling the UNROUTE procedure. The UNROUTE procedure frees p_k by nullifying $f_k(p_k)$ and z_k, and by decreasing the edge variables x_e for the edges along p_k. The freeing of p_k decreases the load along the edges in p_k. As a result of this decrease, it may happen that a path allocated to an alive request might be very heavy compared to a lightest path. In such a case, the request should be rerouted. This is why the MAKEFEASIBLE procedure is invoked after the UNROUTE procedure.

Rerouting is done by the MAKEFEASIBLE procedure. This rerouting is done by freeing a path and then routing the request again. Requests with improved alternative paths are rerouted.

The listing of the online algorithm ALG appears in Algorithm 1.

4   Primal-Dual Analysis of ALG

In this section we prove that the load on every edge is always $O(\log |V|)$, and that each request is rerouted at most $O(\log |V|)$ times. We refer to an input sequence σ as *feasible* if there is an allocation A, such that for all requests that are alive at time t, it holds that $load_t(A) \leq 1$. The following theorem holds under the assumption that the input sequence σ is feasible. Note that the removal of this assumption increases the competitive ratio only by a constant factor by standard doubling techniques [1].

Theorem 1 ([1]). *If the input sequence σ is feasible and assuming that $c_e \geq 1$, then ALG is:*

1. *An $O(\log |V|)$-competitive online algorithm.*
2. *Every request is rerouted at most $O(\log |V|)$ times.*

We point out that the allocation computed by ALG is *nonsplittable* in the sense that at every given time each request is served by a single path. The optimal

Algorithm 1 ALG: Online routing algorithm. The input consists of (1) a graph $G = (V, E)$ where each $e \in E$ has capacity c_e, and (2) a sequence of events $\sigma = \{\sigma_t\}_{t \in [N]}$.

Main(σ_t)
1: $\forall k \in [N] : z_k \leftarrow 0$.
2: $\forall e \in E : x_e \leftarrow \frac{1}{4m}$, where $m = |E|$.
3: $\forall r_k \in [N] \; \forall p : f_k(p) \leftarrow 0$.
4: **Upon** arrival of event σ_t **do**
5:     **if** σ_t is an arrival of request r_k **then Call Route(r_k)**.
6:     **else** (σ_t is an departure of request r_k) **Call Depart(r_k)**.

Route(r_k)
1: Find the "lightest" path: $p_k \leftarrow \operatorname{argmin}\{\sum_{e \in p'} \frac{x_e}{c_e} \mid p' \in \Gamma_k\}$.
2: $z_k \leftarrow 1 - \frac{1}{2} \cdot \sum_{e \in p_k} \frac{x_e}{c_e}$.
3: Route r_k along p_k: $f_k(p_k) \leftarrow 1$.
4: **for all** $e \in p_k$ **do**
5:     $x_e \leftarrow x_e \cdot \lambda_e$ where $\lambda_e \triangleq \left(1 + \frac{1}{4c_e}\right)$. {Update edge "load"}

Depart(r_k)
1: **Call** UNROUTE(r_k).
2: **Call** MAKEFEASIBLE(x, z).

UnRoute(r_k)
1: Free variables: $z_k, f_k(p_k)$.
2: **for all** $e \in p_k$ **do**
3:     $x_e \leftarrow x_e/\lambda_e$ where $\lambda_e \triangleq \left(1 + \frac{1}{4c_e}\right)$. {Update edge "load"}

MakeFeasible(x, z)
1: $\forall r_j \in Alive_t$ **if** $\exists p \in \Gamma_j \; : z_j + \sum_{e \in p} \frac{x_e}{c_e} < 1$ **then**
2:     **Call** UNROUTE(r_j).
3:     **Call** ROUTE(r_j).

allocation, on the other hand, is both totally flexible and *splittable*. Namely, the optimal allocation may reroute all the requests in each time step, and, in addition, may serve a request by a convex combination of paths.

The rest of the proof is as follows. We begin by formulating a packing and covering programs for our problem in Section 4.1. We then prove Lemma 1 in Section 4.2. We conclude the analysis with the proof of Theorem 1 in Section 4.3

4.1   Formulation as an Online Packing Problem

For the sake of analysis, we define for every prefix of events $\{\sigma_j\}_{j=1}^{t}$ a *primal* linear program P-LP(t) and its *dual* linear program D-LP(t). The primal LP is a *covering* LP, and the dual LP is a *packing* LP. The LP's appear in Figure 1.

The variables of the LPs correspond to the variables maintained by ALG, as follows. The covering program P-LP(t) has a variable x_e for every edge $e \in E$, and a variable z_k for every $r_k \in Alive_t$. The packing program D-LP(t) has a variable $f_k(p)$ for every request $r_k \in Alive_t$, and for every path $p \in \Gamma_k$. The variable $f_k(p)$ equals to the fraction of r_k's "demand" that is routed along path $p \in \Gamma_k$.

The dual LP has three types of constraints: capacity constraints, demand constrains, and sign constraints. In the fractional setting the load of an edge is defined by

$$load_t(e) \triangleq \frac{1}{c_e} \cdot \sum_{r_k \in Alive_t} \sum_{\{p | p \in \Gamma_k, e \in p\}} f_k(p) .$$

The capacity constraint in the dual LP requires that the load of each edge is at most one. The demand constraints require that each request r_k that is alive at time t is allocated a convex combination of paths.

If the dual LP is feasible, then the objective function of the dual LP simply equals the number of requests that are alive at time step t, i.e., $|Alive_t|$.

The primal LP has two types of constraints: covering constraints and sign constraints. The covering constraints requires that for every request r_k that is alive and for every path $p \in \Gamma_k$, the sum of z_k and the "weight" of p is at least 1. Note that the sign constraints apply only to the edge variables x_e whereas the request variables z_k are free.

Note that the assumption that σ is feasible is equivalent to requiring that the dual program D-LP(t) is feasible for every t.

$$\underline{\text{P-LP}(t)}: \quad \min \sum_{r_k \in Alive_t} z_k + \sum_{e \in E} x_e \text{ s.t.}$$

$$\forall r_k \in Alive_t \; \forall p \in \Gamma_k : z_k + \sum_{e \in p} \frac{x_e}{c_e} \geq 1 \text{ (Covering Constraints.)}$$

$$x \geq 0$$

(I)

$$\underline{\text{D-LP}(t)}: \quad \max \sum_{r_k \in Alive_t} \sum_{p \in \Gamma_k} f_k(p) \text{ s.t.}$$

$$\forall e \in E : \frac{1}{c_e} \cdot \sum_{r_k \in Alive_t} \sum_{\{p | p \in \Gamma_k, e \in p\}} f_k(p) \leq 1 \text{ (Capacity Constraints.)}$$

$$\forall r_k \in Alive_t : \sum_{p \in \Gamma_k} f_k(p) = 1 \text{ (Demand Constraints.)}$$

$$f \geq 0$$

(II)

Fig. 1. (I) The primal LP, P-LP(t). (II) The dual LP, D-LP(t).

4.2   Bounding the Primal Variables

In this section we prove that the primal variables x_e are bounded by a constant, as formalized in the following Lemma.

Lemma 1. *If σ_t is an original event, then $\forall e \in E : x_e^{(t)} \leq 3$.*

The proof of Lemma 1 is based on a few lemmas that we prove first.

Notation. Let $x_e^{(t)}, z_k^{(t)}$ denote the value of the primal variables x_e, z_k before event σ_t is processed by ALG. Let P_t denote the objective function's value of P-LP(t), formally:

$$P_t \triangleq \sum_{r_k \in Alive_t} z_k^{(t)} + \sum_{e \in E} x_e^{(t)}.$$

Let $\Delta_t P \triangleq P_{t+1} - P_t$.

Note that P_t refers to the value of P-LP(t) at the beginning of time step t. The definition of $Alive_t$ implies that the constraints and variables of P-LP(t) are not influenced by the event σ_t (this happens only for P-LP$(t+1)$). Hence the variables in the definition of P_t are indexed by time step t.

Dummy events. The procedure ROUTE is invoked in two places: (i) in Line 5 of MAIN as a result of an arrival of a request, or (ii) in Line 3 of MAKEFEASIBLE. To simplify the discussion, we create "dummy" events each time the MAKEFEASIBLE procedure reroutes a request. Dummy events come in pairs: first a dummy departure event for request r_k is introduced, and then a dummy arrival event for a "continuation" request r_k is introduced. The combination of original events and dummy events describes the execution of ALG. The augmentation of the original input sequence of events by dummy events does not modify the optimal value of the dual LP at time steps t that correspond to original events. Hence, we analyze the competitive ratio $\rho(\text{ALG}(\sigma))$ by analyzing the competitive ratio with respect to the augmented sequence at time steps t that correspond to original events.

The following lemma follows immediately from the description of the algorithm ALG and the definition of dummy events.

Lemma 2 (Primal Feasibility). *If σ_t is an original event, then the variables $\{x_e^{(t)}\}_{e \in E} \cup \{z_\ell^{(t)}\}_{\ell \in Alive_t}$ constitute a feasible solution for P-LP(t).*

Proof. When an original event $\sigma_{t'}$ occurs, the MAKEFEASIBLE procedure generates dummy events at the end of the time step to guarantee that the primal variables are a feasible solution of the primal LP. Hence, if σ_t is an original event, then the primal variables at the beginning of time step t are a feasible solution for P-LP(t).

Lemma 3. *If σ_t is an arrival of request, then $\Delta_t P < 1$.*

Proof. Assume that σ_t is an event in which request r_k arrives. In Step 2 of the ROUTE algorithm z_k is set to $1 - \frac{1}{2} \cdot \sum_{e \in p_k} \frac{x_e^{(t)}}{c_e}$. In Step 5 of the ROUTE algorithm, for every $e \in p_k$, x_e is increased by $\frac{x_e^{(t)}}{4c_e}$. All the other edge variables x_e remain unchanged. Hence,

$$
\begin{aligned}
\Delta_t P &= 1 - \frac{1}{2} \cdot \sum_{e \in p_k} \frac{x_e^{(t)}}{c_e} + \sum_{e \in p_k} \frac{x_e^{(t)}}{4c_e} \\
&= 1 - \frac{1}{4} \cdot \sum_{e \in p_k} \frac{x_e^{(t)}}{c_e} \tag{1} \\
&< 1,
\end{aligned}
$$

as required.

We refer to the number of requests that are routed along edge e by allocation ALG at time t by $paths_t(e)$.

Lemma 4. *For every t and $e \in E$, $x_e^{(t)} = \frac{1}{4m} \cdot \lambda_e^{paths_t(e)}$.*

Proof. The proof is by induction on t. At time $t = 0$, we have $x_e^{(0)} = \frac{1}{4m}$ and $paths_t(e) = 0$. The proof of the induction basis for $t + 1$ depends on whether at time step t an arrival or a departure occurs. If the event does not affect edge e, then the induction step clearly holds. Assume that the event affects edge e. If a request r_k arrives at time t, then $paths_{t+1}(e) = paths_t(e) + 1$ and $x_e^{(t+1)} = x_e^{(t)} \cdot \lambda_e$. If a request r_k departs at time t, then $paths_{t+1}(e) = paths_t(e) - 1$ and $x_e^{(t+1)} = x_e^{(t)} / \lambda_e$.

Let $Dead_t \triangleq \{r_k \mid b_k < t\}$. In general, it is not true that $\Delta_{a_j} P + \Delta_{b_j} P \leq 0$, however on average it is true, as stated in the following lemma.

Lemma 5. *For every t,*

$$
\sum_{r_j \in Dead_t} \left(\Delta_{a_j} P + \Delta_{b_j} P \right) \leq 0. \tag{2}
$$

Proof. First we prove the following proposition.

Proposition 1. *Consider a set of $I = \{I_j = [\alpha_j, \beta_j]\}_{j=1}^{q}$ such that no two intervals share a common endpoint. Let $cut(t)$ denote the number of intervals that contain t. Then, there is a permutation $\pi : [1, q] \to [1, q]$ such that*

$$
\forall j \in [1, q] : \quad cut(\alpha_j) = cut(\beta_{\pi(j)}). \tag{3}
$$

Proof. The proof is by induction on the number of intervals. The induction basis, for $q = 1$ holds trivially because $cut(\alpha_1) = cut(\beta_1) = 1$. The proof of the induction step is based on the existence of a pair $\alpha_i < \beta_j$ such that the open interval (α_i, β_j) does not contain any endpoint of the intervals in I. For such a pair, we immediately have $cut(\alpha_i) = cut(\beta_j)$ so we define $\pi(i) = j$ and apply the induction hypothesis.

We first show that such a pair $\alpha_i < \beta_j$ exists. We say that an interval I_m is *minimal* if $I_m \cap I_k \neq \emptyset$ implies that $I_m \subseteq I_k$. If there exists a minimal interval I_m, then set $\alpha_i = \alpha_m$ and $\beta_j = \beta_m$. In such a case since $\pi(m) = m$, we can erase I_m and proceed by applying the induction hypothesis to the remaining intervals. Note that equality of cut sizes is preserved when the interval I_m is deleted.

Consider the set of pairs of intersecting intervals without containment defined as follows $A \triangleq \{(i,j) \mid \alpha_j < \alpha_i < \beta_j < \beta_i\}$. If there is no minimal interval, the set A is not empty. Any pair $(i,j) \in A$ that minimizes the difference $(\beta_j - \alpha_i)$ has the property that the interval (α_i, β_j) lacks endpoints of intervals in I.

We can define $\pi(i) = j$. We proceed by applying the induction hypothesis on $(I \setminus \{I_j, I_i\}) \cup I_k$, where $I_k = I_i \cup I_j$. Note that equality of cut sizes is preserved when I_i and I_j are merged into one interval.

The difference $\Delta_{a_j} P$ consists of two parts: $\Delta_{a_j} P = z_j^{(a_j+1)} + \sum_{e \in p_j} \frac{x_e^{(a_j)}}{4c_e}$. The difference $\Delta_{b_j} P$ consists of two parts as well: $\Delta_{b_j} P = -z_j^{(b_j)} - \sum_{e \in p_j} \frac{x_e^{(b_j+1)}}{4c_e}$. It follows that

$$\sum_{r_j \in Dead_t} (\Delta_{a_j} P + \Delta_{b_j} P) = \sum_{r_j \in Dead_t} \sum_e \frac{1}{4c_e} \cdot \left(x_e^{(a_j)} - x_e^{(b_j+1)} \right)$$

$$= \sum_{r_j \in Dead_t} \sum_e \frac{1}{4c_e} \cdot \left(x_e^{(a_j)} - x_e^{(b_{\pi(j)}+1)} \right),$$

where π is any permutation over the set of requests. In fact, we shall use for each edge e, a different permutation $\pi = \pi(e)$ that is a permutation over the requests r_k such that $e \in p_k$.

Assume first that $Alive_t = \emptyset$. We later lift this assumption.

Fix an edge e. For each request r_j such that $e \in p_j$, map the duration $(a_j, b_j]$ of request r_j to the interval $[a_j + 1, b_j]$. The resulting set of intervals satisfies $cut(t) = paths_t(e)$ for every time step t. Let π denote the permutation guaranteed by Prop. 1. Then, it suffices to prove that

$$x_e^{(a_j)} - x_e^{(b_{\pi(j)}+1)} = 0. \tag{4}$$

Indeed, by Lemma 4, $4m \cdot \left(x_e^{(a_j)} - x_e^{(b_{\pi(j)}+1)} \right) = \lambda_e^{paths_{a_j}} - \lambda_e^{paths_{b_{\pi(j)}+1}}$. In addition, the property of permutation π states that $cut(a_j+1) = cut(b_{\pi(j)})$. It follows that $paths_{a_j+1} = paths_{b_{\pi(j)}}$. But, $paths_{a_j} = paths_{a_j+1} - 1$ and $paths_{b_{\pi(j)}+1} = paths_{b_{\pi(j)}} - 1$, and Equation 4 follows.

To complete the proof, consider the requests in $Alive_t$. Because $a_j, b_{\pi(j)} \leq t$, requests in $Alive_t$ do not increase the difference $x_e^{(a_j)} - x_e^{(b_{\pi(j)}+1)}$. Thus $x_e^{(a_j)} - x_e^{(b_{\pi(j)}+1)} \leq 0$, and the lemma follows.

We are now ready to prove Lemma 1. Recall that Lemma 1 states that the primal variables x_e are bounded by a constant. The proof of Lemma 1 is by contradiction. In fact, we reach a contradiction to *weak duality*, that is, we show that the value of the primal solution is strictly smaller than the value of a feasible dual solution.

Proof (Lemma 1). The proof is by contradiction. Assume $x_e^{(t)} > 3$ and σ_t is an original event. Define $t_2 \triangleq \min\{t \mid x_e^{(t)} > 3$ and σ_t is an original event$\}$. Let t_1 be the time step for which $x_e^{(t_1)} < 1$ and $x_e^{t'} \geq 1$ for every $t' \in [t_1 + 1, t_2]$.

Define $Alive_{\in e}(t_1, t_2) \triangleq \{r_j \mid t_1 < a_j < t_2 < b_j, e \in p_j\}$.

Let δ_e denote the difference between the number of arrivals and the number of departures in the time interval $[t_1, t_2)$ among the requests that were routed along e. Clearly $\delta_e \leq |Alive_{\in e}(t_1, t_2)|$.

Lemma 4 implies that $x_e^{(t_2)} = x_e^{(t_1)} \cdot \left(1 + \frac{1}{4c_e}\right)^{\delta_e}$. The assumption that $x_e^{(t_2)} > 3$ and $x_e^{(t_1)} < 1$ imply that $\left(1 + \frac{1}{4c_e}\right)^{\delta_e} \geq 3$. Since $1 + x \leq e^x$, it follows that $\delta_e > 4 \cdot c_e$. Hence,

$$|Alive_{\in e}(t_1, t_2)| > 4 \cdot c_e. \tag{5}$$

By Equation 1, for each $r_j \in Alive_{\in e}(t_1, t_2)$, we have:

$$\Delta_{a_j} P < 1 - \frac{1}{4c_e}. \tag{6}$$

Hence,

$$
\begin{aligned}
P_{t_2} &= \frac{1}{4m} \cdot m + \sum_{t=0}^{t_2-1} \Delta_t P \\
&= \frac{1}{4} + \sum_{r_j \in Dead_{t_2}} (\Delta_{a_j} P + \Delta_{b_j} P) + \sum_{r_j \in Alive_{t_2}} \Delta_{a_j} P \\
&\leq \frac{1}{4} + \sum_{r_j \in Alive_{t_2}} \Delta_{a_j} P \\
&< \frac{1}{4} + |Alive_{t_2}| - \frac{|Alive_{\in e}(t_1, t_2)|}{4c_e} \\
&< |Alive_{t_2}|.
\end{aligned}
\tag{7}
$$

The justification for these lines is as follows. The first line follows from the initialization of the primal variables. The second line follows since every event in time step $t \in [0, t_2 - 1]$ is either an arrival of a request in $Dead_{t_2} \cup Alive_{t_2}$ or a departure of a request in $Dead_{t_2}$. The third inequality is due to Lemma 5. The fourth equation is due to Equation 6. The last inequality follows from Equation 5.

By Lemma 2, the primal variables at time t_2 are a feasible solution of P-LP(t_2). The optimal value of D-LP(t_2) equals $|Alive_{t_2}|$. Hence, Equation 7 contradicts weak duality, and the lemma follows.

4.3   Proof of Theorem 1

We now turn to the proof of the main result. The proof is as follows.

Proof (Theorem 1). We begin by proving the bound on the competitive ratio. Lemma 4 states that $\forall t \ \ \forall e \in E : x_e = \frac{1}{4m} \cdot \left(1 + \frac{1}{4c_e}\right)^{paths_t(e)}$. Hence, by Lemma 1, for each original event σ_t, $\forall e \in E : \frac{1}{4m} \cdot \left(1 + \frac{1}{4c_e}\right)^{paths_t(e)} \leq 3$. Since $2^x \leq 1 + x$ for all $x \in [0, 1]$, it follows that for each original event σ_t

$$\forall e \in E : paths_t(e) \leq c_e \cdot 4 \log(12m) \,,$$

and the first part of the theorem follows.

We now prove the bound on the number of reroutes. Rerouting an alive request r_j occurs if there exists a path $p \in \Gamma_j$ such that $\sum_{e \in p} \frac{x_e}{c_e} < 1 - z_j$. By Line 2 of the ROUTE algorithm, this condition is equivalent to: $\sum_{e \in p} \frac{x_e}{c_e} < \frac{1}{2} \cdot \sum_{e \in p_j} \frac{x_e^{(a_j)}}{c_e}$. Namely, each time a request is rerouted, the weight of the path is at least halved. Note that the halving is with respect to the weight of the path at the time it was allocated.

Let us consider request r_j. Let $p^* \triangleq \text{argmin}_{p \in \Gamma_j} \{\sum_{e \in p} \frac{1}{c_e}\}$. By the choice of a "lightest" path and by Lemma 1, the weight of path p_j is upper bounded by

$$\sum_{e \in p_j} \frac{x_e}{c_e} \leq \sum_{e \in p^*} \frac{x_e}{c_e} \leq 3 \cdot \sum_{e \in p^*} \frac{1}{c_e} \,.$$

By Lemma 4, $x_e \geq 1/(4m)$, hence the weight of path p_j is lower bounded by

$$\sum_{e \in p} \frac{x_e}{c_e} \geq \frac{1}{4m} \cdot \sum_{e \in p} \frac{1}{c_e} \geq \frac{1}{4m} \cdot \sum_{e \in p^*} \frac{1}{c_e} \,.$$

It follows that the number of reroutes each request undergoes is bounded by $\log_2 (12m)$, and the second part of the theorem follows.

Remark 1. *Note that the first routing request will not be rerouted at all, the second routing request will be rerouted at most twice, and so on. In general, a routing request that arrives at time t will be rerouted at most $|Alive_t|$ times.*

5   Discussion

We present a primal-dual analysis of an online algorithm in a nonmonotone setting. Specifically, we analyze the online algorithm by Awerbuch et al. [1] for online routing of virtual circuits with unknown durations. We think that the main advantage of this analysis is that it provides an alternative explanation to the stability condition for rerouting that appears in [1]. According to the primal-dual analysis, rerouting is used simply to preserve the feasibility of the solution of the covering LP.

Our analysis provides a small improvement compared to [1] in the following sense. The optimal solution in our analysis is both totally flexible (i.e., may reroute every request in every time step) and splittable (i.e., may serve a request using a convex combination of paths). The optimal solution in the analysis of Awerbuch et al. [1] is only totally flexible and must allocate a path to each request.

The primal-dual approach of Buchbinder and Naor [5] is based on bounding the change in the value of the primal solution by the change in the dual solution (this is often denoted by $\Delta P \leq \Delta D$). The main technical challenge we encountered was that this bound simply does not hold in our case. Instead, we use an averaging argument to prove an analogous result (see Lemma 5).

References

1. Awerbuch, B., Azar, Y., Plotkin, S., Waarts, O.: Competitive routing of virtual circuits with unknown duration. Journal of Computer and System Sciences 62(3), 385–397 (2001)
2. Awerbuch, B., Azar, Y., Plotkin, S.: Throughput-competitive on-line routing. In: FOCS 1993: Proceedings of the 1993 IEEE 34th Annual Foundations of Computer Science, pp. 32–40. IEEE Computer Society, Washington, DC (1993)
3. Azar, Y., Kalyanasundaram, B., Plotkin, S., Pruhs, K.R., Waarts, O.: On-line load balancing of temporary tasks. Journal of Algorithms 22(1), 93–110 (1997)
4. Aspnes, J., Azar, Y., Fiat, A., Plotkin, S., Waarts, O.: On-line routing of virtual circuits with applications to load balancing and machine scheduling. Journal of the ACM (JACM) 44(3), 486–504 (1997)
5. Buchbinder, N., Naor, J.S.: The design of competitive online algorithms via a primal-dual approach. Foundations and Trends in Theoretical Computer Science 3(2-3), 99–263 (2009)
6. Buchbinder, N., Feldman, M., Ghosh, A., Naor, J(S.): Frequency capping in on-line advertising. In: Dehne, F., Iacono, J., Sack, J.-R. (eds.) WADS 2011. LNCS, vol. 6844, pp. 147–158. Springer, Heidelberg (2011)

Self-organizing Flows in Social Networks

Nidhi Hegde[1,*], Laurent Massoulié[2], and Laurent Viennot[3,**]

[1] Technicolor, France
nidhi.hegde@technicolor.com
[2] Microsoft Research - INRIA Joint Centre, France
laurent.massoulie@inria.fr
[3] INRIA – Paris Diderot University, France
laurent.viennot@inria.fr

Abstract. Social networks offer users new means of accessing information, essentially relying on "social filtering", i.e. propagation and filtering of information by social contacts. The sheer amount of data flowing in these networks, combined with the limited budget of attention of each user, makes it difficult to ensure that social filtering brings relevant content to interested users. Our motivation in this paper is to measure to what extent self-organization of a social network results in efficient social filtering.

To this end we introduce *flow games*, a simple abstraction that models network formation under selfish user dynamics, featuring user-specific interests and budget of attention. In the context of homogeneous user interests, we show that selfish dynamics converge to a stable network structure (namely a pure Nash equilibrium) with close-to-optimal information dissemination. We show that, in contrast, for the more realistic case of heterogeneous interests, selfish dynamics may lead to information dissemination that can be arbitrarily inefficient, as captured by an unbounded "price of anarchy".

Nevertheless the situation differs when user interests exhibit a particular structure, captured by a metric space with low doubling dimension. In that case, natural autonomous dynamics converge to a stable configuration. Moreover, users obtain all the information of interest to them in the corresponding dissemination, provided their budget of attention is logarithmic in the size of their interest set.

Keywords: Network formation, self organisation, budget of attention, price of anarchy, social filtering.

1 Introduction

Information access has been revolutionized by the advent of social networks such as Facebook, Google+ and Twitter. These platforms have brought about the new paradigm of "social filtering", whereby one accesses information by "following" social contacts.

This is especially true for twitter-like microblogging social networks. In such networks the functions of filtering, editing and disseminating news are totally distributed,

---

* This work was partially funded by the European Commission under the FIRE SCAMPI (FP7- IST-258414) project and by the French National Research Agency (ANR) under the PROSE project (ANR-09-VERS-007).
** Supported by the INRIA project-team "Gang".

T. Moscibroda and A.A. Rescigno (Eds.): SIROCCO 2013, LNCS 8179, pp. 116–128, 2013.
© Springer International Publishing Switzerland 2013

in contrast to traditional news channels. The efficiency of social filtering is critically affected by the network topology, as captured by the contact-follower relationships. Today's networks provide recommendations to users for potentially useful contacts to follow, but don't interfere any further with topology formation. In this sense, these networks self-organize, under the selfish decisions of individual users.

This begs the following question: when does such autonomous and selfish self-organizing topology lead to efficient information dissemination? The answer will in turn indicate under what circumstances self-organization is insufficient, and thus when additional mechanisms, such as incentive schemes, should be introduced.

Two parameters play a key role in this problem. On the one hand each user aims to maximize the coverage of the topics of his interest. On the other hand, a user pays with his attention: filtering interesting information from spam (i.e. information that does not fall in his topics of interest) incurs a cost. Users must therefore trade-off topic coverage against attention cost. As pointed out by Simon [23], as information becomes abundant, another resource becomes scarce: attention.

Furthermore, there is an interplay between participants in a social network where filtering by one user may benefit another, inducing complex dependencies in decisions on creating connections. To model this, we introduce a network formation game called *flow game* where some users produce news about specific topics and each user is interested in receiving all news about a set of topics specific to him. Each user is a selfish agent that can choose its incoming connections within a certain budget of attention in order to maximize the coverage of his set of topics of interest.

This model is of interest on its own, as it enriches the class of existing network formation games with a focus on flow dissemination under bounded connections. This model could also be of interest in the context of peer-to-peer streaming and file sharing or publish/subscribe applications.

1.1 Our Results

An important feature in our model is a user's budget of attention for the consumption of content. In previous work [14] the budget of attention was modelled as a limit on the rate with which a user consults a friend, with a different objective of minimizing delay in receiving all content. In the present work we are interested in a more fundamental question, of how efficient social networks are formed in the first place. We consider the model where users are interested in specific subsets of topics and their objective is to maximize the number of flows received corresponding to these topics. As such, we model the budget of attention as a constraint on the number of connections a user may create (rather than a rate of consultation). Our aim is to build a simple model capturing the complexity of the problem. This way of capturing the budget of attention amounts to assuming that each connection consumes the same amount of attention. We discuss in the conclusion how we could tweak our model to more finely model attention consumption.

We capture users' interests in topics through user-specific values for each topic and define the *utility* a user receives to be the sum of values of all received topics. Each user's objective in a *flow game* is then to choose connections so as to maximize his utility. We additionally assume that a user may produce news about one topic at most even if he redistributes other topics. This is coherent with an empirical study of twitter traces [5] where it is shown that ordinary users (as opposed to celebrities or newspapers) can gain influence by concentrating on a single topic.

Our main results relate to the stability and efficiency of the formation of information flows. We derive conditions where selfish dynamics converge to a pure Nash equilibrium.We then give approximation ratios bounding the quality of an equilibrium compared to an optimal solution. This is traditionally measured through the price of anarchy, the ratio of the global welfare (measured as the sum of user utilities) at an optimal solution compared to that at the worst equilibrium.

1.2   Related Work

Information spread in networks has been studied extensively. Much of the past work study the properties of information diffusion on given networks with given sharing protocols. Our goal in this work is to study how networks form when users create connections with the objective of efficient content dissemination in a game-theoretical approach. This work thus follows the large amount of work in network formation games. However, to the best of our knowledge, the objective of efficient information dissemination under edge constraints and interest sets that we consider here is novel. We now discuss some work in those domains that are most relevant to this paper.

Network formation games have been considered in previous work in economics and in the context of the formation of Internet peering relations and peer-to-peer overlay networks. Economic models of network formation [13] use edges to represent social relations and it is typically assumed that the creation of an edge needs bilateral agreement since both users benefit from an edge. Our model is oriented and unilateral agreement is more relevant to the notion of *following* in social networks. A non-cooperative one-way link connection game has been considered been in previous work [3], where each created link incurs a cost and users are interested in connecting to all other users. Our model is richer and more realistic where we consider connections to subsets of information flows that hold user-specific intrinsic values.

Network creation games in the context of the Internet have been considered [19], where distributed formation of undirected edges with a linear cost on each edge formed is studied. In such games, each user's objective is to minimize total formation cost while either minimizing distance to all other users [7], or ensuring connection to a given subset of nodes [2]. We consider a bound on edge costs, in the form of a limit on the number of in-edges at each node, and further, we focus on connections that allow specific flows of information.

Interestingly, bounded budget network formation games have already been considered. Bounded budget connection games [16] consider a bound on each user's budget in creating edges, with the objective being the minimization of the sum of weighted distances to other nodes. A similar model is considered in [4] where each user's objective is to maximize his influence, measured using betweenness centrality. In our work however, rather than minimizing distance to any node, we consider a formation game with the objective of ensuring connections to a subset of flows of interest, without regard to the particular nodes.

The notion of connecting to users that can provide a content flow of interest is similar to peer-to-peer live streaming systems [17]. Unlike peer-to-peer streaming, we do not aim to satisfy flow rates, rather our aim is to connect to as many sets of relevant flows as possible. Moreover, our model allows differing user interests.

To the best of our knowledge the only work considering content dissemination with some game-theoretical approach concerns the b-matching and acyclic preference systems studied in the context of peer-to-peer applications [10]. As a generalization of the stable marriages problem, those systems consider configurations of undirected edges based on mutual acceptance of an edge, whereas unilateral decision is more suitable in our model. Our model is more intricate in the sense that connections are based not only on preferences but also on complementarity of content obtained through various connections.

In Section 5 we model the space of user interests by a metric space with low doubling dimension. Modeling interests of users through a metric space seems a natural approach and bounded growth metrics, or more generally doubling metrics, have shown to be very a general model [20] that can capture general situations, while still providing an algorithmic perspective. The doubling dimension extends the notion of dimension from Euclidean spaces to arbitrary metric spaces. It has proven to be useful in many application domains such as nearest neighbor queries to databases [6], network construction [1], closest server selction [15], etc. Doubling metrics have notably been used to model distances in networks such as Internet [9].

1.3 Organization of the Paper

Section 2 introduces the model. We study the case of homogeneous interests in Section 3. The heterogeneous case in its full generality is considered in Section 4 which details some negative results. Section 5 is dedicated to the specific scenario where users' interests are captured by a doubling metric, enabling some positive results. We finally conclude in Section 6 describing potential extensions of the current work.

2 Model

We consider a social network where users interested in some set of content topics (or subjects) connect to (or *follow* in social networking parlance) other users in order to obtain such contents, materialized by flows of news. Each user may produce news for at most one topic (but may forward news from other topics she is interested in). To distinguish the role of publisher from that of follower, we technically assume that news concerning a given topic (or subject) are produced at a given node called producer which is identified with that topic.

A *flow* game is defined as a tuple (V, P, S, Δ) where V is a set of users, P a set of producers (or subjects or topics) and $S : V \to P$ is a function associating to each user u its interest set $S_u \subseteq P$, and $\Delta : V \to \mathbb{N}$ is a function associating to each user u its budget of attention Δ_u. We let $n = |V|$ and $p = |P|$ denote the number of users and producers respectively. A flow game is *homogeneous* if all users have the same interest set: $S_u = P$ for all $u \in V$. If this is not the case, the game is said to be *heterogeneous*.

A strategy for user u is a subset F_u of $\{(v, u) : v \in V \cup P\}$ such that $|F_u| \leq \Delta_u$ (Δ_u is an upper bound on the in-degree of u). For all $(v, u) \in F_u$, we say that u *follows* v or equivalently that u is connected to v (such a link (v, u) created by u is oriented according to the data flow, that is from v to u). The collection $F = \{F_u : u \in V\}$ forms a network defined by the directed graph $G(F) = (V \cup P, E(F))$ where $E(F) = \cup_{u \in V} F_u$. A user u is *interested* in a subject s if $s \in S_u$. A user u *receives* a subject $s \in P$ if there exists a directed path from s to u in $G(F)$ such that all intermediate nodes are interested in s. This is where *filtering* occurs: a user retransmit only subjects she is interested in. The utility $U_u(F)$ for user u is the number of subjects in S_u she receives. The utility of u is maximized if $U_u(F) = |S_u|$.

We denote by *move*, a shift from a collection F of strategies to a collection F' where a single user u changes her strategy from a set F_u to another F'_u. (We say that u rewires her connections.) The move is *selfish* if $U_u(F') > U_u(F)$. *Selfish dynamics* (or dynamics for short) are the sequences of selfish moves. We say that dynamics *converge* if any sequence of selfish moves is necessarily finite. The network is at equilibrium (or stable) if no selfish move is possible. In standard game-theoretic terminology, this corresponds to a pure Nash equilibrium. The *global welfare* of the system is defined as the overall system utility: $\mathcal{U} = \sum_{u \in V} U_u$. The efficiency of selfish, self-organization of a game is classically captured by the notion of price of anarchy defined as the ratio of the optimal global welfare over the global welfare of the worst equilibrium: $PoA = \frac{\max_{F \in \mathcal{F}} \sum_{u \in V} U_u(F)}{\min_{F \in \mathcal{E}} \sum_{u \in V} U_u(F)}$, where $\mathcal{F}$ denotes the set of possible collection of strategies and $\mathcal{E} \subseteq \mathcal{F}$ denotes the set of equilibria.

In some of our proofs we make use of the notion of potential functions. An ordinal (or general [8]) potential function [18] is a function $f : \mathcal{F} \to \mathbb{R}$ such that $sign(f(F') - f(F)) = sign(U_u(F') - U_u(F))$ for any move from F to F' where user u changes her strategy. If $f(F') - f(F) = U_u(F') - U_u(F)$, f is called an exact potential function. This notion was introduced by Monderer and Shapley [18] who show that it is tightly related to the notion of a congestion game [21]. The use of potential functions is a standard technique to show convergence of dynamics and to bound price of anarchy [8,22].

3   Homogeneous Interests

We first consider the case where all users have identical sets of interests, $S_u = P$, for all $u \in V(G)$. In this context, we first establish an upper bound on the price of anarchy. We will then show convergence of dynamics.

3.1   Price of Anarchy

We first derive a simple upper bound on the overall system utility under an optimal centrally designed configuration. Clearly, any user u cannot achieve utility larger than p, which corresponds to obtaining all the subjects in P. Moreover, he cannot obtain more subjects than the aggregate budget of attention of all users, that is $\sum_{u \in V(G)} \Delta_u = n\overline{\Delta}$, where $\overline{\Delta}$ is the average in-degree per node. We can slightly improve this bound by restricting ourselves to the more interesting case where all users have budget less than p

and where there are at least two users with budget at least 2. One can easily see that the optimal solution then consists in forming an oriented ring between users whose budget is at least 2 and then connecting budget 1 users to some user of the ring. All remaining connections are used to obtain distinct subjects. Each node then receives the same set of subjects. As each node connects to a non-producer, the number of subjects gathered is at most $\sum_{u \in V(G)} \Delta_u - 1$. We thus obtain that the maximal utility U^* a user can get is:

$$U^* = \min\left(p, n(\overline{\Delta} - 1)\right). \tag{1}$$

We now consider a distributed setting where each user selfishly rewires his incoming connections if he can improve his utility, i.e., if this allows him to receive more subjects. The following proposition shows that with homogeneous user interests and budget of attention at least 3, self organization is efficient if dynamics converge, achieving a price of anarchy close to 1.

Proposition 1. *Assume that* $3 \leq \Delta_u < p$ *for every user* $u \in V$ *of a homogeneous flow game. Then under any equilibrium the utility of a user is at least* $\frac{\overline{\Delta}-2}{\overline{\Delta}-1}U^*$ *where* U^* *is his optimal utility. The price of anarchy is thus at most* $1 + 1/(\overline{\Delta} - 2)$, *approaching 1 for large* $\overline{\Delta}$.

We first note that the above proposition is tight in the sense that high price of anarchy can arise when $\Delta_u \leq 2$ for all user u, as shown in Figure 1. In this particular case, a doubly linked chain forms a Nash equilibrium gathering only two subjects in total while an oriented cycle gathers n subjects. The price of anarchy is thus $n/2$.

To establish Proposition 1, we use two lemmas. Due to lack of space, we omit the proofs which are in the extended version of this paper [12]. The first one allows to show the existence of strongly connected components at equilibrium while the second allows to build on the avoidance of redundant links.

Lemma 1. *If an equilibrium is reached such that there exists a path* $x, u_1, \ldots, u_k$ *where* x *is a producer,* u_k *has in-degree bound* $\Delta_{u_k} \geq 3$ *and a producer* y *is not received by* u_k, *then there is a path from* u_k *to* u_1.

Lemma 2. *Consider a strongly connected graph* G *with* n *nodes and* m *arcs (multiple arcs are allowed). If* $m \geq 2n - 1$, *then* G *contains a transitivity arc (i.e. an arc* (s, t) *such that there exists a directed path from* s *to* t).

Proof.[of Proposition 1] Consider any equilibrium. Assume that a user u receives less than p subjects. u must be connected to some producer x by a path $x, u_1, \ldots, u_k = u$. Consider the graph G' induced by users reachable from u_1 that receive less than p subjects. By Lemma 1, G' is strongly connected and all its users receive the same number $p' < p$ of subjects.

We claim that two users u and v of G' cannot follow the same producer y. As there exists a path from u to v, the link (y, v) would be redundant and v would be better off following some unreceived subject instead. Moreover, the fact that users in G' do not receive all subjects implies that they have spent all their budget of attention. We thus conclude that the number of edges in G', $m(G') = \sum_{u \in V(G')} \Delta_u - p'$. As the

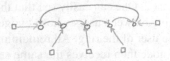

(a) Benchmark configuration

(b) A Nash equilibrium configuration

Fig. 1. Homogeneous interest sets with
degree $\Delta = 2$

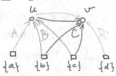

$\{a\}\quad\{b\}\quad\{c\}\quad\{d\}$

Fig. 2. A 4-cycle $(A,C) \to (B,C) \to$
$(B,D) \to (A,D) \to (A,C)$ in the
strategy space

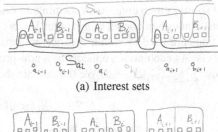

(a) Interest sets

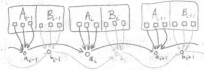

(b) Benchmark configuration

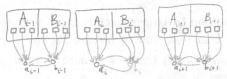

(c) A Nash equilibrium configuration

Fig. 3. Heterogeneous interest sets

network is stable, there is no transitivity arc in G'. Lemma 2 thus implies $m(G') \leq 2n(G') - 2 \leq 2n(G')$, where $n(G')$ is the number of nodes in G'. We thus get $p' \geq \sum_{u \in V(G')} \Delta_u - 2n(G') = \sum_{u \in V(G')} (\Delta_u - 2)$.

First consider the case $p' \leq p - 2$. Suppose there exists a user $w \notin V(G')$. he cannot receive two subjects not received in G' otherwise u_1 would unfollow x and connect to w. As $\Delta_w \geq 3$, w can gather the p' subjects received in G' plus two others by connecting to one node in G' plus the two corresponding producers, a contradiction as this would increase his utility. We thus conclude that G' indeed contains all users, implying $p' \geq n(\overline{\Delta} - 2)$. Using (1), the utility of each user is at least $p' \geq \frac{\overline{\Delta}-2}{\Delta-1}U^*$.

Finally, in all remaining cases to consider, all users receive at least $p - 1$ subjects. The utility of each user is thus at least $\frac{p-1}{p}U^* \geq \frac{\overline{\Delta}-2}{\overline{\Delta}-1}U^*$ as $p \geq \overline{\Delta} - 1$. $\qquad\square$

3.2 Convergence of Dynamics

We have thus shown that stable configurations of self-organizing networks with homogeneous user interests are efficient. However, do network dynamics converge to an equilibrium ? The following proposition answers this question in the affirmative.

Proposition 2. *Any homogeneous flow game has an ordinal potential function, implying that selfish dynamics always converge to an equilibrium in finite time.*

The proof, whose details are in the extended version of this paper [12], demonstrates that the function $-\sum_{0 \leq i \leq p} n_i\, n^{p-i}$ is an ordinal potential function. Our proof yields a very loose bound of n^{p+1} on convergence time. We leave as an open question whether exponential time of convergence can really arise. However, a homogeneous flow game is not equivalent to a congestion game in general as illustrated by Figure 2 where a 4-cycle in the strategy space forbids the existence of an exact potential function. This rules out the possibility of using techniques similar to [8] to find equilibria in polynomial time, and more generally to easily bound convergence time.

Combining Proposition 1 and Proposition 2, we obtain:

Theorem 1. *In a homogeneous flow game where each user has budget of attention at least 3, less than p, and $\overline{\Delta}$ in average, selfish dynamics converge to an equilibrium such that the utility of a user is at least $\frac{\overline{\Delta}-2}{\overline{\Delta}-1}U^*$ where U^* is the optimal utility he can get, implying a price of anarchy of $1 + 1/(\overline{\Delta} - 2)$ at most.*

4 Heterogeneous Interests

We now consider the more realistic case where users have differing sets of interests. Here we assume user u is interested in a subset $S_u \subseteq P$ of topics. As a user may connect to other users whose interest sets differ from his own, he potentially receives subjects out of his interest set. The user may not have the resources to process and store this irrelevant information. We thus assume a natural filtering rule, where a user only retransmits subjects that are in his own interest set.

Price of Anarchy. We now show that the price of anarchy of such a system may be unbounded.

Proposition 3. *In a heterogeneous flow game, the price of anarchy can be arbitrarily large: specific choices yield a PoA of $\Omega\left(\frac{n}{\Delta}\right)$.*

Proof. We show the result through an example, illustrated in Figure 3. For integer k, consider a system with $n = 2k$ users having budget of attention $\Delta \geq 2$ each, and $p = 2(\Delta - 1)k$ producers. We distinguish two set of users $\{a_1, \ldots, a_k\}$ and $\{b_1, \ldots, b_k\}$. Similarly, the producers are partitionned into groups $\{A_1, \ldots, A_k\}$ and $\{B_1, \ldots, B_k\}$ where each A_i (resp. B_i) contains $\Delta - 1$ producers.

As illustrated in Figure 3(a), each user a_i has a value of 1 for each topic in $A_i \cup B_i$ and additionally the first element of each A_j for $j \neq i$. Similarly, each user b_i has a value of 1 for each topic in $A_i \cup B_i$ and additionally the first element of each B_j for $j \neq i$. Users have a value of zero for all other topics.

A benchmark configuration is shown in Figure 3(b), with two oriented rings, one for users a_i, $i = 1, \ldots, k$ and one for users b_i, $i = 1, \ldots, k$. User a_i is connected to a_{i-1} (with a_0 corresponding to a_k) and to all producers in A_i. User b_i is connected to b_{i-1} (with b_0 corresponding to b_k) and to all producers in B_i. The corresponding utility is $n(n/2 + \Delta - 2)$, so that the optimal global welfare $\mathcal{U}^*$ satisfies $\mathcal{U}^* \geq n^2/2$.

The configuration shown in Figure 3(c) is an equilibrium, where each user a_i (resp. b_i) connects to producers in A_i (resp. B_i) and to b_i (resp. a_i). The global utility here is $\mathcal{U} = n(2\Delta - 2) \leq 2n\Delta$, and the price of anarchy is thus at least $\frac{n}{4\Delta}$. □

Convergence of dynamics. We have shown that the price of anarchy can be unbounded with respect to the number of users in some cases. Further, the selfish dynamics do not even guarantee convergence to a Nash Equilibrium. The details of this result are shown in the extended version of the current paper [12].

5 Structured Interest Sets

We now revisit the efficiency of social filtering in an heterogeneous scenario, where interest sets are no longer arbitrary but instead are organized according to a well be-haved geometry. Specifically we assume the following model. A metric d is given on a set $P' \supseteq P$ of subjects. The interest set S_u of each user u then coincides with a *ball* $B(s_u, R_u)$ in this metric, specified by a *central subject* s_u and a *radius of interest* R_u. Without loss of generality, we can assume $P' = \{s_u : u \in V\} \cup P$ and $S_u = B(s_u, R_u) \cap P$. We shall first give conditions on the metric d and the sets S_u under which an efficient configuration exists. We will then introduce modified dynamics and filtering rules which guarantee stability, i.e. convergence to an equilibrium. A flow game where interest sets can be defined in this way is called a *metric flow* game.

The model can easily be generalized to more eclectic user interests where topics a user is interested in correspond to the disjoint union of a constant number of balls. We leave out the details of such generalizations so as to keep the focus of the paper. However, we include a brief discussion later in the section, in the context of Proposition 4.

5.1 Sufficient Conditions for Optimal Utility

Consider the following properties of the interest set geometry.

1. γ-doubling: d is γ-doubling, i.e. for any subject s and radius R, the ball $B(s, R)$ can be covered by γ balls of radius $R/2$: there exists $I \subset S$ such that $|I| \leq \gamma$ and $B(s, R) \subset \cup_{t \in I} B(t, R/2)$.
2. r-covering: r is a covering radius, i.e. any subject $s \in P$ is at distance at most r from the central subject s_u of some user u with interest radius $R_u \geq r$.
3. (r, δ)-sparsity: there are at most δ subjects within distance r: $|B(s, r)| \leq \delta, \forall s$.
4. r-interest-radius regularity: for any users u, v with $d(s_u, s_v) < 3R_u/2 + r$, we have $R_v \geq R_u/2 + r$ (users with similar interests have comparable interest radii).

Property (1) is a classical generalization of dimension from Euclidean geometry to abstract metric spaces (an Euclidean space with dimension k is $2^{\Theta(k)}$-doubling). This is a natural assumption if user interests can be modeled by proximity in a hidden low-dimensional space. Property (2) states that all subjects are within distance r from some user's center of interest and can thus be seen as an assumption of minimum density of users' interests over the whole set P of available subjects. Property (3) puts an upper bound on the density of subjects. In other words, we assume a level of granularity under which we do not distinguish subjects. Property (4) is another form of regularity assumption, requiring some smoothness in the radii of interests of nearby users. This may be the most debatable assumption, for instance if we consider the case of an expert next to an amateur. However, if we assume that a topic is split into several subjects according to the level of expertise required to understand the corresponding news, the assumption becomes more natural as an expert is still interested in related subjects (with lower level of understanding) and an amateur still has some focus if the correct number of levels is considered.

We now show that an optimal solution exists, i.e. one in which each user receives all subjects in his interest set, as soon as his budget of attention is at least $\gamma\delta + \gamma^2 \log \frac{R_m}{r}$ where R_m is the maximum radius of interest over all users. This will be a direct consequence of the following proposition.

Proposition 4. *Consider a metric flow game satisfying the γ-doubling, r-covering, (r, δ)-sparsity and r-interest-radius regularity assumptions. If in addition each user u has a budget of attention at least $\gamma\delta + \gamma^2 \log \frac{R_u}{r}$, then there exists a collection of user strategies allowing each user u to receive all subjects in S_u.*

This result can easily be extended to the case where each user interest set is given by a disjoint union of balls (the number of balls being at most a constant b). It suffices to repeat the construction of the proof for each ball, resulting in a factor b in the resulting required budget of attention. The assumptions have to be slightly modified so that any subject is covered by some ball of a user (in the covering assumption) and that two nearby balls have comparable radii (in the regularity assumption). The details of the proof can be found in the extended version of the paper [12].

The core of the construction consists in covering a a given ball radius of $2^i r$ with a set of γ balls of radius $2^{i-1}r$. As a covering set of γ^2 balls can be computed through a simple greedy covering algorithm [11], a solution where the required budget of attention is within a factor γ from the bound of Proposition 4 can thus be computed in polynomial time.

As previously mentioned, a budget of attention of $\Delta = \gamma\delta + \gamma^2 \log \frac{R_m}{r}$ per user is thus enough for maximum utility. This scales logarithmically in R_m, while under the assumptions of the theorem one can arrange interest sets to have size polynomial in R_m (take for example interests to be regularly placed on a lattice). Thus this configuration gives substantial savings in comparison to one where users would connect directly to all their subjects.

Clearly the configuration graph identified in this theorem is an equilibrium: as maximum utility is reached, no user can increase its utility by reconnecting. We now study conditions that guarantee convergence of dynamics.

5.2 Sufficient Conditions for Stability

We first define two rules regarding republication of subjects received and reconnections.

1. Expertise-filtering rule: when a user u is connected to a user v, u only receives subjects s such that $d(s_v, s) \leq d(s_u, s)$.
2. Nearest-subject rule for re-connection: when reconnecting, each user u gives priority to subjects that are closer to s_u: a new subject s is gained by u so that no subject t with $d(s_u, t) < d(s_u, s)$ is lost. (On the other hand, any subject t with $d(s_u, t) > d(s_u, s)$ can be lost.)

Rule 1 can be interpreted as follows. The center of expertise of a user is the same as its center of interest, and the distance d also captures expertise of users about subjects, in that u is more expert than v on subject s if and only if $d(s_u, s) \leq d(s_v, s)$. The rule then amounts to a sanity check where u discards news from sources that have less expertise than himself on the subject. We capture with the following slight variation of the model. A flow game *with expertise-filtering* is a flow game where reception of a subject s by user u occurs only when there exists a directed path $s = u_0, \ldots, u_k = u$ from s to u such that for each $1 \leq i < k$, $s \in S_{u_i}$ (i.e. $d(s_{u_i}, s) \leq R_{u_i}$) and $d(s_{u_i}, s) \leq d(s_{u_{i+1}}, s)$.

Rule 2 states that a user u prefers to receive a subject he is more interested in (i.e. closer to s_u) rather than any number of subjects that are less interesting. A flow game is denoted to be *with nearest-subject priority* if the utility function of each user u is defined by $U_u(F) = \max\{R : u \text{ receives all } s \in B(s_u, R)\}$.

Proposition 5. *Any metric flow game with expertise-filtering and nearest-subject priority has an ordinal potential function, implying that selfish dynamics always converge to an equilibrium in finite time.*

The proof shows that the function $\sum_{0 \leq i \leq m} n_i (n+p)^{2(m-i)}$ is an ordinal potential function (details in the extended version [12]). As in the previous section, the bound on convergence time implied by the above proof is very loose. We leave open the question of determining better bounds or faster convergence conditions.

We are now ready to prove the following:

Theorem 2. *Consider a metric flow game with expertise-filtering and nearest-subject priority that satisfies the γ-doubling, r-covering, (r, δ)-sparsity and r-interest-radius regularity assumptions. If in addition each user u has budget of attention at least $\gamma\delta + \gamma^2 \log \frac{R_u}{r}$, selfish dynamics converge to an equilibrium where each user u receives all subjects in S_u, implying that the price of anarchy is then 1.*

Proof. Consider a configuration where some users do not receive some subject in their interest ball. Let (u, s) be a user-subject unsatisfied pair such that $d(s_u, s)$ is minimal. Consider the smallest integer i such that $d(s_u, s) \leq 2^i r$ holds. According to the construction of Proposition 4, user u can receive all subjects in $B_{u,i} = B(s_u, \min(R_u, 2^i r))$ as long as every user v receives all subjects in his ball of radius $\min(R_v, 2^{i-1}r)$ which is the case according to the choice of the pair (u, s). Note that this construction follows the expertise filtering rule as each subject at distance greater than $2^{i-1}r$ is retrieved through a user at distance at most $2^{i-1}r$ from the subject. User u can retrieve $B_{u,i}$ using at most $\gamma\delta + \gamma^2(i-1)$ connections. The configuration is thus unstable as long as $\Delta_u \geq \gamma\delta + \gamma^2(i-1)$ which is the case for $\Delta_u \geq \gamma\delta + \gamma^2 \log \frac{R_u}{r}$. Since the system must stabilize to some equilibrium according to Proposition 5, every user u must receive all news about subjects in S_u in that stable configuration. □

Interestingly, the above proof implies that the convergence is fast: as soon as all users receive their ball of radius $2^{i-1}r$, one reconnection by each user will allow him to receive his ball of radius $2^i r$ (expertise-filtering and nearest-subject priority ensure that other users will not lose subjects at distance less than $2^i r$). Convergence is thus achieved after $\log \frac{R_m}{r}$ rounds where each round consists in letting each user reconnect once (or more).

6   Concluding Remarks

We have shown that a flow game can have complex dynamics that may not converge. However, we can prove convergence to efficient equilibrium for both homogeneous flow games (with very weak assumptions) and metric flow games (with more technical assumptions). While our proofs give exponential bounds on convergence time in general, we get linear convergence time (up to a logarithmic factor) for structured interest set with expertise-filtering and nearest-subject priority, showing that understanding the structure of interests and its relation to forwarding mechanisms is a key aspect of information flow in social networks. Direct follow up of this work concerns the study of the speed of convergence in general and the characterization of flow games having pure Nash equilibria.

Our model makes several simplifying assumptions. A natural generalization would be to consider a real-valued cost of attention for establishing a link (v, u) instead of a unitary cost. The cost of establishing link (v, u) could typically be a function of S_u and S_v. A natural cost taking into account the attention required to filter out uninteresting content would then be $c(v, u) = \frac{|S_v|}{|S_u \cap S_v|}$, for example.

A dual variant of our model could be to consider that every user gathers all the subjects he is interested in while he tries to minimize the required cost of attention. We could also mix both models, using utility functions combining coverage of interest set and cost of attention (the function being increasing in the number of interesting subjects received and decreasing in the costs of attention of the formed links).

In that context, we believe the two following directions are promising for efficient social dissemination. First, incentive mechanisms, e.g. reputation counters maintained by users, or payments between users, may considerably augment the performance of self-organizing social flows. Second, more elaborate content filtering between contact-follower pairs may also lead to substantial improvements. We have already introduced expertise filtering, which could translate into implementable mechanisms in existing social networking platforms. More generally there appears to be a rich design space of filtering rules based on combinations of interests and expertise.

References

1. Abraham, I., Malkhi, D., Dobzinski, O.: LAND: stretch $(1 + \epsilon)$ locality-aware networks for DHTs. In: Munro, J.I. (ed.) SODA, pp. 550–559. SIAM (2004)
2. Anshelevich, E., Dasgupta, A., Kleinberg, J., Tardos, E., Wexler, T., Roughgarden, T.: The price of stability for network design with fair cost allocation. In: Proceedings of the 45th Annual IEEE Symposium on Foundations of Computer Science, FOCS 2004, pp. 295–304. IEEE Computer Society, Washington, DC (2004)
3. Bala, V., Goyal, S.: A noncooperative model of network formation. Econometrica 68(5), 1181–1229 (2000)
4. Bei, X., Chen, W., Teng, S.-H., Zhang, J., Zhu, J.: Bounded budget betweenness centrality game for strategic network formations. Theor. Comput. Sci. 412(52), 7147–7168 (2011)
5. Cha, M., Haddadi, H., Benevenuto, F., Gummadi, P.K.: Measuring user influence in twitter: The million follower fallacy. In: Cohen, W.W., Gosling, S. (eds.) ICWSM. The AAAI Press (2010)

6. Clarkson, K.: Nearest neighbor queries in metric spaces. Discrete & Computational Geometry 22(1), 63–93 (1999)
7. Fabrikant, A., Luthra, A., Maneva, E., Papadimitriou, C.H., Shenker, S.: On a network creation game. In: Proc. ACM PODC, pp. 347–351 (2003)
8. Fabrikant, A., Papadimitriou, C.H., Talwar, K.: The complexity of pure nash equilibria. In: Babai, L. (ed.) STOC, pp. 604–612. ACM (2004)
9. Fraigniaud, P., Lebhar, E., Viennot, L.: The inframetric model for the internet. In: Proceedings of the 27th IEEE International Conference on Computer Communications (INFOCOM), Phoenix, pp. 1085–1093 (2008)
10. Gai, A.-T., Lebedev, D., Mathieu, F., de Montgolfier, F., Reynier, J., Viennot, L.: Acyclic Preference Systems in P2P Networks. In: Kermarrec, A.-M., Bougé, L., Priol, T. (eds.) Euro-Par 2007. LNCS, vol. 4641, pp. 825–834. Springer, Heidelberg (2007)
11. Har-Peled, S., Mendel, M.: Fast construction of nets in low dimensional metrics, and their applications. In: Mitchell, J.S.B., Rote, G. (eds.) Symposium on Computational Geometry, pp. 150–158. ACM (2005)
12. Hegde, N., Massoulié, L., Viennot, L.: Self-organising flows in social networks. arXiv e-print, arXiv:1212.0952 (2012)
13. Jackson, M.: Social and Economic Networks. Princeton University Press (2010)
14. Jiang, B., Hegde, N., Massoulié, L., Towsley, D.: How to optimally allocate your budget of attention in social networks. In: Proc. IEEE Infocom (2013)
15. Karger, D.R., Ruhl, M.: Finding nearest neighbors in growth-restricted metrics. In: Reif, J.H. (ed.) STOC, pp. 741–750. ACM (2002)
16. Laoutaris, N., Poplawski, L.J., Rajaraman, R., Sundaram, R., Teng, S.-H.: Bounded budget connection (BBC) games or how to make friends and influence people, on a budget. In: Proceedings of the Twenty-seventh ACM Symposium on Principles of Distributed Computing, PODC 2008, pp. 165–174. ACM, New York (2008)
17. Massoulié, L., Twigg, A.: Rate-optimal schemes for peer-to-peer live streaming. In: Journal of Performance Analysis (2008)
18. Monderer, D., Shapley, L.: Potential games. In: Games and Economic Behavior, pp. 124–143 (1996)
19. Nisan, N., Roughgarden, T., Tardos, É., Vazirani, V.V. (eds.): Algorithmic Game Theory. Cambridge Univ. Press (2007)
20. Plaxton, C.G., Rajaraman, R., Richa, A.W.: Accessing nearby copies of replicated objects in a distributed environment. Theory Comput. Syst. 32(3), 241–280 (1999)
21. Rosenthal, R.W.: A class of games possessing pure-strategy Nash equilibria. International Journal of Game Theory 2, 65–67 (1973)
22. Roughgarden, T.: Potential functions and the inefficiency of equilibria. In: Proceedings of the International Congress of Mathematicians (ICM), vol. 3, pp. 1071–1094 (2006)
23. Simon, H.A.: Designing organizations for an information rich world. In: Greenberger, M. (ed.) Computers, Communications, and the Public Interest, pp. 37–72. The Johns Hopkins Press, Baltimore (1971)

Performance/Security Tradeoffs for Content-Based Routing Supported by Bloom Filters

Hugues Mercier*, Emanuel Onica, Etienne Rivière, and Pascal Felber

Institute of Computer Science, Université de Neuchâtel
2000 Neuchâtel, Switzerland
first.last@unine.ch

Abstract. Content-based routing is widely used in large-scale distributed systems as it provides a loosely-coupled yet expressive form of communication: consumers of information register their interests by the means of subscriptions, which are subsequently used to determine the set of recipients of every message published in the system. A major challenge of content-based routing is security. Although some techniques have been proposed to perform matching of encrypted subscriptions against encrypted messages, their computational cost is very high. To speed up that process, it was recently proposed to embed Bloom filters in both subscriptions and messages to reduce the space of subscriptions that need to be tested. In this article, we provide a comprehensive analysis of the information leaked by Bloom filters when implementing such a "prefiltering" strategy. The main result is that although there is a fundamental trade-off between prefiltering efficiency and information leakage, it is practically possible to obtain good prefiltering while securing the scheme against leakages with some simple randomization techniques.

1 Introduction

Content-based publish/subscribe [1] is an efficient and powerful communication paradigm for the development of large-scale applications. It supports decoupled communication between the producers and the consumers of information, respectively called *publishers* and *subscribers*. Users interested in specific data register subscriptions, which consist of predicates on the content of messages called *publications* sent by the publishers. Data is routed through the network by comparing the content of messages against the predicates and is delivered to all users with at least one matching subscription. For instance, as illustrated at the right of Figure 1, a user monitoring the stock market might register a subscription $s_1 = \{symbol = \texttt{ACME} \land price \geq 100 \land currency = \texttt{USD}\}$ to be notified of quotes for company ACME with share price greater than or equal to \$100. A publication quote with content $p_1 = \{symbol = \texttt{ACME}, price = $

* This work is partially sponsored by European Commission's Seventh Framework Program (FP7) under grant agreement No. 257843.

T. Moscibroda and A.A. Rescigno (Eds.): SIROCCO 2013, LNCS 8179, pp. 129–140, 2013.
© Springer International Publishing Switzerland 2013

102.23, $volume$ = 210425, $currency$ = USD}, shown at the left of Figure 1, matches the subscription s_1 and is forwarded to the corresponding subscriber. The publication also matches Subscription s_2 = {$symbol$ = ACME $\wedge$ $price$ $\geq$ 80}, but not s_3 = {$symbol$ = ACME $\wedge$ $price$ = 90}.

Security is a major hurdle to the wide adoption of content-based routing. Indeed, as predicates must be compared against the content of messages, the entity performing the routing must be trusted by all parties. The nature of subscriptions may in particular reveal sensitive information about users (e.g., stock portfolios, whether two users have the same interests, etc.). Therefore, both the publications and subscriptions must be encrypted ($p_i \rightarrow E(p_i)$ and $s_i \rightarrow E(s_i)$ in Figure 1), and the routing operation must be performed without access to the decryption keys. Unfortunately, existing encrypted filtering techniques, like *asymmetric scalar-product preserving encryption* (ASPE) [2], have a high computational cost that makes them inadequate for high-throughput content-based routing.

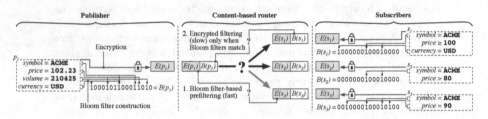

Fig. 1. Encrypted content-based routing with prefiltering using Bloom filters

A promising approach to tackle both the security and performance challenges of content-based routing is to use an efficient "prefiltering" technique, described in [3], to significantly reduce the space of subscriptions that must be tested by the encrypted filtering engine during routing. In a nutshell, the principle is to embed Bloom filters [4] inside publications and subscriptions; these Bloom filters encode the values carried by the publications and the equality constraints of the subscriptions. The simplified example illustrated in Figure 1 considers Bloom filters of 16 bits with 2 hash functions. The values ACME, 102.23, 210425 and USD are encoded in the Bloom filter $B(p_1)$ of p_1, ACME and USD are encoded in $B(s_1)$, ACME in $B(s_2)$, and ACME and 90 in $B(s_3)$. By testing the Bloom filters of subscriptions for inclusion in those of publications, one can efficiently determine the possibility for a message to match a subscription: if the test is negative, the message is guaranteed *not* to match; otherwise, it *might* match. In Figure 1, $B(s_1)$ and $B(s_2)$ are included in $B(p_1)$, but $B(s_3)$ is not. After the prefiltering, only the subscriptions with matching Bloom filters need to be tested using the computationally intensive encrypted filtering operation. Since most workloads are dominated by equality constraints, prefiltering can significantly improve the performance of high-throughput content-based routing systems.

Despite their hash-based construction and their probabilistic nature, Bloom filters may still convey sensitive information. In particular, *containment* produces

Bloom filters that are included in one another.[1] For security reasons, most encrypted filtering schemes are designed such that containment relationships between subscriptions cannot be derived from the encrypted matching mechanism. However, an attacker can use containment relationships from the Bloom filters to build groups of subscriptions whose size depends on the popularity of their values. The attacker can then try to derive values based on the size of the groups, compromising the overall system security. This problem can be mitigated by randomly removing bits from the subscriptions' Bloom filters and adding bits to those of publications, but this degrades the prefiltering accuracy. This performance/security tradeoff of Bloom filter-based prefiltering for content-based routing is the focus of this article.

1.1   Our Contributions

Our main insight is to introduce randomness in the prefiltering process by randomly choosing a small number of hash functions from a larger set for each subscription value encoded in the Bloom filters. Using mathematical analysis and simulations, we show that there is a fundamental tradeoff between the information that can be leaked to an attacker from the Bloom filters and the performance of prefiltering. Although this might seem like a discouraging result, we prove that it is practically possible to obtain excellent prefiltering performance while making the containment extremely challenging to derive from the Bloom filters, henceforth securing the prefiltering scheme against leakages.

The main attack we analyze is an attacker trying to derive subscriptions containment relationships based on Bloom filter containment. The major difficulty is the analysis of the dependencies introduced by hash functions collisions, which are cumbersome to model mathematically. In practice, we use small Bloom filters and the domain of possible subscriptions can be small and nonuniform, resulting in a large number of collisions in the Bloom filters. We provide evidence, both mathematically and using numerical simulations, that efficient and secure prefiltering can be achieved for real-life practical systems.

The article is organized as follows. We start by formally stating the problem and its assumptions in Section 2. The complete theoretical analysis of the performance-security tradeoff of the prefiltering scheme is presented in Section 3. A discussion of the results is presented in Section 4. In Section 5, we study other possible ways for an attacker to obtain containment relationships from subscriptions' Bloom filters and prove that our prefiltering scheme is secure against them. We conclude the paper in Section 6.

1.2   Related Work

Bloom filters are widely used in distributed systems for both performance and security purposes. The prefiltering technique for content-based publish/subscribe

---

[1] We say that a subscription s_i *contains* another subscription s_j if any event that matches s_j also matches s_i, in other words if s_i is more general than s_j. At the right of Figure 1, s_2 contains s_1 and s_3.

systems described in the introduction was introduced by Barazzutti et al. [3]. Kerschbaum [5] used Bloom filters to protect supply chains information against, for instance, malicious suppliers counterfeiting products; Bloom filters are used to encode suppliers IDs attached to items as they pass through the supply chain. Shikfa et al. [6] used counting Bloom filters to securely compute a weighted matching ratio in a broker-based scenario. Jerzak and Fetzer [7] used Bloom filters to speed up matching in publish/subscribe systems by constructing efficient routing tables for the broker network. Perl et al. [8] used of Bloom filters for a sort of prefiltering of DNA search using homomorphic encryptions. They used pollution to obfuscate the filters, which is complementary to the pruning approach we use in this article. Goh [9] presented a secure indexing technique allowing word searches over a collection of encrypted documents. Both documents and queries are encoded in Bloom filters, and additional bits in the documents filters are randomly set to 1. Bellovin and Cheswick [10] briefly mentioned that randomizing the inputs could be beneficial for using Bloom filters for encrypted search, but they did not develop the idea further. Kuzu et al. [11] perform a cryptanalysis on Bloom filters used in private record linkage applications. This applies to keywords n-grams which are encoded in Bloom filters. Their work shares some characteristics with ours: their Bloom filters are not encrypted, and they make similar assumptions on the information available to attackers (domain probability space and number of hash functions). Finally, following the seminal work of Bloom [4], a large body of research was also done on Bloom filters themselves and is described is the extended version of this manuscript.

2   Problem Statement and Assumptions

2.1   Notation and Assumptions

Consider a subscription s_1 with $c_1 \geq 0$ different equality constraints encoded in a Bloom filter $B_\alpha(s_1)$ of size n using α out of k hash functions per constraint. Furthermore, let us consider a second subscription s_2 with $c_2 \geq c_1$ different equality constraints also encoded in a Bloom filter $B_\alpha(s_2)$ of size n with α out of k hash functions per constraint. We also consider a publication p with $c_p \geq c_1$ different values. The publication is encoded in a Bloom filter $B_k(p)$, that is, all k hash functions are used. We write $B_\alpha(s_1) \sqsubseteq B_\alpha(s_2)$ if and only if all the nonzero bits in the Bloom filter of s_1 are also nonzero in the Bloom filter of s_2. Likewise, we write $B_\alpha(s_1) \sqsubseteq B_k(p)$ if the Bloom filter of a subscription s_1 is included in the Bloom filter of a publication p.

We only consider the number of distinguishable values for subscriptions and publications since it is assumed that the same value appearing for multiple attributes is only hashed once. As mentioned in the introduction, the subscriptions may also contain "nonequality" constraints ($\neq, <, \geq, \dots$). However, nonequality constraints are not encoded in the Bloom filters and cannot be prefiltered; we therefore assume that they are not present in our analysis. Thus, s_1, s_2 and p can formally be defined as sets of cardinality c_1, c_2 and c_p, respectively, and we use the standard notation $A \subseteq B$ to denote that A is a subset of B.

2.2 Hashing Strategy

It is assumed that the Bloom filters use a set of k independent and perfectly random hash functions. Our hashing strategy $\mathcal{H}_\alpha$ consists, for a fixed $\alpha \in \{1, 2, \ldots, k\}$, of randomly selecting α among the k hash functions for each subscription value to be encoded. We emphasize again that the hashing strategy is only applied to subscriptions; all the k hash functions must be used for publications, otherwise false negatives can occur, i.e., matching publication-subscription pairs can be rejected during the prefiltering operation.

2.3 Subscription/Publication Domains

We assume that there are a possible attributes $A_1, A_2, \ldots, A_a$ with equality constraints that can be used for each subscription and in publications, and that the set of v possible values for attribute A_i is $V_i = \{v_1^i, v_2^i, \ldots, v_v^i\}$. It is assumed that when a subscriber or publisher selects an attribute to put in a subscription or publication, it chooses any of the a attributes with probability $\frac{1}{a}$. It is also assumed that the attribute takes each of its possible values with probability $\frac{1}{v}$. The domain we consider, noted $\mathcal{D}_{\neq}$, assumes that $V_i \cap V_j = \emptyset$ for $i \neq j$. Non-disjoint value sets can always be made disjoint by encoding each value with a small prefix representing its attribute. Other domains are considered in the extended version of this work.

It is assumed that an attacker knows the domain and its probability distribution and also k and α. We assume that when an attacker examines a subscription Bloom filter, it knows the number of different equality constraints that were encoded in it. This increases its power, but only slightly because the number of nonzero bits in a Bloom filter is highly correlated with the number of encoded values. Furthermore, this allows us to do the analysis without any assumption on the distribution of the number of equality attributes per subscription.

The domain uniformity is not a realistic assumption for most content-based systems, although two remarks must be made. Firstly, the domain uniformity allows us to analyze rigorously how an attacker can infer subscription containment from the Bloom filters as well as the performance of the prefiltering scheme. Secondly, the analysis can be carried out numerically for any content-based system for which an estimate of the statistics at the subscriber, publisher, or system level is available. One such example is presented in Section 4.

2.4 Problem Statement

The objective of this work is to study, using our hashing strategy, the tradeoff between the amount of information about subscriptions that can be inferred by an attacker from a set of leaked Bloom filters, and the performance of the prefiltering process.

Leaked Information from the Bloom Filters. We want to prove that randomly selecting a small number of hash functions for each coded attribute of the Bloom filters restricts the attacker capacity to derive useful subscription information based on their Bloom filters. The expression of interest is

$$P_\alpha \triangleq \Pr[s_1 \subseteq s_2 \mid B_\alpha(s_1) \sqsubseteq B_\alpha(s_2)]. \tag{1}$$

The higher this probability is, the easier it is for an attacker to build an accurate containment graph when having access to a large number of subscription Bloom filters included in other subscription Bloom filters. Using Bayes' law, we can write

$$P_\alpha = \Pr[s_1 \subseteq s_2] \cdot \frac{\Pr[B_\alpha(s_1) \sqsubseteq B_\alpha(s_2) \mid s_1 \subseteq s_2]}{\Pr[B_\alpha(s_1) \sqsubseteq B_\alpha(s_2)]}. \tag{2}$$

When all the k hash functions are chosen, $\Pr[B_k(s_1) \sqsubseteq B_k(s_2) \mid s_1 \subseteq s_2] = 1$.

Prefiltering Efficiency. Decreasing the amount of information to potential attackers is useless if it results in a prefiltering operation that fails to discard a significant fraction of the subscriptions with equality constraints that do not match an incoming publication. The expression of interest is the probability of false positive

$$P_{\text{f\_pos}} \triangleq \Pr[B_\alpha(s_1) \sqsubseteq B_k(p) \mid s_1 \not\subseteq p]. \tag{3}$$

Simply, it is the probability, when the equality constraints of a subscription are not included in the values set of a publication, that the Bloom filter of the subscription is included in the Bloom filter of the publication.

3   Exact Analysis with Collisions

In this section, we derive exact expressions for P_α and $P_{\text{f\_pos}}$. We first state preliminary results whose proofs are in the extended version of this work.

Consider the number of common values between a pair of subscriptions. Let $\Pr_{\text{common}}[\gamma]$ be the probability that there are γ common values between subscriptions s_1 and s_2 for $0 \leq \gamma \leq c_1$.

Lemma 1

$$\Pr_{\text{common}}[\gamma] = \frac{\binom{c_2}{\gamma} \cdot \sum_{\delta=0}^{c_1-\gamma} \left[\binom{c_2-\gamma}{\delta} \binom{a-c_2}{c_1-\gamma-\delta} (v-1)^\delta v^{c_1-\gamma-\delta} \right]}{\binom{a}{c_1} v^{c_1}}.$$

Corollary 1

$$\Pr[s_1 \subseteq s_2] = \Pr_{\text{common}}[c_1] = \frac{\binom{c_2}{c_1}}{\binom{a}{c_1} v^{c_1}}.$$

We now focus on $\Pr[B_\alpha(s_1) \sqsubseteq B_\alpha(s_2) \mid s_1 \subseteq s_2]$ and $\Pr[B_\alpha(s_1) \sqsubseteq B_\alpha(s_2)]$. Our main insight is to split the hashes into two categories. First, consider an attribute value common to subscriptions s_1 and s_2. If, for this value, there is a hash function h which is randomly chosen for both Bloom filters $B_\alpha(s_1)$ and

$B_\alpha(s_2)$, we say there is a *good hash* between s_1 and s_2. Second, it is also possible that unrelated hash functions selected by both s_1 and s_2 randomly hash at the same position in the Bloom filters; this is called a *lucky hash*. Since we want $B_\alpha(s_1) \sqsubseteq B_\alpha(s_2)$, all the hashes of s_1 in $B_\alpha(s_1)$ must be covered by the hashes of s_2 in $B_\alpha(s_2)$ by *good* and/or *lucky hashes*.

Let $\Pr_{good}[g, \gamma]$ be the probability that there are g *good hashes* between s_1 and s_2, where γ is the number of common values between both subscriptions. $\Pr_{good}[g, \gamma]$ is difficult to calculate for $\alpha > 1$ because the probability that there are l good hashes for one common value between s_1 and s_2 is different from the probability that there are l common values with one good hash each. Thus, to evaluate $\Pr_{good}[g, \gamma]$ we must consider the integer partitions of g [12].

Let IntPart$(g, \gamma, \alpha_{min}, \alpha)$ be the set of all integer partitions of g into exactly γ parts with the additional constraint that the size of each part is at least α_{min} and at most α. We write $\lambda \vdash$ IntPart$(g, \gamma, \alpha_{min}, \alpha)$ to denote that λ is an integer partition of g with the desired properties. The frequency representation of a partition $\lambda \vdash$ IntPart$(g, \gamma, \alpha_{min}, \alpha)$ is $(\alpha_{min}{}^{p_{\alpha_{min}}} \alpha_{min+1}{}^{p_{\alpha_{min+1}}} \ldots \alpha^{p_\alpha})$, meaning that the value α_i appears p_{α_i} times in the partition.

Lemma 2

$$\Pr_{good}[g, \gamma] = \sum_{\lambda \vdash IntPart(g, \gamma, \alpha_{min}, \alpha)} \left[\binom{\gamma}{p_{\alpha_{min}}, p_{\alpha_{min+1}}, \ldots, p_\alpha} \prod_{\beta=\alpha_{min}}^{\alpha} \left[\frac{\binom{\alpha}{\beta}\binom{k-\alpha}{\alpha-\beta}}{\binom{k}{\alpha}} \right]^{p_\beta} \right]$$

where $\alpha_{min} = \max(0, 2\alpha - k)$.

The formula is simpler when $\alpha = 1$, since in that case here is only one integer partition of g into γ parts when each part is either 0 or 1. The following corollary describes this.

Corollary 2   When $\alpha = 1$,

$$\Pr_{good}[g, \gamma] = \binom{\gamma}{g} \left(\frac{1}{k} \right)^g \left(\frac{k-1}{k} \right)^{\gamma - g}.$$

If $B_\alpha(s_1) \sqsubseteq B_\alpha(s_2)$, all the Bloom filter bits derived from the hash functions randomly chosen for s_1 must somehow be covered by some of the hash functions randomly chosen for the Bloom filter of s_2. If we have g *good hashes* between s_1 and s_2, that leaves $c_1\alpha - g$ *lucky hashes* of s_1 that must be covered by some of the hash functions of s_2 randomly hashing at the same positions in the Bloom filter. Let $\Pr_{lucky}[h_1, h_2]$ be the probability that h_1 hashes of a subscription or publication are covered by some of the h_2 hashes of a second subscription or publication randomly hashing at the same positions in the Bloom filter.

Lemma 3

$$\Pr_{lucky}[h_1, h_2] = \frac{1}{n^{h_1+h_2}} \sum_{i=0}^{h_2} \left[\binom{n}{i} i! \left\{ {h_2 \atop i} \right\} \sum_{j=0}^{i} \binom{i}{j} j! \left\{ {h_1 \atop j} \right\} \right].$$

where $\left\{ {k_1 \atop k_2} \right\}$ represent the Stirling numbers of the second kind [13].

Lemma 4

$$\Pr[B_\alpha(s_1) \sqsubseteq B_\alpha(s_2) \mid s_1 \subseteq s_2] = \sum_{g=0}^{c_1\alpha} \left(\Pr_{good}[g,c_1] \Pr_{lucky}[c_1\alpha - g, c_2\alpha] \right).$$

Lemma 5

$$\Pr[B_\alpha[s_1] \sqsubseteq B_\alpha(s_2)] = \sum_{\gamma=0}^{c_1} \left(\Pr_{common}[\gamma] \sum_{g=0}^{\gamma\alpha} \left(\Pr_{good}[g,\gamma] \Pr_{lucky}[c_1\alpha - g, c_2\alpha] \right) \right).$$

We now present the main results of this work.

Theorem 1

$$P_\alpha = \frac{\binom{c_2}{c_1}}{\binom{a}{c_1} v^{c_1}} \cdot \frac{\sum\limits_{g=0}^{c_1\alpha} \left(\Pr\limits_{good}[g,c_1] \Pr\limits_{lucky}[c_1\alpha - g, c_2\alpha] \right)}{\sum\limits_{\gamma=0}^{c_1} \left(\Pr\limits_{common}[\gamma] \sum\limits_{g=0}^{\gamma\alpha} \left(\Pr\limits_{good}[g,\gamma] \Pr\limits_{lucky}[c_1\alpha - g, c_2\alpha] \right) \right)}.$$

Proof The theorem can be derived by merging (2), Lemmas 1, 2, 3, 4, 5, and Corollaries 1 and 2.

Theorem 2

$$P_{f\_pos} = \sum_{\gamma=0}^{c_1-1} \left(\Pr_{common}[\gamma] \Pr_{lucky}[\alpha(c_1 - \gamma), c_p k] \right).$$

Proof From (3), $P_{f\_pos} \triangleq \Pr[B_\alpha(s_1) \sqsubseteq B_k(p) \mid s_1 \not\subseteq p]$. Suppose that there are $\gamma < c_1$ common attribute values between s_1 and p. Since the publication uses all the k hash functions for each value in its Bloom filter, it follows that there are $\alpha\gamma$ *good hashes* for s_1, leaving $\alpha(c_1 - \gamma)$ *lucky hashes* that must be covered by the $c_p k$ hashes of p randomly hashing at the same positions in the Bloom filters. The probability of this occurring is given by $\Pr\limits_{lucky}[\alpha(c_1 - \gamma), c_p k]$. The result can be obtained by summing, for all values of γ between 0 and $c_1 - 1$ ($s_1 \subseteq p$ if $\gamma = c_1$), the probability that there are γ common attribute values between s_1 and p times the probability of $\alpha(c_1 - \gamma)$ *lucky hashes*.

4   Results and Discussion

In this section, we use Theorems 1 and 2 to study the fundamental tradeoffs between the efficiency and the security of our prefiltering scheme. We first consider as an example the uniform domain $\mathcal{D}_{\neq}$ with $a = 10$ attributes and $v = 100$ values per attribute. It is further assumed that all the subscriptions contain

$c_1 = c_2 = 3$ equality constraints, that all the publications contain $c_p = 5$ values, and that a set with $k = 5$ hash functions is used. Figure 2 shows how P_α and P_{f_pos} vary with the Bloom filter size. For both P_α and P_{f_pos}, the five curves correspond to $\alpha \in \{1, 2, 3, 4, 5\}$, where α is the number of hash functions randomly selected. Figure 3 uses the same domain as Figure 2, except that it shows how P_α and P_{f_pos} vary with k with Bloom filters of 128 bits. The main observation from both figures is that in order to minimize the amount of information to an attacker, we can choose a small Bloom filter and select one hash function randomly out of a very large set. In fact, we can show from (2) and Theorem 1 that $\lim_{k\to\infty} P_\alpha = \Pr[s_1 \subseteq s_2]$. With a very large k, there is no information leaked to an attacker who knows the domain probability space. However in this case, since all the hash functions must be encoded in the Bloom filters for publications, the publications Bloom filters have all their bits set to one, thus the probability of false positive approaches 1 and the prefiltering scheme is absolutely useless.

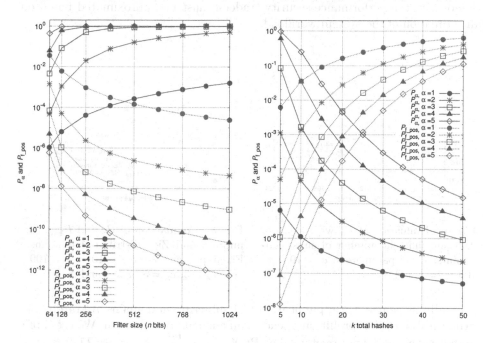

Fig. 2. P_α/P_{f_pos} tradeoff as n increases. Fixed parameters: $a = 10$, $v = 100$, $c_1 = c_2 = 3$, $c_p = 5$ and $k = 5$.

Fig. 3. P_α/P_{f_pos} tradeoff as k increases. Fixed parameters: $a = 10$, $v = 100$, $c_1 = c_2 = 3$, $c_p = 5$ and $n = 128$.

Figure 4 further emphasizes that there is no free lunch. In Figure 4, we choose k such that the false positive rate is always 3% ($P_{f\_pos} \approx 0.03$) and study how it affects P_α for $\alpha = 1$ as the Bloom filter size varies. The conclusion is that for a fixed false positive rate, the security of the prefiltering is essentially fixed no matter how we vary the size of the Bloom filters and the cardinality of the hash functions set. One observation that can be drawn from this is that for complexity reasons we should set the Bloom filter size as small as we can get away with (for very small Bloom filters it is not possible to set the probability of false positive low enough). Typically, Bloom filters of size 64 or 128 are more than sufficient. A similar behavior can be observed with $\alpha > 1$, but the tradeoff generally becomes worst as α increases. Figure 4 also shows that the performance-security tradeoff is better when the number of subscription equality constraints approaches the number of publication values. Despite the observed tradeoffs, our truncation hashing strategy is a valuable mechanism. For instance, it can be observed from Figure 4 that for a 3% false positive rate, which is excellent, P_α is approximately $5 \cdot 10^{-11}$ when $c_p = c_1 = c_2 = 5$. This makes it extremely challenging for an attacker to build an accurate containment graph and run statistical attacks using a set of encrypted subscriptions. Of course in practical systems, subscribers do not necessarily generate subscriptions with a fixed number of equality attributes, nor do publishers generate publications with a fixed number of elements. When this occurs, the performance-security tradeoff must be approximated based on an estimation of the system workload.

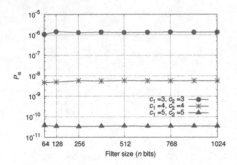 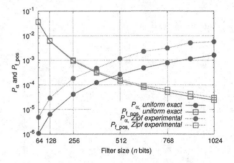

Fig. 4. Evaluation of P_α with $P_{\text{f\_pos}} \approx 0.03$ and different number of subs. constraints. Fixed parameters: $a = 10$, $v = 100$, $c_p = 5$ and $\alpha = 1$.

Fig. 5. Comparison of P_α and $P_{f_{pos}}$ for uniform and Zipf distributed domains. Fixed parameters: $a = 10$, $v = 100$, $c_1 = c_2 = 3$, $c_p = 5$, $k = 5$ and $\alpha = 1$.

Finally, since real-world publication and subscription domains are rarely uniform, we also test our prefiltering scheme on a nonuniform domain. We use Zipf's law with exponential parameter 0, i.e., $\Pr[X = x] = \frac{\Pr[X=1]}{x}$ for $x = \{1, 2, 3, \dots\}$. Zipf's law and other power laws appear in a wide range of disciplines like finance, demography, computer science, biology, ... [14] We consider that the domain attributes follow Zipf's law; likewise, we assume that the possible values for each attribute also follow Zipf's law. We use $a = 10$, $v = 100$, 5 values per publication, 3 equality constraints per subscription, $k = 5$, and $\alpha = 1$. We compute P_α and $P_{\text{f\_pos}}$ using numerical simulations. The results for P_α and $P_{\text{f\_pos}}$ comparing Zipf's law and a uniform domain are shown in Figure 5. It can be observed that P_α is slightly higher when the domain follows Zipf's law, which is not surprising since the domain entropy is lower and the probability that the subscriptions share common values increases. Furthermore, $P_{\text{f\_pos}}$ is almost equal for small

Bloom filters. This example provides evidence that our prefiltering scheme can be made secure and efficient with nonuniform real-life practical workloads.

5   Other Containment and Noncontainment Attacks

The attacker can also try to derive containment relationships based on mismatched Bloom filters, i.e., Bloom filters $B_\alpha(s_1)$ and $B_\alpha(s_2)$ such that $B_\alpha(s_1) \not\sqsubseteq B_\alpha(s_2)$ and $B_\alpha(s_2) \not\sqsubseteq B_\alpha(s_1)$. Let us consider the probability

$$\overline{P}_\alpha \triangleq \Pr\left[s_1 \subseteq s_2 \mid B_\alpha(s_1) \not\sqsubseteq B_\alpha(s_2)\right].$$

$\overline{P}_\alpha$ can be evaluated using the formulas derived in Section 3. It should be clear that $\overline{P}_k = 0$ and that $\overline{P}_\alpha > 0$ for $\alpha < k$, so at first sight randomly choosing some of the hash functions appears to provide more information to an attacker than selecting all the hash functions. This is misleading, since $\overline{P}_\alpha$ increases from zero to $\Pr[s_1 \subseteq s_2]$, providing no more information to an attacker than from random subscriptions. This is a welcome tradeoff considering how our hashing strategy decreases the amount of information leaked from contained subscriptions.

Finally, an attacker could try to derive "noncontainment" attributes about pairs of encrypted subscriptions. The probability of interest is

$$\Pr\left[s_1 \not\subseteq s_2 \mid B_\alpha(s_1) \not\sqsubseteq B_\alpha(s_2)\right],$$

which is equal to 1 with $\alpha = k$ and smaller than 1 when $\alpha < k$. This means that when $\alpha < k$, the attacker cannot even conclude with certainty that two subscriptions are not included into each other.

6   Conclusion and Future Work

Content-based routing middleware yields multiple advantages in terms of flexibility, scalability, and simplicity for the development distributed applications. A major impairment to wider adoption of these techniques is their security: messages, individual subscriptions, and containment relationships among a set of subscriptions may reveal important information about a user or group of users. While encrypted routing hides the content of messages and subscriptions, it requires costly encrypted processing algorithms that make the filtering operation computationally prohibitive for high-throughput systems. Prefiltering allows us to greatly reduce the performance gap between non-encrypted and encrypted filtering. By adding Bloom filters that encode publication values and subscription equality constraints, we can discard a large fraction of subscriptions before reaching the costly encrypted filtering operation. Furthermore, when selecting a subset of the hash functions for encoding subscription equality constraint in the Bloom filters, we showed that it is possible prefilter a large fraction of the subscriptions while leaking very little information to potential attackers.

We believe that our results can be applied to many other problems in distributed systems where set inclusions are represented using a compact structure such as Bloom filters, and where the security and confidentiality requirements are stringent. One such application is encrypted distributed databases [15]. The mathematical analysis presented in this article could also be applied to other techniques studied empirically, for instance filter obfuscation [8].

References

1. Eugster, P.T., Felber, P.A., Guerraoui, R., Kermarrec, A.M.: The many faces of publish/subscribe. ACM Computing Surveys 35(2), 114–131 (2003)
2. Choi, S., Ghinita, G., Bertino, E.: A privacy-enhancing content-based publish/Subscribe system using scalar product preserving transformations. In: Bringas, P.G., Hameurlain, A., Quirchmayr, G. (eds.) DEXA 2010, Part I. LNCS, vol. 6261, pp. 368–384. Springer, Heidelberg (2010)
3. Barazzutti, R., Felber, P., Mercier, H., Onica, E., Rivière, E.: Thrifty privacy: efficient support for privacy-preserving publish/subscribe. In: Proceedings of the 6th ACM International Conference on Distributed Event-Based Systems, DEBS 2012, pp. 225–236. ACM (2012)
4. Bloom, B.H.: Space/time trade-offs in hash coding with allowable errors. Communications of the ACM 13(7), 422–426 (1970)
5. Kerschbaum, F.: Public-key encrypted Bloom filters with applications to supply chain integrity. In: Li, Y. (ed.) DBSec. LNCS, vol. 6818, pp. 60–75. Springer, Heidelberg (2011)
6. Shikfa, A., Önen, M., Molva, R.: Broker-based private matching. In: Fischer-Hübner, S., Hopper, N. (eds.) PETS 2011. LNCS, vol. 6794, pp. 264–284. Springer, Heidelberg (2011)
7. Jerzak, Z., Fetzer, C.: Bloom filter based routing for content-based publish/subscribe. In: Proceedings of the Second International Conference on Distributed Event-Based Systems, DEBS 2008, pp. 71–81. ACM, New York (2008)
8. Perl, H., Mohammed, Y., Brenner, M., Smith, M.: Fast confidential search for biomedical data using Bloom filters and homomorphic cryptography. In: 2012 IEEE 8th International Conference on E-Science (e-Science), pp. 1–8 (2012)
9. Goh, E.J.: Secure indexes. Cryptology ePrint Archive, Report 2003/216 (2003)
10. Bellovin, S.M., Cheswick, W.R.: Privacy-enhanced searches using encrypted Bloom filters. Cryptology ePrint Archive, Report 2004/022 (2004)
11. Kuzu, M., Kantarcioglu, M., Durham, E., Malin, B.: A constraint satisfaction cryptanalysis of Bloom filters in private record linkage. In: Fischer-Hübner, S., Hopper, N. (eds.) PETS 2011. LNCS, vol. 6794, pp. 226–245. Springer, Heidelberg (2011)
12. Andrews, G.E.: The Theory of Partitions. Cambridge Mathematical Library (1998)
13. Graham, R.L., Knuth, D.E., Patashnik, O.: Concrete Mathematics: A Foundation for Computer Science, 2nd edn. Addison-Wesley Longman Publishing (1994)
14. Newman, M.E.J.: Power laws, Pareto distributions and Zipf's law. Contemporary Physics 46, 323–351 (2005)
15. Popa, R.A., Redfield, C.M.S., Zeldovich, N., Balakrishnan, H.: CryptDB: protecting confidentiality with encrypted query processing. In: Proceedings of the Twenty-Third ACM Symposium on Operating Systems Principles, SOSP 2011, pp. 85–100. ACM, New York (2011)

Influence Diffusion in Social Networks under Time Window Constraints*

Luisa Gargano[1], Pavol Hell[2], Joseph Peters[2], and Ugo Vaccaro[1]

[1] Dipartimento di Informatica, University of Salerno, Italy
[2] School of Computing Science, Simon Fraser University, Canada

Abstract. We study a combinatorial model of the spread of influence in networks that generalizes existing schemata recently proposed in the literature. In our model agents change behaviors/opinions on the basis of information collected from their neighbors in a time interval of *bounded size* whereas agents are assumed to have *unbounded memory* in previously studied scenarios. In our mathematical framework, one is given a network $G = (V, E)$, an integer value $t(v)$ for each node $v \in V$, and a time window size λ. The goal is to determine a small set of nodes (*target set*) that influences the whole graph. The spread of influence proceeds in rounds as follows: initially all nodes in the target set are influenced; subsequently, in each round, any uninfluenced node v becomes influenced if the number of its neighbors that have been influenced in the previous λ rounds is greater than or equal to $t(v)$. We prove that the problem of finding a minimum cardinality target set that influences the whole network G is hard to approximate within a polylogarithmic factor. On the positive side, we design exact polynomial time algorithms for paths, rings, trees, and complete graphs.

1 Introduction

Many phenomena can be represented by dynamical processes on networks. Examples include cascading failures in physical infrastructure networks [21], information cascades in social and economic systems [8], spreads of infectious diseases [2], and the spreading of ideas, fashions, or behaviors among people [12, 40]. Therefore, it comes as no surprise that the study of dynamical processes on complex networks is an active area of research, crossing a variety of different disciplines. Epidemiologists, social scientists, physicists, and computer scientists have studied diffusion phenomena using very similar models to describe the spreading of diseases, knowledge, behaviors, and innovations among individuals of a population (see [4, 9, 24] for surveys of the area).

A particularly important diffusion process is that of *viral marketing* [30], which refers to the spread of information about products and behaviors and their adoption by people. Recently, it has also become an important tool in the communication strategies of politicians [31, 39]. Although there are many similarities

---

* This research was supported in part by the Ebco Eppich Endowment Fund at Simon Fraser University and by NSERC of Canada.

T. Moscibroda and A.A. Rescigno (Eds.): SIROCCO 2013, LNCS 8179, pp. 141–152, 2013.

between social and epidemiological contagion [23], social contagion is usually an intentional act on the part of the transmitter and/or the adopter, unlike a pathogen contagion. The spread of ideas requires extra mechanisms in addition to mere exposure, e.g., some kind of "social pressure". More importantly, in the marketing scenario one is interested in *maximizing* the spread of information [22], while this is not likely to happen in the spread of pathogenic viruses. The intent of maximizing the spread of viral information across a network naturally suggests many optimization problems. Some of them were first articulated in the seminal papers [27, 28], under various adoption paradigms. In the next section, we will explain and motivate our model of information diffusion, state the problem that we are investigating, describe our results, and discuss how they relate to the existing literature in the area.

Due to space constraints, proofs are omitted from this extended abstract.

2   The Model, the Context, and the Results

The network is represented by a pair (G, t), where $G = (V, E)$ is an undirected graph and $t : V \longrightarrow \mathbb{N} = \{1, 2, \ldots, \}$ is a function assigning integer thresholds to nodes. We assume that $1 \leq t(v) \leq \deg(v)$ for each $v \in V$, where $deg(v)$ is the degree of v. For a given set $S \subseteq V$ and a time window size $\lambda \in \mathbb{N}$, we consider a dynamical process of influence diffusion in G defined by two sequences of node subsets, $\texttt{Influenced}[S, r]$ and $\texttt{Active}[S, r]$, $r = 0, 1, \ldots$, where $\texttt{Influenced}[S, 0] = S$, $\texttt{Active}[S, 0] = \emptyset$, and for any $r \geq 1$ it holds that

$$\texttt{Influenced}[S, r] = \texttt{Influenced}[S, r-1] \cup \{v : |N(v) \cap \texttt{Active}[S, r]| \geq t(v)\} \tag{1}$$

$$\texttt{Active}[S, r] = \begin{cases} \texttt{Influenced}[S, r-1] & \text{if } r \leq \lambda \\ \texttt{Influenced}[S, r-1] \setminus \texttt{Influenced}[S, r-1-\lambda] & \text{if } r > \lambda \end{cases}$$

Intuitively, the set S might represent a group of people who are initially influenced/convinced to adopt a product or an idea. Then the cascade proceeds in rounds. In each round r, the set of influenced nodes is augmented by including each node v that has a number of influenced and *still active* neighbors greater than or equal to its threshold $t(v)$. A node is active for λ rounds after it becomes influenced and then it becomes inactive.

Our model is based on the models in [20, 33] which assume that people can be divided into three classes at any time instant. *Ignorants* are those not aware of a rumor/not yet influenced, *spreaders* are those who are spreading it, and *stiflers* are those who know the rumor/have been influenced but have ceased to spread the rumor/influence.[1] Several rules have been proposed to govern the transition from ignorants to spreaders and from spreaders to stiflers, and many papers have studied the dynamics of these systems, mostly in stochastic scenarios

---

[1] The reader will notice an analogy with the SIR model of mathematical epidemiology [2], in which individuals can be classified as Susceptible, Infected, and Recovered.

(see [7, 34] and references quoted therein). Here, we posit that any ignorant node becomes a spreader if the number of its neighbors who are spreaders is above a certain threshold (i.e., the node is subject to a large enough amount of "social pressure"), and any spreader becomes a stifler after λ rounds (because the spreader loses interest in the rumor, for instance). Other papers have studied information diffusion under similar assumptions [19, 26].

Our model also captures another important characteristic of influence diffusion. It is well known (e.g. [3]) that people are more inclined to react to pieces of information cumulatively heard during a "short" time interval than to information heard during a considerably longer period of time. In other words, one is more likely to be convinced of an opinion heard from a certain number of friends during the last few days than by an opinion heard sporadically during the last year from the *same* number of people. Therefore, it seems reasonable to study diffusion processes in which people have *bounded memory*, and only the number of spreaders heard during the last λ rounds may contribute to the change of status of an ignorant node.[2] Formally, one has a dynamical process of influence diffusion on G described by the sequence of node subsets $\texttt{Influenced}'[S, r]$, $r = 0, 1, \ldots$, where $\texttt{Influenced}'[S, 0] = S$, and for any $r \geq 1$ it holds that

$$\texttt{Influenced}'[S, r] =$$
$$= \texttt{Influenced}[S, r-1] \cup \{v : |N(v) \cap \texttt{Influenced}'[S, r-1]| \geq t(v)\} \quad (2)$$

if $r \leq \lambda$, and

$$\texttt{Influenced}'[S, r] = \texttt{Influenced}'[S, r-1] \quad (3)$$
$$\cup \{v : |N(v) \cap (\texttt{Influenced}'[S, r-1] \setminus \texttt{Influenced}'[S, r-1-\lambda])| \geq t(v)\}$$

if $r > \lambda$. It is immediate that (2) and (3) are an equivalent way to write (1) and (2): for any $S \subseteq V$ and $r \geq 1$, $\texttt{Influenced}'[S, r] = \texttt{Influenced}[S, r]$, so we get that the spreading process with "stiflers" also describes the spreading process with "bounded memory" governed by (2) and (3).

Summarizing, the problem that we shall study in this paper is the following:

TIME WINDOW CONSTRAINED TARGET SET SELECTION (TWC–TSS)
Input: A graph $G = (V, E)$, a threshold function $t : V \longrightarrow \mathbb{N}$, and a time window size λ.
Output: A minimum size $S \subseteq V$ s.t. $\texttt{Influenced}[S, r] = V$, for some $r \geq 0$.

When λ is large enough, for instance equal to the number n of nodes, our TIME WINDOW CONSTRAINED TARGET SET SELECTION problem is equivalent to the classical TARGET SET SELECTION problem studied in [1, 5, 6, 10, 13–18, 37, 41]. In terms of our second formulation of the TWC–TSS problem, the classical TARGET SET SELECTION problem can be viewed as an extreme case in which it is assumed that people have unbounded memory. In general, the TWC–TSS and the TSS problems are quite different. One of the main difficulties of the new TWC–TSS problem is that the sequence of sets $\texttt{Active}[S, r]$, $r = 0, 1, \ldots$ is not necessarily monotonically non-decreasing: it is possible that $\texttt{Active}[S, r]$

---

[2] Another model in which individuals carry a memory of the "amount of influence" received during a bounded time interval has been studied in [23].

is larger than Active$[S, r + 1]$ for some values of r. When $\lambda = n$, we have Active$[S, r] = $ Influenced$[S, r - 1]$ for any r, and monotonicity is restored. At the other extreme, when $\lambda = 1$, a node v becomes influenced at time r only if at least $t(v)$ of its neighbors become influenced at *exactly* time $r - 1$. This sort of synchronization in the propagation of influence poses new challenges, both in the assessment of the computational complexity of the TWC–TSS problem and, especially, in the design of algorithms for its solution. The example in which the graph G is a path is particularly illuminating. As we shall see in Section 4.1, the TARGET SET SELECTION problem is trivial to solve on a path; it is far from being so when there is a fixed time window size λ.

Our Results. In Section 3, we prove a polylogarithmic inapproximability result for the TWC–TSS problem under a plausible computational complexity assumption. The result is obtained by a modification of the very clever proof of the inapproximability of TSS by Chen [13]. In view of the strong inapproximability of the TWC–TSS problem, we then turn our attention to special cases of the problem. In Section 4 we present the main results of the paper: exact polynomial time algorithms for paths, rings, complete graphs, and trees. The algorithms for paths and rings are based on dynamic programming, and the algorithm for complete graphs is greedy. The algorithm for trees is also based on dynamic programming and requires the solution of polynomially many integer linear programs. The polynomial time solvability of each integer linear program is guaranteed by the unimodularity of the associated matrix of coefficients.

3   Hardness of TWC–TSS

In general, our optimization problem TWC–TSS is unlikely to be efficiently approximable, as the following result shows.

Theorem 1. *For any fixed value of the time window size λ, the TWC–TSS problem cannot be approximated within a ratio of $O(2^{\log^{1-\epsilon} n})$ for any fixed $\epsilon > 0$, unless $NP \subseteq DTIME(n^{polylog(n)})$.*

Theorem 1 is a generalization of a similar inapproximability result given in [13] for the TARGET SET SELECTION problem that, as said before, corresponds to our TIME WINDOW CONSTRAINED TARGET SET SELECTION problem when the time window size λ is unbounded. Our result holds for any fixed value of λ. The proof details are presented in the Appendix; here we sketch the main idea. We prove Theorem 1 by a polynomial time reduction from the same MIN REP problem used in [13].

Let $H = (V_A \cup V_B, E)$ be a bipartite graph, where $V_A \cap V_B = \emptyset$ and $E \subseteq V_A \times V_B$. Let $\mathcal{A}$ be a family of subsets of V_A that partitions V_A into $|\mathcal{A}|$ equally sized subsets, and analogously let the family $\mathcal{B}$ be a partition of V_B into $|\mathcal{B}|$ equally sized subsets. Given graph H and partitions $\mathcal{A}$, $\mathcal{B}$, the MIN REP problem asks for a subset $U \subseteq V$ of minimum size such that for each $A \in \mathcal{A}$ and $B \in \mathcal{B}$

$$E \cap (A \times B) \neq \emptyset \text{ implies } [E \cap (A \times B)] \cap (U \times U) \neq \emptyset. \tag{4}$$

Theorem 2. [13] *The* MIN REP *problem cannot be approximated within a ratio of* $O(2^{\log^{1-\epsilon} n})$ *for any fixed* $\epsilon > 0$, *unless* $NP \subseteq DTIME(n^{polylog(n)})$.

Given an instance of MIN REP consisting of the bipartite graph $H = (V_A \cup V_B, E)$ and the pair of partitions $(\mathcal{A}, \mathcal{B})$, we construct an instance $\mathcal{I}$ for the TWC–TSS problem. More precisely, for the instance $\mathcal{I}$ we will only specify a suitable graph $G = (V, E)$ and threshold function $t : V \longrightarrow \mathbb{N} = \{1, 2, \ldots, \}$, since our aim is to prove inapproximability for *any* value of λ. We denote by Γ_ℓ the gadget shown in Figure 1(a), which consists of ℓ paths of length 2 connecting the same pair of nodes. If $\lambda \leq 8$, we need another gadget Γ_ℓ^λ shown in Figure 1(b); it consists of ℓ paths, each having length $11 - \lambda$ and connecting the same pair of extremal nodes. All internal nodes of the gadgets have threshold 1.

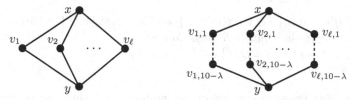

Fig. 1. (a) The gadget Γ_ℓ consisting of ℓ paths of length 2 sharing the extremal nodes. (b) The gadget Γ_ℓ^λ consisting of ℓ paths of length $11 - \lambda$ sharing the extremal nodes.

Let $N = |V| + |E|$. The graph G has node set $V_1 \cup V_2 \cup V_3 \cup V_4$ where

- $V_1 = V$ and each node has threshold N^2,
- $V_2 = \{x_{(a,b)} : (a,b) \in E\}$; each node $x_{(a,b)} \in V_2$ has threshold $2N^5$. The node $x_{(a,b)}$ is connected to both $a \in V_1$ and $b \in V_1$ by a gadget Γ_{N^5}; moreover, if $\lambda \leq 8$ then $x_{(a,b)}$ is also connected to both a and b by a gadget $\Gamma_{N^5}^\lambda$.
- $V_3 = \{y_{A,B} : (A \times B) \cap E \neq \emptyset\}$; each node $y_{A,B} \in V_3$ has threshold N^4 and is connected by a gadget Γ_{N^4} to each $x_{(a,b)} \in V_2$ with $a \in A$ and $b \in B$, and
- $V_4 = \{z_1, \ldots, z_N\}$; each node $z \in V_4$ has threshold $|V_3| \times N^2$ and is connected by a gadget Γ_{N^2} to each node in V_3 and by a gadget Γ_N to each node in V_1.

Theorem 1 follows by showing that any optimal solution U of the MIN REP instance gives rise to a solution $U \subseteq V_1$ to the TWC–TSS problem with input instance (G, t, λ). Vice versa, if S is a solution to the TWC–TSS istance, then in polynomial time one can construct a MIN REP solution of size at most $2|S|$.

4 Polynomially Solvable Cases of TWC–TSS

We now present exact polynomial time algorithms to solve the TWC–TSS problem in several classes of graphs.

4.1 Paths

Let $L^n = (V, E)$ be a path on n nodes, with $V = \{0, \ldots, n-1\}$ and $E = \{(v, v + 1) : 0 \leq v \leq n - 2\}$. Since the threshold of each node cannot exceed its degree, we have that $t(0) = t(n - 1) = 1$ and $t(v) \in \{1, 2\}$, for each $v = 1, \ldots, n - 2$.

The TWC–TSS problem is trivial to solve in case λ is unbounded. Letting $\{v_{i_1}, v_{i_2}, \ldots, v_{i_k}\}$ be the nodes of L^n having threshold equal to 2, one can see that $\{v_{i_1}, v_{i_3}, \ldots, v_{i_{k-2}}, v_{i_k}\}$ is an optimal solution when k is odd, whereas the subset $\{v_{i_1}, v_{i_3}, \ldots, v_{i_{k-1}}, v_{i_k}\}$ is optimal when k is even. In case λ has some fixed value, the situation is much more complicated. Indeed, because of the time window constraint, one must judiciously choose the initial target set in such a way that, for every node with threshold 2 that does not belong to the initial target set, its two neighbors become influenced at the correct times.

To avoid trivialities, we assume that L^n has at least two nodes with threshold equal to 2. Should it be otherwise, for instance all nodes have threshold 1, then any subset S of V with $|S| = 1$ is an optimal solution. If exactly one node, say v, has threshold 2, then $\{v\}$ is an optimal solution.

Lemma 1. *If $\ell = \min\{v \in V : t(v) = 2\}$ and $s = \max\{v \in V : t(v) = 2\}$, then there exists an optimal solution S such that*
   i) $S \cap \{0, \ldots \ell - 1\} = \emptyset = S \cap \{s + 1, \ldots n - 1\}$;
   ii) $\ell, s \in S$.

Lemma 1 implies that we can ignore all nodes in L^n that are to the left of the lowest numbered node with threshold 2, and to the right of the highest numbered node with threshold 2. Equivalently, from now on we can assume that $t(0) = t(n-1) = 2$. Define the array $D[0 \ldots (n-1)]$, where $D[n-1] = n-1$ and

$$D[i] = \min\{j : i < j \leq n-1 \text{ and } t(j) = 2\}, \tag{5}$$

for $i = 0, \ldots, n-2$. Since $t(n-1) = 2$, value $D[i]$ is always well defined. One can check that the following algorithm computes an array D satisfying (5).

Algorithm ARRAY(L^n) *[Input: A path L^n with threshold function $t(\cdot)$]*
   1. Set $D[n-1] = n-1$ and $j = n-1$
   2. for $i = (n-2)$ **down to** 0 **do**
   3.   set $D[i] = j$
   4.   **if** $t(i) = 2$ **then** set $j = i$

For each $i = 0, \ldots, n-1$, let L_i^n denote the sub-path consisting of the last $n-i$ nodes $\{i, i+1, \ldots, n-1\}$ of L^n. We denote by $\mathbf{s}(i)$ the minimum size of a TWC target set for L_i^n that contains both the extreme nodes, that is, i and $n-1$. Our first goal is to compute $\mathbf{s}(0)$, the size of an optimal solution for $L_0^n = L^n$.

Lemma 2. *Fix the time window size λ and consider the family of all TWC target sets for L_i^n that include both i and $n-1$. If $i < n-1$, such a family contains a minimum size TWC target set whose second smallest element is in*

$$\{D[i]\} \cup \{x : \max\{D[i]+1, 2D[i]-i-\lambda+1\} \leq x \leq \min\{2D[i]-i+\lambda-1, D[D[i]]\}\}. \tag{6}$$

From Lemma 2, we have $\mathbf{s}(n-1) = 1$ and, for each $i = 0, \ldots, n-2$,

$$\mathbf{s}(i) = 1 + \min\left\{\mathbf{s}(D[i]), \min_j \mathbf{s}(j)\right\} \tag{7}$$

where j satisfies $\max \left\{ D[i] + 1,\, 2D[i] - i - \lambda + 1 \right\} \leq j \leq \min \left\{ 2D[i] - i + \lambda - 1,\, D[D[i]] \right\}$. The size of an optimal target set for L_n can be computed as $\mathsf{s}(0)$.

The actual TWC target set of optimal size $\mathsf{s}(0)$ can be constructed using standard backtracking techniques.

Theorem 3. *For any time window size λ, an optimal TWC target set for the path L^n can be computed in time $O(n)$.*

4.2 Rings

We can use Theorem 3 above to design an algorithm for the TWC–TSS problem on rings. Let R^n denote the ring on n nodes $\{0, \ldots, n-1\}$ with edges $(i, (i+1) \bmod n)$ and thresholds $t(i)$, for $i = 0, \ldots, n-1$.

We first notice that if all nodes have threshold 2, then an optimal TWC target set for R^n trivially has size $\lceil n/2 \rceil$, so let us now assume that there exists a node j that has threshold $t(j) = 1$. Either j is in an optimal TWC target set for R^n or it is not. Consider the path $R^n_{j,2}$ obtained by "breaking" the ring R^n at node j, duplicating node j into j and j', and assigning threshold 2 to both j and j' (regardless of the original threshold value $t(j) = 1$ in R^n). Therefore, the edges of $R^n_{j,2}$ are $(j, j+1), (j+1, j+2), \ldots, (n-2, n-1), (n-1, 0), \ldots (j-2, j-1), (j-1, j')$. The thresholds of $R^n_{j,2}$ are

$$t_{j,2}(i) = \begin{cases} t(i) & \text{if } 0 \leq i \leq n-1 \text{ and } i \neq j \\ 2 & \text{if } i = j \text{ or } i = j'. \end{cases}$$

We can use the algorithm of Section 4.1 to compute the size of an optimal TWC target set $S_{j,2}$ for the path $R^n_{j,2}$. Notice that both j and j' must be in $S_{j,2}$, so $S_{j,2} - \{j'\}$ is a TWC target set for the ring R^n, optimal among all TWC target sets that include node j.

Now we want to compute a TWC target set for the ring R^n that is optimal among all TWC target sets that do not include node j. To do this, consider the path $R^n_{j,1}$ that has the same nodes and edges as $R^n_{j,2}$ but has thresholds

$$t_{j,1}(i) = \begin{cases} t(i) & \text{if } 0 \leq i \leq n-1 \text{ and } i \neq j \\ 1 & \text{if } i = j \text{ or } i = j'. \end{cases}$$

In particular, the endpoints of $R^n_{j,1}$ have thresholds $t_{j,1}(j) = t_{j,1}(j') = 1$. First, we apply Lemma 1 to $R^n_{j,1}$ and then we use the algorithm of Section 4.1 to compute (the size of) an optimal TWC target set $S_{j,1}$. Since $j, j' \notin S_{j,1}$, we have that $S_{j,1}$ is a TWC target set for the ring R^n, optimal among all TWC target sets that do not include node j.

An optimal solution for the ring R^n is then obtained by choosing the smaller of $S_{j,2} - \{j'\}$ and $S_{j,1}$. In conclusion we have the following result.

Theorem 4. *For any value of the time window size λ, an optimal TWC target set for the ring R^n can be computed in time $O(n)$.*

4.3 Trees

Let $T = (V, E)$ be a tree with threshold function $t : V \longrightarrow \mathbb{N}$, and let $\lambda \geq 1$ be a fixed value of the time window size. We consider T to be rooted at some arbitrary node $p \in V$. For each node $v \in V$, we denote by $T_v = (V_v, E_v)$ the subtree of T rooted at v. Moreover, we denote by $\mathrm{Ch}(v)$ the set of all children of node v in T_v.

Definition 1. *Given node $v \in V$ and integers t, r, with $t \in \{t(v), t(v) - 1\}$ and $r \geq 0$, we denote by $\mathrm{s}(v, t, r)$ the minimum size of a TWC target set $S \subseteq V_v$ for subtree T_v that influences node v in round r (that is, $v \in \mathrm{Influenced}[S, r] \setminus \mathrm{Influenced}[S, r - 1])$, under the assumption that v has threshold t in T_v. The threshold of each other node $w \neq v$ in T_v is the original one $t(w)$.*

The size of an optimal TWC target set for the tree T can be computed as

$$\min_r \mathrm{s}(p, t(p), r), \tag{8}$$

where r ranges between 0 and the maximum possible number of rounds needed to complete the influence diffusion process. The number of rounds is always upper bounded by the number of nodes in the graph (since at least one new node must be influenced in each round before the diffusion process stops). However, for a tree T, this value is upper bounded by the length of the longest path in T. In other words, the parameter r in Definition 1 is bounded by the diameter $diam(T)$ of T.

We use a dynamic programming approach to compute the value in (8). Then, the corresponding optimal TWC target set S can be built using standard backtracking techniques. In our dynamic programming algorithm we compute all of the values

$\mathrm{s}(v, t, r)$ for each $v \in V$, $t \in \{t(v), t(v) - 1\}$ and $r = 0, \ldots, diam(T)$,

and the computation is performed according to a breadth-first search (BFS) reverse ordering of the nodes of T, so that each node v is considered only when all of the values $\mathrm{s}(\cdot, \cdot, \cdot)$ for all of its children are known. The rationale behind the computation of both $\mathrm{s}(v, t(v), r)$ and $\mathrm{s}(v, t(v) - 1, r)$ is the following:

i) $\mathrm{s}(v, t(v), r)$ corresponds to the case of a target set S for tree T such that
   - $v \in \mathrm{Influenced}[S, r] \setminus \mathrm{Influenced}[S, r - 1]$ and
   - at least $t(v)$ of v's children belong to $\mathrm{Active}[S \cap V_v, r] \subseteq \mathrm{Active}[S, r]$;
ii) $\mathrm{s}(v, t(v) - 1, r)$ is the size of an optimal target set S for T satisfying
   - $v \in \mathrm{Influenced}[S, r] \setminus \mathrm{Influenced}[S, r - 1]$,
   - $\mathrm{Active}[S, r]$ contains v's parent in T, and
   - at least $t(v) - 1$ of v's children belong to $\mathrm{Active}[S \cap V_v, r] \subseteq \mathrm{Active}[S, r]$.

In the following, we show how to compute the above values $\mathrm{s}(\cdot, \cdot, \cdot)$. The procedure is summarized in algorithm TREE.

First, consider the computation of $\mathrm{s}(v, t, r)$ when v is a leaf of T. In this case we have $t(v) = deg(v) = 1$.

– If $r = 0$, v trivially must belong to the target set since v needs to be active at time 0; hence $\mathrm{s}(v, t, 0) = 1$.

– If $r > 0$ and $t = t(v) = 1$, we observe that any TWC target set that influences leaf v at time *exactly* r cannot contain v and, therefore, must influence v's parent at time $r - 1$. To do so, we set $\mathrm{s}(v, 1, r) = \infty$ in the algorithm; this forces the minimum at line **14** or **18** to be reached with threshold $t(v) - 1 = 0$, thus forcing

v's parent to be active in round $r - 1$.
- If $r > 0$ and $t = t(v) - 1 = 0$ then, trivially, $s(v, 0, r) = 0$.

Algorithm TREE(T, p, λ, t)
　　　　　　[Input: Tree T rooted at p, time window size λ, threshold function t.]
1. **For** each $v \in T$ in reverse order to a BFS of T
2. 　　　　　*[We compute $s(v, t, r)$ for each $t \in \{t(v), t(v) - 1\}$ and $0 \leq r \leq diam(T)$]*
3. 　**If** v is a leaf **then** *[here $t(v) = 1$]*
4. 　　**For** $r = 0, \ldots, diam(T)$
5. 　　　　Set $s(v, 0, r) = 0$ and $s(v, 1, r) = \begin{cases} 1 & \text{if } r = 0 \\ \infty & \text{otherwise} \end{cases}$
6. 　**If** v is NOT a leaf in T **then**
7. 　　**For** each $\left(r = 0, \ldots, diam(T) \text{ AND } t \in \{t(v), t(v) - 1\} \text{ (only } t = t(v) \text{ if } v = p) \right)$
8. 　　　**If** $r = 0$ **then**
9. 　　　　Set $s(v, t, 0) = 1 + \sum_{w \in Ch(v)} \min \{\min_{1 \leq j \leq \lambda} s(w, t(w) - 1, j), \min_{j \geq 0} s(w, t(w), j)\}$
10. 　　　**If** $r \geq 1$ and $t = 0$ **then**
11. 　　　　Set $s(v, 0, r) = \sum_{w \in Ch(v)} \min \{\min_{r+1 \leq j \leq r+\lambda} s(w, t(w) - 1, j), \min_{j \geq r-1} s(w, t(w), j)\}$
12. 　　　**If** $r \geq 1$ and $t = 1$ **then**
13. 　　　　**For** each $w \in Ch(v)$
14. 　　　　　Compute $m(w) = \min \{\min_{r+1 \leq j \leq r+\lambda} s(w, t(w) - 1, j), \min_{j \geq r-1} s(w, t(w), j)\}$
15. 　　　　Set $z = \text{argmin}_{w \in Ch(v)} \{s(w, t(w), r - 1) - m(w)\}$
16. 　　　　Set $s(v, 1, r) = \sum_{w \in Ch(v) \backslash \{z\}} m(w) + s(z, t(z), r - 1)$
17. 　　　**If** $r \geq 1$ and $t > 1$ **then**
18. 　　　　Set $s(v, t, r) = \min \sum_{w \in Ch(v)} m(w)$, where
19. 　　　　$m(w) \in \{s(w, t, j) : (t = t(w) \text{ AND } j \geq 0) \text{ OR } (t = t(w) - 1 \text{ AND } r < j \leq r + \lambda)\}$
20. 　　　　$|\{w : m(w) = s(w, t(w), j), r - \lambda \leq j \leq r - 1\}| \geq t$
21. 　　　　$|\{w : m(w) = s(w, t(w), j), \ell - \lambda \leq j \leq \ell - 1\}| < t, \quad \forall \ell = 1, \ldots, r - 1$

Now consider an arbitrary internal node v. Since we process nodes in a BFS reverse order, each child of v has already been processed when the algorithm processes v. If $r = 0$, then v must necessarily be in the target set and any $w \in Ch(v)$ can benefit from this. Therefore, the size $s(v, t, 0)$ of an optimal solution for the subtree T_v is equal to

$$s(v, t, 0) = 1 + \sum_{w \in Ch(v)} \min \left\{ \min_{1 \leq j \leq \lambda} s(w, t(w) - 1, j), \min_{0 \leq j \leq diam(T)} s(w, t(w), j) \right\}.$$

Notice that we have constrained j to be in the range $1, \ldots, \lambda$ in the formula above when w's threshold is $t(w) - 1$. This is correct since v is active and able to influence w only in rounds $j = 1, \ldots, \lambda$.

Now, let us consider the computation of $s(v, t, r)$ with $r \geq 1$, that is, when v is not part of the target set and v is influenced at time r by t of its children (plus its parent if $t = t(v) - 1$). To determine the optimal solution, we need to know the best among the values $s(w, \tau, j)$ for each $w \in Ch(v)$ and for all possible values of parameters τ and j, subject to the following two constraints:

1) if $\tau = t(w) - 1$, then $r + 1 \leq j \leq r + \lambda$ (indeed v is active and can influence w only during the λ rounds after it has become influenced, that is, in rounds $j = r + 1, \ldots, r + \lambda$),

2) at least t nodes in $\text{Ch}(v)$ are active in round r but at most $t-1$ are active in any previous round $j \leq r - 1$ (otherwise v would become influenced before the required round r).

The special case $t = 0$ can hold only if $t(v) = 1$ and $t = t(v) - 1$; hence node v must be influenced by its parent at round r and none of its children can be active before round r.

Lemma 3. *The computation at lines 18–21 of the algorithm TREE(T, p, λ) can be done in polynomial time.*

Theorem 5. *For any tree T, the optimal TWC target set can be computed in polynomial time.*

4.4 Complete Graphs

Let $K_n = (V, E)$ denote the complete graph on n nodes. The following observation was made in [35] for target set selection without a time window constraint; it is easy to see that it also holds in our scenario.

Lemma 4. *[35] If the optimal TWC target set for K_n has size k, then there exists an optimal TWC target set consisting of k nodes with the largest thresholds.*

Lemma 4 follows from the observation that in any target set S for K_n, if there exist $v \in S$ and $u \in V - S$ with $t(v) < t(u)$, then $S \setminus \{v\} \cup \{u\}$ is also a target set for K_n. Lemma 4 implies that we only need to determine the size of an optimal TWC target set. The following algorithm $MAX(n, k)$ determines the largest number of nodes that can be influenced using a TWC target set of k nodes. The algorithm assumes that the thresholds have been sorted in non-decreasing order. Moreover, it assumes the precomputation of the integer vector $A[1..n-1]$ such that $A[\ell] = |\{v \in V \mid t(v) \leq \ell\}|$, for $\ell = 1, \ldots, n - 1$. Notice that both sorting the thresholds, by counting sort, and computing A can be done in linear time.

Algorithm $MAX(n, k)$ *[Input: vector A[1..n-1], parameters λ and k]*
  1. Set $\ell = k$
  2. If $A[\ell] > 0$ **then** *[at least one node outside the target set can be influenced]*
  3.   **For** $j = 0, \ldots, \lambda - 2$
  4.   Set $X[j] = -k$
  5.   Set $X[\lambda - 1] = 0$, Set $j = 0$;
  6.   **Repeat**
  7.   Set $y = A[\ell]$, $\ell = A[\ell] - X[j]$, $X[j] = y$, $j = (j + 1) \bmod \lambda$
  8.   **Until** $(A[\ell] - X[j] \leq \ell$ OR $A[\ell] + k \geq n)$
  9. Output $\min\{n, k + A[\ell]\}$

We can show that the algorithm $MAX(n, k)$ requires $O(n)$ time to compute the largest number of nodes that can be influenced in K_n using a TWC target set of size k. Using a binary search for the optimal value of k we obtain the following result.

Theorem 6. *The optimal TWC target set in a complete graph K_n can be computed in time $O(n \log n)$.*

References

1. Ackerman, E., Ben-Zwi, O., Wolfovitz, G.: Combinatorial model and bounds for target set selection. Theoretical Computer Science 411, 4017–4022 (2010)
2. Anderson, R.M., May, R.M.: Infectious Diseases of Humans: Dynamics and Control. Oxford University Press (1991)
3. Banos, R.A., Borge-Holthoefer, J., Moreno, Y.: The role of hidden influentials in the diffusion of online information cascades. arXiv:1303.4629v1 [physics.soc-ph] (2013)
4. Barrat, A., Barthelemy, M., Vespignani, A.: Dynamical Processes on Complex Networks. Cambridge University Press (2012)
5. Bazgan, C., Chopin, M., Nichterlein, A., Sikora, F.: Parameterized Approximability of Maximizing the Spread of Influence in Networks. arXiv:1303.6907 [cs.DS]
6. Ben-Zwi, O., Hermelin, D., Lokshtanov, D., Newman, I.: Treewidth governs the complexity of target set selection. Discrete Optimization 8, 87–96 (2011)
7. Bettencourt, L.M.A., Cintron-Arias, A., Kaiser, D.I., Castillo-Chavez, C.: The power of a good idea: Quantitative modeling of the spread of ideas from epidemiological models. Physica A 364, 513–536 (2006)
8. Bikhchandani, S., Hirshleifer, D., Welch, I.: A theory of fads, customs, and cultural change as informational cascades. Journal of Political Economy 100, 992–1026 (1992)
9. Boccaletti, S., Latora, V., Moreno, Y., Chavez, M., Hwang, D.-U.: Complex Networks: Structure and Dynamics. Physics Reports 472, 175–308 (2006)
10. Centeno, C.C., Dourado, M.C., Penso, L.D., Rautenbach, D., Szwarcfiter, J.L.: Irreversible conversion of graphs. Theoretical Computer Science 412, 3693–3700 (2011)
11. Centola, D.: The spread of behavior in an online social network experiment. Science 329, 1194–1197 (2010)
12. Centola, D., Macy, M.: Complex contagions and the weakness of long ties. American Journal of Sociology 113, 702–734 (2007)
13. Chen, N.: On the approximability of influence in social networks. SIAM J. Discrete Math. 23, 1400–1415 (2009)
14. Chiang, C.Y., Huang, L.H., Huang, W.T., Yeh, H.G.: The Target Set Selection Problem on Cycle Permutation Graphs, Generalized Petersen Graphs and Torus Cordalis. arXiv:1112.1313 (2011)
15. Chopin, M., Nichterlein, A., Niedermeier, R., Weller, M.: Constant Thresholds Can Make Target Set Selection Tractable. In: Even, G., Rawitz, D. (eds.) MedAlg 2012. LNCS, vol. 7659, pp. 120–133. Springer, Heidelberg (2012)
16. Chiang, C.-Y., Huang, L.-H., Li, B.-J., Wu, J., Yeh, H.-G.: Some results on the target set selection problem. Journal of Combinatorial Optimization (to appear)
17. Chiang, C.-Y., Huang, L.-H., Yeh, H.-G.: Target Set Selection Problem for Honeycomb Networks. SIAM J. Discrete Math. 27(1), 310–328 (2013)
18. Cicalese, F., Cordasco, G., Gargano, L., Milanič, M., Vaccaro, U.: Latency-Bounded Target Set Selection in Social Networks. In: Bonizzoni, P., Brattka, V., Löwe, B. (eds.) CiE 2013. LNCS, vol. 7921, pp. 65–77. Springer, Heidelberg (2013)
19. Comellas, F., Mitjana, M., Peters, J.G.: Broadcasting in Small-world Communication Networks. In: SIROCCO. Proceedings in Informatics, Carleton Scientific, vol. 13, pp. 73–85 (2002)
20. Daley, D.J., Kendall, D.G.: Epidemics and rumours. Nature 204, 1118 (1964)

21. Duenas-Osorio, L., Vemuru, S.M.: Cascading failures in complex infrastructure systems. Structural Safety 31(2), 157–167 (2009)
22. Domingos, P., Richardson, M.: Mining the network value of customers. In: Proc. of the Seventh ACM SIGKDD International Conference on Knowledge Discovery and Data Mining, pp. 57–66 (2001)
23. Dodds, P.S., Watts, D.J.: A generalized model of social and biological contagion. J. Theoretical Biology 232, 587–604 (2005)
24. Easley, D., Kleinberg, J.: Networks, Crowds, and Markets: Reasoning About a Highly Connected World. Cambridge University Press (2010)
25. Flocchini, P., Královic, R., Ruzicka, P., Roncato, A., Santoro, N.: On time versus size for monotone dynamic monopolies in regular topologies. J. Discrete Algorithms 1, 129–150 (2003)
26. Karimi, F., Holme, P.: Threshold model of cascades in temporal networks. arXiv:1207.1206v2 [physics.soc-ph] (2012)
27. Kempe, D., Kleinberg, J.M., Tardos, E.: Maximizing the spread of influence through a social network. In: Proc. of the Ninth ACM SIGKDD International Conference on Knowledge, Discovery and Data Mining, pp. 137–146 (2003)
28. Kempe, D., Kleinberg, J.M., Tardos, É.: Influential Nodes in a Diffusion Model for Social Networks. In: Caires, L., Italiano, G.F., Monteiro, L., Palamidessi, C., Yung, M. (eds.) ICALP 2005. LNCS, vol. 3580, pp. 1127–1138. Springer, Heidelberg (2005)
29. Kephart, J.O., White, S.R., Chess, D.M.: Computers and epidemiology. IEEE Spectrum 30, 20–26 (1993)
30. Leskovic, H., Adamic, L.A., Huberman, B.A.: The dynamic of viral marketing. ACM Transactions on the WEB 1 (2007)
31. Leppaniemi, M., Karjaluoto, H., Lehto, H., Goman, A.: Targeting Young Voters in a Political Campaign: Empirical Insights into an Interactive Digital Marketing Campaign in the 2007 Finnish General Election. Journal of Nonprofit & Public Sector Marketing 22, 14–37 (2010)
32. Lopez-Pintado, D.: Diffusion in complex social networks. Games and Economic Behavior 62, 573–590 (2008)
33. Maki, D.P., Thompson, M.: Mathematical Models and Applications, with Emphasis on the Social, Life and Management Sciences. Prentice Hall (1973)
34. Nekovee, M., Moreno, Y., Bianconi, G., Marsili, M.: Theory of rumour spreading in complex social networks. Physica A: Statistical Mechanics and its Applications 374, 467–470 (2007)
35. Nichterlein, A., Niedermeier, R., Uhlmann, J., Weller, M.: On Tractable Cases of Target Set Selection. Social Network Analysis and Mining, 1–24 (2012)
36. Peleg, D.: Local majorities, coalitions and monopolies in graphs: a review. Theoretical Computer Science 282, 231–257 (2002)
37. Reddy, T.V.T., Rangan, C.P.: Variants of spreading messages. J. Graph Algorithms Appl. 15(5), 683–699 (2011)
38. Schrijver, A.: Combinatorial Optimization: Polyhedra and Efficiency. Springer (2004)
39. Tumulty, K.: Obama's Viral Marketing Campaign. TIME Magazine (July 5, 2007)
40. Ugander, J., Backstrom, L., Marlow, C., Kleinberg, J.: Structural Diversity in Social Contagion. Proc. of the Nat'l Academy of Sciences (PNAS) 109, 5962–5966 (2012)
41. Zaker, M.: On dynamic monopolies of graphs with general thresholds. Discrete Mathematics 312(6), 1136–1143 (2012)

Analysis of Fully Distributed Splitting and Naming Probabilistic Procedures and Applications

(Extended Abstract)

Yves Métivier, John Michael Robson, and Akka Zemmari

Université de Bordeaux, LaBRI, UMR CNRS 5800
351 cours de la Libération, 33405 Talence, France
{metivier,robson,zemmari}@labri.fr

Abstract. This paper proposes and analyses two fully distributed probabilistic splitting and naming procedures which assign a label to each vertex of a given anonymous graph G without any initial knowledge. We prove, in particular, that with probability $1 - o(n^{-1})$ (resp. with probability $1 - o(n^{-c})$ for any $c \geq 1$) there is a unique vertex with the maximal label in the graph G having n vertices. In the first case, the size of labels is $O(\log n)$ with probability $1 - o(n^{-1})$ and the expected value of the size of labels is also $O(\log n)$. In the second case, the size of labels is $O\left((\log n)(\log^* n)^2\right)$ with probability $1 - o(n^{-c})$ for any $c \geq 1$; their expected size is $O\left((\log n)(\log^* n)\right)$.

We analyse a basic simple maximum broadcasting algorithm and prove that if vertices of a graph G use the same probabilistic distribution to choose a label then, for broadcasting the maximal label over the labelled graph, each vertex sends $O(\log n)$ messages with probability $1 - o(n^{-1})$.

From these probabilistic procedures we deduce Monte Carlo algorithms for electing or computing a spanning tree in anonymous graphs without any initial knowledge and for counting vertices of an anonymous ring; these algorithms are correct with probability $1 - o(n^{-1})$ or with probability $1 - o(n^{-c})$ for any $c \geq 1$. The size of messages has the same value as the size of labels. The number of messages is $O(m \log n)$ for electing and computing a spanning tree; it is $O(n \log n)$ for counting the vertices of a ring.

We illustrate the power of the splitting procedure by giving a probabilistic election algorithm for rings having n vertices with identities which is correct and always terminates; its message complexity is equal to $O(n \log n)$ with probability $1 - o(n^{-1})$. (Proofs are omitted for lack of space).

1 Introduction

The Problem. We consider anonymous, and, more generally, partially anonymous networks: unique identities are not available to distinguish the processes (or we cannot guarantee that they are distinct). We do not assume any global knowledge of the network, not even its size or an upper bound on its size. The

T. Moscibroda and A.A. Rescigno (Eds.): SIROCCO 2013, LNCS 8179, pp. 153–164, 2013.

processes have no knowledge on position or distance. In this context, solutions for classical distributed problems, such as the construction of spanning trees, counting or election, must use probabilistic algorithms. This paper presents and studies splitting and naming procedures which provide solutions to these problems. The question of anonymity is often considered when processes must not divulge their identities during execution, due to privacy concerns or security policy issues [GR05]. In addition, each process may be built in large scale quantities from which it is quite infeasible to ensure uniqueness. Therefore, each process must execute the same finite algorithm in the same way, regardless of its identity, as explained in [AAER07].

The Model. Our model is the usual asynchronous message passing model [AW04, Tel00]. A network is represented by a simple connected graph $G = (V(G), E(G)) = (V, E)$ where vertices correspond to processes and edges to direct communication links. Each process can distinguish different incident edges, i.e., for each $u \in V$ there exists a bijection between the neighbours of u in G and $[1, \deg_G(u)]$ (where $\deg_G(u)$ is the number of neighbours of u in G). The numbers associated by each vertex to its neighbours are called *port-numbers*. Each process v in the network represents an entity that is capable of performing computation steps, sending messages via some ports and receiving any message via some port that was sent by the corresponding neighbour. We consider asynchronous systems, i.e., each computation may take an unpredictable (but finite) amount of time. Note that we consider only reliable systems: no fault can occur on processes or communication links. In this model, a distributed algorithm is given by a local algorithm that all processes should execute; thus the processes having the same degree have the same algorithm. A local algorithm consists of a sequence of computation steps interspersed with instructions to send and to receive messages. As Tel [Tel00] (p. 71), we define the time complexity by supposing that internal events need zero time units and that the transmission time (i.e., the time between sending and receiving a message) is at most one time unit. This corresponds to the number of rounds needed by a synchronous execution of the algorithm.

A probabilistic algorithm is an algorithm which makes some random choices based on some given probability distributions; non-probabilistic algorithms are called deterministic. A distributed probabilistic algorithm is a collection of local probabilistic algorithms. Since our networks are anonymous, if two processes have the same degree their local probabilistic algorithms are identical and have the same probability distribution. A Las Vegas algorithm is a probabilistic algorithm which terminates with a positive probability (in general 1) and always produces a correct result. A Monte Carlo algorithm is a probabilistic algorithm which always terminates; nevertheless the result may be incorrect with a certain probability.

Distributed algorithms presented in this paper are message terminating. This means (see [Tel00], Chapter 8) that algorithms reach a terminal configuration (a configuration in which no further steps are applicable) and processes are not aware that the computation has terminated. We speak of process termination if, when algorithms reach a terminal configuration, processes are in a terminal state (a state in which there is no event of the process applicable).

Some results on graphs having n vertices are expressed with high probability, meaning with probability $1 - o(n^{-1})$ (w.h.p. for short) or with very high probability, meaning with probability $1 - o(n^{-c})$ for any $c \geq 1$ (w.v.h.p. for short). We recall that $\log^* n = min\{i | \log^{(i)} n \leq 2\}$, where $\log^{(1)} n = \log n$ and $\log^{(i+1)} n = \log(\log^{(i)} n)$.

Our Contribution. Let $G = (V, E)$ be an anonymous connected graph having n vertices. We assume no knowledge on G. In the first part of this paper, we provide and analyse the following procedure by which each vertex builds its label. Each vertex v of G draws a bit b_v uniformly at random. Let t_v be the number of random draws of b_v on the vertex v until $b_v = 1$; it is called the lifetime of the vertex v. Each vertex v uses its lifetime to draw at random a number id_v in the set $\{0, ..., 2^{t_v + 3\log_2(t_v)} - 1\}$; finally, v is labelled with the couple (t_v, id_v). Let T be the maximal value in the set $\{t_v | v \in V(G)\}$. We prove that w.h.p.:

$$\log_2 n - \log_2(2\log n) < T < 2\log_2 n + \log^* n.$$

We prove that, w.h.p., there exists exactly one vertex v such that $t_v = T$ and $id_v > id_w$ for any vertex w different from v such that $t_w = T$. The size of labels is $O(\log n)$ w.h.p. and the expected value of the size of labels is also $O(\log n)$. We also prove that w.v.h.p.: $\frac{1}{2}\log_2 n < T < (\log^* n)\log_2 n$. If each vertex v draws id_v uniformly at random in the set: $\{0, ..., 2^{t_v \log^* t_v} - 1\}$ then, w.v.h.p., there exists exactly one vertex v such that $t_v = T$ and $id_v > id_w$ for any vertex w different from v such that $t_w = T$. In this case the size of labels is $O((\log n)(\log^* n)^2)$ w.v.h.p.; their expected size is $O((\log n)(\log^* n))$.

We analyse a very simple maximum broadcasting algorithm and prove that if vertices of a graph G use the same distribution to choose their label then, for broadcasting the maximal label over G, each vertex sends $O(\log n)$ messages w.h.p.

In the second part of this paper, we apply these procedures to classical problems, in anonymous graphs without any knowledge, such as spanning tree construction, counting the number of vertices of a ring or electing. In this way, we obtain:

- Monte Carlo spanning tree algorithms correct w.h.p. (resp. w.v.h.p.),
- Monte Carlo counting ring size algorithms correct w.h.p. (resp. w.v.h.p.),
- Monte Carlo election algorithms correct w.h.p. (resp. w.v.h.p.).

The size of messages used by these algorithms is the same as the size of labels generated by splitting and naming procedures. We prove also that the message complexities (the number of messages through the graph) is $O(m \log n)$ for the spanning tree computation and election; it is $O(n \log n)$ for counting the vertices of a ring.

We illustrate the power of the splitting procedure by giving a probabilistic election algorithm for ring graphs with identities which is correct and always terminates; its message complexity is equal to $O(n \log n)$ w.h.p. (where n is the number of vertices).

A conclusion explains how to obtain a Monte Carlo algorithm which solves the counting problem of a ring w.h.p. and which ensures an error probability bounded by ϵ where ϵ is the smallest value among a set of error probabilities fixed by vertices of the ring R (each vertex knows only its own error probability).

Related Work: Comparisons and Comments. Chapter 9 of [Tel00] and [Lav95] give a survey of what can be done and of impossibility results in anonymous networks. In particular, no deterministic algorithm can elect (see Angluin [Ang80], Attiya et al. [ASW88] and Yamashita and Kameda [YK88]); furthermore, with no knowledge on the network, there exists no Las Vegas election algorithm [IR90]. In fact, [IR90] proves that, in this context, the best we can achieve for electing, counting or spanning tree computing are message terminating Monte Carlo algorithms; there are no such algorithms which are process terminating.

Message terminating Monte Carlo election algorithms for anonymous graphs without knowledge are presented in [IR90, AM94, SS94].

The idea that each vertex draws at random a bit until it gets 1 (or 0) is used in two different contexts in [Pro93, FMS96, KMW11] and in [AM94].

In [Pro93, FMS96, KMW11], typically, as explained in [FMS96], processors have identification numbers and know the number of processors. A group of n processors play a game to identify a winner by tossing fair coins; the winner is the elected vertex. All processors that throw heads are eliminated; those that throw tails remain candidate and flip their coins again. The process is repeated among candidate winners until a single winner is identified. If at any stage all remaining candidate winners throw heads then all remaining players participate again as candidate winners in the next round of coin tossing. The main parameters studied are the number of rounds till termination and the total number of coin flips.

In the case where no knowledge on the network is available, Monte Carlo election algorithms presented in [AM94, SS94] are correct with probability $1 - \epsilon$, where ϵ is fixed and known to all vertices. In [AM94] executing the algorithm, presented in Section 2 (Networks with unknown size), each vertex tosses a fair coin until it gets a head for the first time. The number of these tosses is used only to have a small number of vertices which compete for the election. To obtain its number, each vertex v selects at random an element id_v of $[1, ..., d]$ where $d = 36 \log 4r$, $r = 1/\epsilon$ and ϵ is fixed, given a priori (see [AM94], p. 315 Fig. 2.). The time complexity is $O(D)$ where D is the diameter of the network. Let $M = max\{id_v | v \in V\}$, then there is a unique vertex u such that t_u is maximal and $id_u = M$ with probability greater than or equal to $1 - \epsilon$. The expected size of each message is $O(\log \log n + \log \epsilon^{-1})$. (The idea of eliminating some vertices before the election appears also in [RFJ+07]; the goal is to reduce the number of messages for the election.) The algorithm presented in [SS94], maintains a rooted spanning forest of G. Each vertex belongs to a tree (initially it is alone). In the course of the algorithm trees expand by merging with adjacent trees. The level of a tree T, denoted $Level(T)$, is the integer part of the (base two) logarithm of the estimated number of nodes of T. The label of T (and thus of its root) is redrawn from the domain $d2^{level(T)}$, where $d = \lceil 2/\epsilon \rceil$ and ϵ is fixed ([SS94] p. 90). Upon termination, the forest consists of one tree with probability $1 - \epsilon$; the root of this tree is taken to be the elected vertex. The size of messages is $O(\log(n/\epsilon))$. The time complexity is $O(n)$. A message terminating Monte Carlo counting algorithm for rings is presented in [IR90]. Each vertex generates messages formed by: an estimation (initially 2) of the size of the ring and a random bit. Next the vertex sends the message along the ring. If a vertex receives a message which indicates that the estimation is not correct then it incrementes its estimation and generates another message. Finally, for each ϵ, [IR90] provides a counting algorithm correct with probability $1 - \epsilon$; its message complexity is $O(n^3)$ and its time complexity is $O(n)$.

Summary. This paper is organised as follows. Sections 2 and 3 present and analyse two fully distributed splitting and naming probabilistic procedures which generate labels on vertices of graphs. Section 4 analyses a classical broadcasting algorithm. Sections 5, 6 and 7 present applications to spanning tree construction, counting and electing for anonymous graphs without any knowledge; more precisely they present Monte Carlo algorithms which solve these problems and which are correct w.h.p. or w.v.h.p. Section 8 applies the splitting procedure to obtain an efficient deterministic election algorithm for named rings.

2   Analysis of a Splitting and Naming Probabilistic Procedure

This section presents and analyses a fully distributed probabilistic procedure, denoted Splitting-Naming-whp (see Algorithm 1), which assigns to each vertex v of a graph G a label (t_v, id_v) defined as follows. A vertex v draws uniformly at random (u.a.r. for short) a bit b_v. We denote by t_v the number of bits generated by the vertex v until $b_v = 1$; it is called the lifetime of v. A vertex v is said to be alive at time t if $t < t_v$. The number id_v is obtained by generating a number choosen u.a.r. in the set $\{0, ..., 2^{t_v+3\log_2(t_v)} - 1\}$.

Algorithm 1: Procedure Splitting-Naming-whp(v)

1: $t_v := 0$
2: **repeat**
3:     draw uniformly at random a bit $b(v)$;
4:     $t_v := t_v + 1$
5: **until** $b(v) = 1$
6: choose uniformly at random a number id_v in the set $\{0, \cdots, 2^{t_v+3\log_2(t_v)} - 1\}$;

We define the order, denoted $<$, on couples by: $(t_v, id_v) < (t'_v, id'_v)$ if: either $t_v < t'_v$ or ($t_v = t'_v$ and $id_v < id'_v$).

The sequel of this section analyses Procedure Splitting-Naming-whp(v).

Analysis of the Maximal Number of Bits Drawn by Each Vertex. We first analyse the value of t_v for any vertex v and the maximal value of t_v among vertices of the graph. Any vertex has probability $1/2$ to draw the bit 1. Then t_v, for any vertex v, is a geometric random variable (r.v. for short) with parameter $1/2$. The maximal lifetime is simply $\max_{v \in V} t_v$, the maximum of n independent identically distributed (i.i.d. for short) geometric r.v. and hence:

Proposition 1. *Let $G = (V, E)$ be a graph having $n > 0$ vertices. We consider a run of Procedure Splitting-Naming-w.h.p. on vertices of G. Let T denote the maximal value of the set $\{t_v | v \in V\}$; T satisfies the following inequalities: 1. $T < 2\log_2 n + \log^* n$ w.h.p., 2. $T > \log_2 n - \log_2(2\log n)$ w.h.p.*

Note that, in Proposition 1, the term $\log^* n$ can be replaced by any slowly-growing non bounded function $g(n)$.

From Proposition 1, we can deduce the following:

Corollary 1. *The expected value of T is equal to $\Theta(\log n)$.*

Analysis of the Number of Vertices That Share the Same Maximum LifeTime. For any $t \geq 0$, we denote by X_t the number of vertices still alive at time t. For any $i > 0$, we have:

$$\Pr\left(X_{t+1} = 0 \mid X_t = i\right) = \frac{1}{2^i}. \tag{1}$$

This yields the following claim:

Claim 1.

$$\Pr\left(X_{t+1} = 0 \mid X_t \geq 2\log_2 n\right) \leq \frac{1}{n^2}. \tag{2}$$

Then, we obtain the following proposition:

Proposition 2. *The number of vertices which have the same maximum lifetime is, with high probability, at most $2\log_2 n$.*

Analysis of the Number of Vertices That Have the Same Maximum LifeTime and the Same Maximum Number. At the end of the initialisation phase, each vertex v obtains an integer t_v. Then it chooses u.a.r. a number

$$id_v \in \left\{0, \cdots, 2^{t_v + 3\log_2(t_v)} - 1\right\}.$$

We have the following proposition :

Proposition 3. *With high probability, there exists a unique vertex v with label (t_v, id_v) such that for any $w \in V \setminus \{v\}$: $(t_v, id_v) > (t_w, id_w)$.*

Analysis of the Size of the Random Numbers. A vertex v chooses u.a.r. a number id_v from the set:

$$\left\{0, \cdots, 2^{t_v + 3\log_2(t_v)} - 1\right\}.$$

This implies that this random number has a size of at most $2\log_2 n + O\left(\log_2 \log_2 n\right)$ bits w.h.p. Furthermore, from Corollary 1 we deduce directly that the expected value of the size of id_v is $O(\log n)$.

3   Analysis of a Variant of the Splitting and Naming Probabilistic Procedure

This section presents and analyses a variant of Procedure Splitting-Naming-whp, it is denoted Splitting-Naming-wvhp (see Algorithm 2). Now, the label of the vertex v is the couple (t_v, id_v) where t_v is still the lifetime of v. The difference is in the drawing of id_v: the number id_v is obtained by generating a number choosen u.a.r. in the set $\{0, ..., 2^{t_v \log^* t_v} - 1\}$.

Algorithm 2: Procedure Splitting-Naming-wvhp(v)

1: $t_v := 0$
2: **repeat**
3:     draw uniformly at random a bit $b(v)$;
4:     $t_v := t_v + 1$
5: **until** $b(v) = 1$
6: choose uniformly at random a number id_v in the set $\{0, ..., 2^{t_v \log^* t_v} - 1\}$;

The remainder of the section is devoted to the analysis of Procedure Splitting-Naming-wvhp.

Analysis of the Maximal Number of Bits Drawn by Each Vertex.

Proposition 4. *Let $G = (V, E)$ be a graph with $n > 0$ vertices. We consider a run of Procedure Splitting-Naming-wvhp. Let T denote the maximal value of the set $\{t_v | v \in V\}$; T satisfies the following inequalities: 1. $T < (\log_2 n) \log^* n$. w.v.h.p., 2. $T > \frac{1}{2} \log_2 n$ w.v.h.p.*

Remark 1. Note that, in Proposition 4, as in Proposition 1, the term $\log^* n$ can be replaced by any slowly-growing non bounded function $g(n)$.

Analysis of the Number of Vertices That Have the Same Maximum LifeTime and the Same Maximum Number

Proposition 5. *With very high probability, there exists a unique vertex v with the label (t_v, id_v) such that for any $w \in V \setminus \{v\}$: $(t_v, id_v) > (t_w, id_w)$.*

Analysis of the Size of the Random Numbers

Proposition 6. *The size of numbers id_v is w.v.h.p. $O\left((\log n)(\log^* n)^2\right)$. Its expected value is $O\left((\log n)(\log^* n)\right)$.*

4   Analysis of the Message Complexity of a Maximum Broadcasting Algorithm

This section analyses the message complexity of a maximum broadcasting algorithm, denoted Algorithm Broadcasting-Max. More precisely, it analyses the number of messages exchanged in a labelled graph for broadcasting the maximal label. This analysis will be useful in next sections. Let G be a graph and L a set of labels totally ordered by a relation $>$. Each vertex v of G chooses a label from L (all the vertices use the same distribution to choose a label), memorises its value in max_v and sends it to neighbours. If vertex v receives a label l greater than max_v then it memorises it in max_v and sends it to neighbours. We have the following proposition:

Proposition 7. *Let G be a graph and let L be a totally ordered set of labels. Each vertex v of G chooses a label from L; all the vertices use the same distribution to choose a label. For each run of Algorithm Broadcasting-Max the number of messages sent by each vertex of G is w.h.p. $O(\log n)$.*

Corollary 2. *Let G be a graph of bounded degree having n vertices and let L be a totally ordered set of labels. Each vertex v of G chooses a label from L; all the vertices use the same distribution to choose a label. Procedure Broadcasting-Max has a message complexity equal to $O(n \log n)$.*

5   A Monte Carlo Spanning Tree Algorithm Correct w.h.p. (resp. w.v.h.p.)

This section presents message terminating Monte Carlo algorithms, denoted Algorithm Spanning-Tree-whp and a variant, denoted Algorithm Spanning-Tree-wvhp, for computing a spanning tree of an anonymous graph without any initial knowledge and with no distinguished vertex; they are correct w.h.p. (resp. w.v.h.p.). Each vertex v is initially labelled with the label $label_v = (t_v, id_v)$ generated by the probabilistic procedure Splitting-naming-whp (resp. Splitting-naming-wvhp). Each vertex attempts to build a tree considering that it is the root. If two vertices are competing to capture a third vertex w then w will join the tree whose root has the higher label; this label is indicated in max_v. The father of vertex v is indicated by the port number $father_v$ (by convention, if $root_v = true$ then $father_v = 0$). The children of v correspond to the set of port numbers $children_v$, neighbours of v which are neither children nor the father of v are indicated in the set of port numbers $other_v$.

Claim 2. *Let G be a graph. For each run of Spanning-Tree-whp, eventually, the network reaches a terminal configuration in which there is no message in transition and no action can happen. The time complexity is $O(D)$, where D is the diameter of the graph.*

Let (t_{max}, id_{max}) be the maximal couple among labels of the vertices of G. When the network reaches a terminal configuration the set of $father_v$ encodes a spanning forest, the root of each tree is a vertex labelled (t_{max}, id_{max}). If there is a unique vertex v such that $(t_v, id_v) = (t_{max}, id_{max})$ then the spanning forest is a spanning tree whose root is v.

Algorithm Spanning-Tree-wvhp is obtained by substituting Splitting-Naming-wvhp for Splitting-Naming-whp in Algorithm Spanning-Tree-whp. Claim 2 is still valid for Algorithm Spanning-Tree-wvhp. Concerning the analysis of complexities, results of previous sections imply:

Proposition 8. *Let G be a graph having n vertices and m edges. Algorithm Spanning-Tree-whp (resp. Algorithm Spanning-Tree-wvhp) computes a spanning tree w.h.p. (resp. w.v.h.p.).*

The size of messages of Algorithm Spanning-Tree-whp (resp. Spanning-Tree-wvhp) is $O(\log n)$ w.h.p., it is also the expected value (resp. $O\left((\log n)(\log^* n)^2\right)$ w.v.h.p. and the expected value is $O\left((\log n)(\log^* n)\right)$). The message complexity of these two algorithms is w.h.p. $O(m \log n)$.

6   A Monte Carlo Counting Algorithm for Rings Correct w.h.p. (resp. w.v.h.p.)

This section presents a message terminating Monte Carlo algorithm, denoted Algorithm Counting-Ring-whp, for computing the size of an anonymous ring without any initial knowledge correct w.h.p. It also presents a variant correct w.v.h.p., denoted Algorithm Counting-Ring-wvhp. The main idea is very simple: each vertex v generates a label $label_v = (t_v, id_v)$ with the probabilistic procedure Splitting-naming-whp; this label is memorised in max_v. Then v sends over the ring a message containing this label and a counter equal to 1, denoted $(label_v, 1)$. Each vertex v memorises in max_v the largest value among labels it has received. A message is rejected if it is received by a vertex u such that max_u is greater than the label it contains. If not, the label is memorised in max_u, the counter is incremented and the message is transmitted to the next vertex until the message is received by a vertex w having the same label as the label of the message. In this case w considers that the message is its own message and the value of the counter is the size of the ring; therefore it sends over the ring a new message to indicate this fact to each vertex which memorises this size and the associated label.

Claim 3. *Let G be a ring. For each run of Counting-Ring-whp, eventually, the network reaches a terminal configuration in which there is no message in transition and no action can happen. The time complexity is $O(n)$, where n is the number of vertices of the ring. Let (t_{max}, id_{max}) be the maximal couple among labels of the vertices of G. If there is a unique vertex v such that $(t_v, id_v) = (t_{max}, id_{max})$ then for each vertex v of G, $size_v$ is equal to the number of vertices of G and $max_v = (t_{max}, id_{max})$.*

Algorithm Counting-Ring-wvhp is obtained by substituting Splitting-Naming-wvhp for Splitting-Naming-whp in Algorithm Counting-Ring-whp. Claim 3 is still valid for Algorithm Counting-Ring-wvhp. As for the spanning tree computation, complexities concerning Algorithm Counting-Ring-whp (resp. Algorithm Counting-Ring-wvhp) are deduced immediately from previous sections and summarised by:

Proposition 9. *Let G be a ring graph having n vertices. Algorithm Counting-Ring-whp (resp. Algorithm Counting-Ring-wvhp) is correct w.h.p. (resp. w.v.h.p.). The size of messages of Algorithm Counting-Ring-whp (resp. Counting-Ring-wvhp) is $O(\log n)$ w.h.p., it is also the expected value (resp. $O\left((\log n)(\log^* n)^2\right)$ w.v.h.p. and the expected value is $O\left((\log n)(\log^* n)\right)$). The message complexity of these two algorithms is w.h.p. $O(n \log n)$.*

7   A Monte Carlo Election Algorithm Correct w.h.p. (resp. w.v.h.p.)

This section presents a message terminating Monte Carlo election for anonymous graphs without any initial knowledge correct w.h.p. (resp. w.v.h.p.). They differ from Algorithm ELECT in [AM94] p. 315 only by the choice of labels of vertices. The aim of Algorithm *Elect-whp* is to choose as elected the unique vertex v (if there exists a unique), such that: $\forall \; w \neq v \quad (t_w, id_w) < (t_v, id_v)$.

Algorithm 3: Elect-whp

I : {If (t_v, id_v) is not defined}
begin
    call Splitting-Naming-whp(v);
    $max_v := (t_v, id_v)$;
    $leader_v := true$;
    send $< (t_v, id_v) >$ to all neighbours
end
D : {A Message (t, id) has arrived at v through port l}
begin
    if (t_v, id_v) *is not defined* **then**
        call Splitting-Naming-whp(v);
        $max_v := (t_v, id_v)$;
        $leader_v := true$;
        send $< (t_v, id_v) >$ to all neighbours
    if $(t, id) > max_v$ **then**
        $max_v := (t, id)$;
        $leader_v := false$;
        send $< (t, id) >$ to all neighbours except through l
end

Claim 4. *Let G be a graph. For each run of Elect-whp, eventually, the network reaches a terminal configuration in which there is no message in transition and no action can happen. The time complexity is $O(D)$, where D is the diameter of G. Let (t_{max}, id_{max}) be the maximal couple among labels of the vertices of G. If there is a unique vertex v such that $(t_v, id_v) = (t_{max}, id_{max})$ then there is a unique vertex such that leader is true, the others have leader equal to false.*

Remark 2. Algorithm Elect-wvhp is obtained by substituting Splitting-Naming-wvhp for Splitting-Naming-whp in Algorithm Elect-whp. Claim 4 is still valid for Algorithm Elect-wvhp.

Proposition 10. *Let G be a graph having n vertices and m edges. Algorithm Elect-whp (resp. Algorithm Elect-wvhp) is an election algorithm w.h.p. (resp.*

w.v.h.p.). The size of messages of Algorithm Elect-whp (resp. Elect-wvhp) is $O(\log n)$ *w.h.p., it is also the expected value (resp.* $O\left((\log n)(\log^* n)^2\right)$ *w.v.h.p. and the expected value is* $O\left((\log n)(\log^* n)\right)$*). The message complexity of these two algorithms is w.h.p.* $O(m \log n)$*.*

Complexities of Algorithm Elect-whp (resp. Elect-wvhp) are the same as algorithms for spanning tree constructions.

8   Ring with Identities + Splitting Procedure = $O(n \log n)$ Message Complexity w.h.p. Election Algorithm

This section gives an illustration of the power of the splitting procedure studied in Section 2 thanks to Proposition 7. Let R be a ring having n vertices. We assume that vertices of R are aware of n, and each vertex v has a unique identity denoted $ident_v$. We consider the order on couples defined in Section 2. For electing a vertex of R, first of all each vertex v of R applies Procedure Splitting obtaining t_v, and then it broadcasts over the ring the triple $(t_v, ident_v, hop)$, where $ident_v$ is the initial identity of v and hop is an integer (its initial value is 1) which is incremented as it moves over the ring. Each vertex memorises the maximal value that it sees. A message is stopped as soon as it reaches a vertex w having a maximal value greater than or equal to the couple of the message. The elected vertex is the vertex u which receives a message $(t, ident, h)$ with $t = t_u$, $ident = ident_u$ and $h = n$. Initially, for each vertex v $leader_v$ is undefined. Proposition 7 implies:

Proposition 11. *Let R be a ring having n vertices such that each vertex has a unique identity and knows n. Algorithm Elect-ring terminates and elects a vertex with messages of size $O(\log n)$ w.h.p. and the number of messages through the ring is w.h.p. $O(n \log n)$.*

In a certain sense our result is optimal since any decentralised wave algorithm for ring networks exchanges $\Omega(n \log n)$ messages, on the average as well as the worst case (see [Tel00] Corollary 7.14).

9   Conclusion

One may wonder whether it is possible to obtain a Monte Carlo algorithm which solves the counting problem of a ring w.h.p. and also ensures an error probability bounded by a constant ϵ as is done by [IR90, AM94, SS94]. The answer is positive. Let R be a ring, v be a vertex of R and let ϵ_v be a constant ($\epsilon_v <$ 1) known by v. The vertex v wants to ensure that the size computed by the Monte Carlo algorithm is correct with probability $1 - \epsilon_v$. From the proof of Proposition 3 we deduce a size K of a ring such that the probability of having two vertices with the maximum lifetime and with the maximum number is bounded by ϵ_v. Now, instead of drawing a bit uniformly at random until it gets 1, the vertex v does this operation K times and it memorises the maximal lifetime, denoted $t_v^{(K)}$, among the K lifetimes it obtains. The sequel of the algorithm is the same as Counting-Ring-whp: the vertex v draws at random a number id_v in the set $\{0, ..., 2^{t_v^{(K)} + 3\log_2(t_v^{(K)})} - 1\}$ etc. In this way, we obtain a Monte Carlo algorithm which solves the counting problem of a ring w.h.p. and ensures an error probability bounded by a constant ϵ where ϵ is the smallest value among the set of error probabilities fixed by vertices of the ring R. The time complexity is $O(D)$ (where D is the diameter of the ring) and the message complexity is w.h.p. $O(n \log n)$. The same constructions and results can be obtained for spanning tree construction and election. This will be developed in a full paper version.

References

[AAER07] Angluin, D., Aspnes, J., Eisenstat, D., Ruppert, E.: The computational power of population protocols. Distributed Computing 20(4), 279–304 (2007)

[AM94] Afek, Y., Matias, Y.: Elections in anonymous networks. Inf. Comput. 113(2), 312–330 (1994)

[Ang80] Angluin, D.: Local and global properties in networks of processors. In: Proceedings of the 12th Symposium on Theory of Computing, pp. 82–93 (1980)

[ASW88] Attiya, H., Snir, M., Warmuth, M.K.: Computing on an anonymous ring. J. ACM 35(4), 845–875 (1988)

[AW04] Attiya, H., Welch, J.: Distributed computing: fundamentals, simulations, and advanced topics. John Wiley & Sons (2004)

[FMS96] Fill, J.A., Mahmoud, H.M., Szpankowski, W.: On the distribution for the duration of a randomized leader election algorithm. Ann. Appl. Probab. 6, 1260–1283 (1996)

[GR05] Guerraoui, R., Ruppert, E.: What can be implemented anonymously? In: Fraigniaud, P. (ed.) DISC 2005. LNCS, vol. 3724, pp. 244–259. Springer, Heidelberg (2005)

[IR90] Itai, A., Rodeh, M.: Symmetry breaking in distributed networks. Inf. Comput. 88(1), 60–87 (1990)

[KMW11] Kalpathy, R., Mahmoud, H.M., Ward, M.D.: Asymptotic properties of a leader election algorithm. Journal of Applied Probability 48(2), 569–575 (2011)

[Lav95] Lavault, C.: Evaluation des algorithmes distribués, Hermès, Paris (1995)

[Pro93] Prodinger, H.: How to select a loser. Discrete Mathematics 120(1-3), 149–159 (1993)

[RFJ+07] Ramanathan, M.K., Ferreira, R.A., Jagannathan, S., Grama, A., Szpankowski, W.: Randomized leader election. Distributed Computing 19(5-6), 403–418 (2007)

[SS94] Schieber, B., Snir, M.: Calling names on nameless networks. Inf. Comput. 113(1), 80–101 (1994)

[Tel00] Tel, G.: Introduction to distributed algorithms. Cambridge University Press (2000)

[YK88] Yamashita, M., Kameda, T.: Computing on an anonymous network. In: PODC, pp. 117–130 (1988)

A Deterministic Worst-Case Message Complexity Optimal Solution for Resource Discovery*

Sebastian Kniesburges, Andreas Koutsopoulos, and Christian Scheideler

Department of Computer Science, University of Paderborn, Germany
seppel@upb.de, {koutsopo,scheideler}@mail.upb.de

Abstract. We consider the problem of resource discovery in distributed systems. In particular we give an algorithm, such that each node in a network discovers the address of any other node in the network. We model the knowledge of the nodes as a virtual overlay network given by a directed graph such that complete knowledge of all nodes corresponds to a complete graph in the overlay network. Although there are several solutions for resource discovery, our solution is the first that achieves worst-case optimal work for each node, i.e. the number of addresses ($\mathcal{O}(n)$) or bits ($\mathcal{O}(n \log n)$) a node receives or sends coincides with the lower bound, while ensuring only a linear runtime ($\mathcal{O}(n)$) on the number of rounds.

Keywords: distributed algorithms, resource discovery, self-stabilization, clique network.

1 Introduction

To perform cooperative tasks in distributed systems the network nodes have to know which other nodes are participating. Examples for such cooperative tasks range from fundamental problems such as group-based cryptography [14], verifiable secret sharing [6], distributed consensus [17]to peer-to-peer(P2P) applications like distributed storage, multiplayer online gaming, and various social network applications such as chat groups. To perform these tasks efficiently knowledge of the complete network for each node is assumed. Considering large-scale, real-world networks this complete knowledge has to be maintained despite high dynamics, such as joining or leaving nodes, that lead to changing topologies. Therefore the nodes in a network need to learn about all other nodes currently in the network. This problem called *resource discovery*, i.e. the discovery of the addresses of all nodes in the network by every single node, is a well studied problem and was firstly introduced by Harchol-Balter, Leighton and Lewin in [23].

1.1 Resource Discovery

As mentioned in [23] the resource discovery problem can be solved by a simple swamping algorithm also known as *pointer doubling*: in each round, every node informs all of

---

* This work was partially supported by the German Research Foundation (DFG) within the Collaborative Research Centre On-The-Fly Computing (SFB 901).

T. Moscibroda and A.A. Rescigno (Eds.): SIROCCO 2013, LNCS 8179, pp. 165–176, 2013.

its neighbors about its entire neighborhood. While this just needs $O(\log n)$ communication rounds, the work spent by the nodes can be very high. We measure the work of a node as the number of addresses each node receives or sends while executing the algorithm. Moreover, in the stable state (i.e., each node has complete knowledge) the work spent by every node in a single round is $\Theta(n^2)$, which is certainly not useful for large-scale systems. Alternatively, each node may just introduce a single neighbor to all of its neighbors in a round-robin fashion. However, it is easy to construct initial situations in which this strategy is not better than pointer doubling. In [23] a randomized algorithm called the *Name-Dropper* is presented that solves the resource discovery problem within $\mathcal{O}(\log^2 n)$ rounds w.h.p. and work of $\mathcal{O}(n^2 \log^2 n)$. In [24] a deterministic solution for resource discovery in distributed networks was proposed by Kutten et al. which takes $\mathcal{O}(\log n)$ rounds and $\mathcal{O}(n^2 \log n)$ amount of work. Konwar et al. presented solutions for the resource discovery problem considering different models, i.e. multicast or unicast abilities and messages of different sizes, where the upper bound for the work is $O(n^2 \log^2 n)$. Recently resource discovery has been studied by Haeupler et. al. in [21], in which they present two simple randomized algorithms based on gossiping that need $\Omega(n \log n)$ time and $\Omega(n^2 \log n)$ work per node on expectation. They only allow nodes to send a single message containing at most one address of size $\log n$ in each round. We present a deterministic solution that follows the idea of [21] and limits the number of messages each node has to send and the number of addresses transmitted in one message. Our goal is to reduce the number of messages sent and received by each node such that we avoid nodes to be overloaded. In detail we show that resource discovery can be solved in $\mathcal{O}(n)$ rounds and it suffices that each node sends and receives $\mathcal{O}(n)$ messages in total, each message containing $\mathcal{O}(1)$ addresses. Our solution is the first solution for resource discovery that not only considers the total number of messages but also the number of messages a single node has to send or receive. Note that $\Omega(n)$ is a trivial lower bound for the work of each node to gain complete knowledge. So our algorithm is worst case optimal in terms of message complexity. Furthermore our algorithm can handle the deletion of edges and joining or leaving nodes, as long as the graph remains weakly connected. Modeling the current knowledge of all nodes as a directed graph, i.e. there is an edge (u, v) iff u knows v's ID, one can think of resource discovery as building and maintaining a complete graph, a clique, as a virtual overlay network. If the overlay can be recovered out of any (weakly connected) initial graph, the corresponding algorithm can be considered to be a *self-stabilizing* algorithm. More precisely, an algorithm is considered as self-stabilizing if it reaches a legal state when started in an arbitrary initial state (*convergence*) and stays in a legal state when started in a legal state (*closure*).

1.2 Topological Self-stabilization

There is a large body of literature on how to efficiently maintain overlay networks, e.g., [1,2,19,13,15,18,22]. While many results are already known on how to keep an overlay network in a legal state, far less is known about self-stabilizing overlay networks. The idea of self-stabilization in distributed computing first appeared in a classical paper by E.W. Dijkstra in 1974 [8] in which he looked at the problem of self-stabilization in a token ring. In order to recover certain network topologies from any weakly connected network, researchers have started with simple line and ring networks, [7,20]. In [16], Onus et al. present a local-control strategy called linearization for converting an arbitrary connected graph into a sorted list. Various self-stabilzing algorithms for different network overlay structures have been considered over the years [12,11,9]. In [9] the authors use a self-stabilizing algorithm in which they collect snapshots of the network along a spanning tree, which could also be used to form a complete graph. However, the authors give no bounds on the message complexity of their algorithm. In [3] the authors present a general framework for the self-stabilizing construction of overlay networks, which may involves the construction of the clique. However, the work in order to do that when using this method is too high.

One could use the distributed algorithms for self-stabilizing lists and rings to form a complete graph, but all algorithms proposed so far for these topologies involve a worst-case work of $\Omega(n^2)$ per node.

Alternatively, a self-stabilizing spanning tree algorithm could be used. A large number of self-stabilizing distributed algorithms has already been proposed for the formation of spanning trees in static network topologies, [5], [4], [10], [10]. However, these spanning trees are either expensive to maintain or the amount of work in these algorithms is not being considered.

In summary, no self-stabilizing algorithm has been presented for the formation of a bounded degree spanning tree if the network topology is under the control of the nodes and there are no outside services for the introduction of nodes.

1.3   Our Model

We use the network model used in [23,24,21]. We model the network as a directed graph $G = (V, E)$ where $|V| = n$. The nodes have unique identifiers with a total order, and these identifiers are assumed to be immutable (for example, we may use the IP addresses of the nodes). We are using a standard synchronous message-passing model: time proceeds in synchronous rounds, and all messages generated in round i are delivered at the end of round i. In order to deliver a message, a node may use any address stored in its local variables. No a priori information about the size or diameter of the network can be assumed by a node and there cannot be made use of some outside rendezvous service to get introduced to other nodes. Hence, the *state* of a node is fully determined by its local variables. Like in [23,24,21] we assume that a node can verify its neighborhood without extra work, such that there are no false identifiers in the network. Only local topology changes are allowed, i.e. a node may decide to cut a link to a neighbor (by deleting its address) or introduce a link to one of its neighbors (by sending it an address). We model the decisions to cut or establish links and to send messages as actions. An action has the form $<$ *guard* $> \rightarrow <$ *commands* $>$. A guard is a Boolean expression over the state of the node. The commands are executed if the guard is true. Any action whose guard is true is said to be *enabled*. We assume that a node can execute all of its enabled actions in the current round.

The *state* of the system is the combination of the states of all nodes in the system. and contains all the information available in the system. A *computation* is a sequence of system states such that for each state s_i at the beginning of round i, the next state s_{i+1} is obtained after executing all actions that are enabled at the beginning of round

i and receiving all messages that they generated. We call a distributed algorithm *self-stabilizing* if from any initial state in which the overlay network is weakly connected, it eventually reaches a legal state and stays in a legal state afterwards. In our case, the legal state is the clique topology. We distinguish between two types of work. The *stabilization work* of a node v is defined as the total number of addresses sent and received by v during the stabilization process. The *maintenance work* of a node v is defined as the maximum number of addresses sent and received by v during a single round of the stable state.

1.4 Our Contributions

In this paper we present a distributed algorithm for resource discovery. We will describe the algorithm as a self-stabilizing algorithm that forms and maintains a clique as a virtual overlay network, and show that our algorithm is worst-case optimal in terms of message complexity.

Theorem 1. *For any initial state in which the network is weakly connected, our algorithm requires at most $\mathcal{O}(n)$ rounds and $\mathcal{O}(n)$ work per node until the network reaches a legal state in which it forms a clique.*

We further show that the maintenance cost per round is $\mathcal{O}(1)$ for each node once a legal state has been reached. We also consider topology updates caused by a single joining or leaving node and show that the network recovers in $\mathcal{O}(n)$ rounds with at most $\mathcal{O}(n)$ messages over all nodes besides the maintenance work. Note that we use a synchronous message passing model to give bounds on the message complexity of our algorithm, but our correctness analysis can also be applied to an asynchronous setting. A detailed version of this paper containing pseudo code, all missing proofs and further details can be found in [26].

The paper is structured as follows: In Section 2 we give a description of our algorithm. In Section 3 we prove that the algorithm is self-stabilizing. We consider the stabilization work and maintenance work in Section 4. In Section 5 we analyze the steps needed for the network to recover after a node joins or leaves the network. Finally, in Section 6 we end with a conclusion.

2   A Distributed Self-stabilizing Algorithm for the Clique

In this section we give a general description of our algorithm. First we introduce the variables being used, and then the actions the nodes take, according to our rules. Each node x has a buffer $B(x)$ for incoming messages from the previous round. We assume that the buffer capacity is unbounded and no messages are lost. We do not require any particular order in which the messages are processed in $B(x)$. Moreover, each node x stores the following internal variables: its predecessor $p(x)$, its successor $s(x)$, its current neighborhood $N(x)$ in a circular list, the nodes received by messages from the predecessor in another circular list $L(x)$, the set of nodes $S(x)$ that are received through scanning messages (defined below), its own identifier $id(x)$ and its status $status(x)$, which is by default set to 'inactive' and can be changed to 'active'.

A message in general consists of the following parts: a *sender id*, which is the id of the node sending the message, an optional *additional id*, if the sender wants to inform the receiving node about another node, and the *type* of the message.

Each node has two different kinds of actions that we call *receive* actions and *periodic* actions. A receive action is enabled if there is an incoming message of the corresponding type in the buffer $B(x)$. A periodic action is enabled in every state, as its guard is simply *true*. Each enabled action is executed once every step.

In order to describe the algorithm formally and prove its correctness later on, we need the definitions given below. In this paper we assume that a predecessor of a node is a node with the next larger identifier. Therefore for all $p(x)$ links, $p(x) > x$. Then all nodes in a connected component considering only $p(x)$ links form a rooted tree, where for each tree the root has the largest identifier. Note here that the *heap H* (defined below) is not a data structure or variable stored by any node. It is a notion used just for the purpose of the analysis.

Definition 1. *We call such a rooted tree formed by $p(x)$ links a* heap H. *We further call the root of the tree the* head h *of the heap H. We further denote with* $heap(x)$ *the heap H such that $x \in H$.*

Definition 2. *A sorted list is a heap H with head h, such that $\forall v \in H - \{h\} : p(v) > v$ and $\forall v \in H - \{h\} : s(p(v)) = v$. We call a heap linearized w.r.t. a node $u \in H$, if $\forall v \in H - \{h\} : p(v) > v$ and $\forall v \in H - \{h\} \wedge v \geq u : s(p(v)) = v$. We further call the time until a heap is linearized w.r.t. a node u the* linearization time *of u. We say that two heaps H_i and H_j are merged if all nodes in H_i and H_j form one heap H.*

2.1 Description of Our Algorithm

We only present the intuition behind our algorithm. Our primary goal is to collect the addresses of all nodes in the system at the node of maximum id, which we also call the *root*. In order to efficiently distribute the addresses from this root to all other nodes in the system we aim at organizing them into a spanning tree of constant degree, which in our case is a sorted list, ordered in descending ids. The root would then be the head of the list. In order to reach a sorted list, we first organize the nodes in rooted trees satisfying the max-heap property, i.e. a parent (also called *predecessor* in the following) of a node has a higher id than the node itself. The rooted trees will then be merged and linearized over time so that they ultimately form a single sorted list.

In our protocol, in order to minimize the amount of messages sent by the nodes, we allow a node in each round to share information only with its immediate *successor* $s(x)$ (which is one of the nodes that considers it as its predecessor) and *predecessor* $p(x)$. More precisely, in each round a node forwards one of its neighbors (i.e. the nodes it knows about) in a round-robin manner to its predecessor.

Moreover, each node chooses the smallest node in its neighborhood that is larger than itself as its predecessor and requests from it to accept it as successor ($pred - request$ message). Each node also looks at the nodes which requested to be its successor, assigns the largest of them as its successor ($pred - accept$) and forwards the rest of these nodes to it ($new - predecessor$).

We also need to ensure that there exists a path of successors from the root to all other nodes so that the information can be forwarded to all. This is initially not the case since there exist many nodes that are the largest in their known neighborhood, thinking they are the root. We call these nodes *heads*. All the nodes having the same head as an ancestor form a *heap*. The challenge is to *merge* all heaps into one. In order to enable the merging of the heaps, the heads continuously scan their neighborhood. A node that receives a *scan* message responds by sending the largest node in its neighborhood through a *scanack* message to the node that sent that *scan* message (could be possibly more than one).

To avoid accumulation of unsent ids in the lists (which would have an effect on the time and message complexity) maintained by the nodes, the following rules are used. When x has no predecessor that it can send a *forward-from-successor* message to, it changes its status to *inactive*, and then (with the help of some special messages) all the nodes in the succesor-line of x are informed in order not to forward any messages to predeseccors, and are activated again if the information flow is possible again.

3   Correctness

In this section we show the correctness of our approach for the self-stabilizing clique.

At first we show some basic lemmas. We then show that in linear time all nodes belong to the same heap. Then we show that the head of this heap (node with the maximal id) is connected with every node and vice versa after an additional time of $\mathcal{O}(n)$. From this state it takes $\mathcal{O}(n)$ more time until every node is connected to every other node and the clique is formed. We give a formal definition of the legal state.

Definition 3. *Let G be a network with node set V and $max = \max\{v \in V\}$ be the node with the maximum id. Then G is in a legal state iff $\forall v \in V : N(v) = V - \{v\}$ and $\forall v \in V - \{max\} : p(v) > v$ and $\forall v \in V - \{max\} : s(p(v)) = v$.*

Note that the legal state contains the clique and also a sorted list over the nodes. In this section we will prove the following theorem.

Theorem 2. *After $\mathcal{O}(n)$ rounds the network stabilizes to a legal state.*

3.1   Phase 0: Recovery to a Valid State

In this phase we show that the network can recover if the internal variables $p(x)$ and $s(x)$ are undefined or set to invalid values, e.g $p(x) < x$.

Theorem 3. *It takes at most 2 rounds until the network is in a valid state.*

3.2   Phase 1: Connect All Heaps by S-Edges

In this phase we show that starting from a valid state all existing heaps will eventually be connected by s-edges (defined below), so that they will merge afterwards.

Definition 4. *We distinguish between two different kinds of edges that can exist at any time in our network, the edges in the set E and the ones in the set E_s. We say that (x, y) is in E, if $y \in N(x)$ and (x, y) in E_s if $y \in S(x)$, resulting from a scan from y. We will call the latter ones* s-edges *and denote them by* $(x, y)_s$.

Definition 5. *In the directed graph we define an* undirected path *as a sequence of edges* $(v_0, v_1), (v_1, v_2), \cdots, (v_{k-1}, v_k)$, *such that* $\forall i \in \{1, \cdots, k\} : (v_i, v_{i-1}) \in E \vee (v_{i-1}, v_i) \in E$. *We say that two heaps H_1 and H_2 are* s-connected *if there exists at least one undirected path from one node in H_1 to one node in H_2 and this path consists of either s-edges or edges having both nodes in the same heap.*

We say that a subset of s-edges $E'_s \subseteq E_s$ is a s-connectivity set *at round t if all heaps in the graph are s-connected to each other through edges in E'_s at round t.*

In the first phase we will show that after $\mathcal{O}(n)$ rounds all heaps have been connected by s-edges. Let E^0 be the set of edges $(u, v) \in E$ at time $t = 0$. We then show that all these edges are scanned in $\mathcal{O}(n)$ rounds, giving us the connections via s-edges.

Theorem 4. *After $\mathcal{O}(n)$ rounds the heaps H_i and H_j connected by $(u, v) \in E^0$ have either merged or been connected by s-edges .*

To prove the theorem we firstly show some basic lemmas needed in the analysis.

Lemma 1. *Let $u_1, \cdots u_{|H|}$ be the elements in a heap H in descending order. Then it takes at most i rounds till H is linearized w.r.t u_i.*

Lemma 2. *Once one head learns about the existence of another head, two heaps are merged.*

In case of a merging of two heaps H_i, H_j, the time it takes until the new heap H is linearized w.r.t. a node u can increase with respect to the linearization time of u in the heap before the merging.

Lemma 3. *If two heaps H_i and H_j merge to one heap H, the linearization time of a node $u \in H_i$ (resp. $u \in H_j$) can increase by at most $|H_j|$ (resp. $|H_i|$).*

From Lemma 1 and Lemma 3 we immediately get via an inductive argument:

Corollary 1. *For any heap H of size $|H|$ in round t it takes at most $|H| - t$ rounds until it forms a sorted list.*

Lemma 4. *If a node sends an id with a* forward-from-successor *message, the id will not be delayed by other* forward-from-successor *messages on its way to the head.*

As a consequence of the observation of Lemma 2 we introduce some additional notation to estimate the time it takes until any id is scanned by a head of a heap.

For any edge $(u, v) \in E^0$ with $u \in H_i$ and $v \in H_j$, where h_i and h_j denote the corresponding heads of the heaps, we define the following notation in a round t: Let $P^t(u)$ be the length of the path from u to h_i, once H_i is linearized w.r.t. u. Let $ID^t(u, v)$ be the number of ids u forwards or scans before sending or scanning v the first time. Let $LT^t(u)$ be the time it takes until the heap is linearized w.r.t. u, i.e. on the path from the head h_i to u each node has exactly one predecessor and successor. Corollary 1 shows that $LT^t(u)$ is bounded by $|H_i|$.

Let $\phi^t(u,v) = P^t(u) + ID^t(u,v) + LT^t(u)$. We call $\phi^t(u,v)$ the *delivery time* of an id v because if $\phi^t(u,v) = 0$, the id is scanned in round t or has already been scanned by h_i. We further denote by $\Phi^t(u,v) = \min\{\phi^t(w,v) : heap(u) = heap(w)\}$ the minimal delivery time of v for the heap containing u.

For any edge $(u,v) \in E^0$, with $u \in H_i$ and $v \in H_j$, (i.e. u and v are in different heaps) and $\Phi^t(u,v) = 0$ the head of H_i scans or has scanned $v \in H_j$ resulting in the s-edge $(v, h_i)_s$. The following holds:

Lemma 5. *If $(u,v) \in E^0$ is an edge between two heaps H_i and H_j, then $\Phi^t(u,v) \leq \max\{2|H_i| + n - t, 0\} \leq \max\{3n - t, 0\}$ for all rounds t.*

Proof. We will show the lemma by induction on the number of rounds. For the analysis we divide each round $t \to t+1$ into two parts: in the first step $t \to t'$ all actions are executed and in the second step $t' \to t+1$ all network changes are considered. Thus, we assume that all actions are performed before the network changes. This is reasonable as a node is aware of changes in its neighborhood only in the next round, when receiving the messages. By network changes we mean the new edges that could be created in the network. These new edges could possibly lead to the merging of some heaps at time $t+1$.

Induction base($t = 0$): For any edge $(u,v) \in E^0$ between H_i and H_j let $x \in H_i$ be the node such that $\Phi^0(u,v) = \phi^0(x,v)$. Then $P^0(x) \leq H_i$ as the path length is limited by the number of nodes in the heap, $ID^0(x,v) \leq n$ as not more than n ids are in the system, and following from Lemma 1, $LT(x) \leq |H_i|$. Then $\Phi^0(u,v) \leq \phi^0(x,v) \leq 2|H_i| + n \leq 3n$.

Induction step($t \to t'$): For any edge $(u,v) \in E^0$ between H_i and H_j let $x \in H_i$ be the node such that $\Phi^t(u,v) = \phi^t(x,v)$.

Then in round t the following actions can be executed.

- x is inactive and can not forward an id. Then the heap is not linearized w.r.t. x, which implies that the linearization time decreases by one, i.e. $LT^{t'}(x) = LT^t(x) - 1$ and $\phi^{t'}(x,v) = \phi^t(x,v) - 1 \leq 2|H_i| + n - t - 1$ as all other values are not affected.
- u is active, but does not send v by a *forward-from-successor* message, then the number of ids that u is sending before v decreases by 1. Note that according to Lemma 4, x hasn't sent a *forward-from-successor* message with v in a round before, as then there would be another node $y \in H_i$ with $\phi^t(y,v) < \phi^t(x,v)$. Then $ID^{t'}(x,v) \leq ID^t(x,v) - 1$ and $\phi^{t'}(x,v)) = \phi^t(x,v) - 1 \leq 2|H_i| + n - t - 1$.
- u sends a *forward-from-successor* message with v, then the length of the path for v to the head h_i decreases by 1 and $\phi^{t+1}(p(x),v) \leq P^t(x) - 1 + ID^t(x,v) + LT^t(x) = \phi^t(x,v) - 1 \leq 2|H_i| + n - t - 1$

Thus, in total $\Phi^{t'}(u,v) \leq \Phi^t(u,v) - 1 \leq 2|H_i| + n - t - 1 \leq 3n - (t+1)$.

Induction step($t' \to t+1$): Now we consider the possible network changes and their effects on the potential $\Phi^{t+1}(u,v)$. Let again $x \in H_i$ be the node such that

$\Phi^t(u, v) = \phi^t(x, v)$ for an edge $(u, v) \in E^0$ between H_i and H_j. The following network changes might occur:

- some heaps H_k and H_l with $k \neq i$ and $l \neq i$ merge. This has no effect on $\Phi^{t'}(u, v)$. Thus, $\Phi^{t+1}(u, v) = \Phi^{t'}(u, v) \leq 2|H_i| + n - t - 1 \leq 3n - (t + 1)$.
- Heaps H_i and H_k merge to H'_i. Obviously the length of the path of x can increase and $P^{t+1}(x) \leq P^{t'}(x) + |H_k|$. According to Lemma 3 also the linearization time of x can increase and $LT^{t+1}(x) \leq LT^{t'}(x) + |H_k|$. In total $\Phi^{t+1}(u, v) \leq \Phi^{t'}(u, v) + 2|H_k| \leq 2|H'_i| + n - t - 1 \leq 3n - (t + 1)$.

Thus, in round $t + 1$, $\Phi^{t+1}(u, v) \leq 2|H_i| + n - t - 1 \leq 3n - (t + 1)$.

Hence for every edge $(u, v) \in E^0$ with $u \in H_i$ and $v \in H_j$, $\Phi^t(u, v) = 0$ after $3n$ rounds, which means that the head of H_i scans or has scanned $v \in H_j$ resulting in the s-edge (v, h_i). Thus, we immediately get Theorem 4.

3.3   Phase 2: Towards One Heap

Based on the results of Phase 1, we will prove that after $O(n)$ further rounds a clique is formed. For the purpose of the analysis below, we use the following definitions:

Definition 6. *Let $ord(x)$ be the order of a node x, i.e. the ranking of the node if we sort all n nodes in the network according to their id (i.e. the node with the largest id m has $ord(m) = 0$, the second largest has order 1, and so on). Moreover, we define the potential $\lambda(x, y)$ of a pair of nodes x and y to be the positive integer equal to $\omega(x, y) = 2 \cdot ord(x) + 2 \cdot ord(y) + K(x, y)$, where $K(x, y) = 1$ if $x > y$ and 0 otherwise. Also, let for a set of edges $E' \subseteq E$, $\Lambda(E') = \max_{(u,v) \in E'} \{\omega(u, v)\}$, if $E' \neq \emptyset$ and 0 otherwise.*

Lemma 6. *Two heaps H_i, H_j that are connected by an s-edge $(x, y)_s$ at time t will either stay connected via s-edges $(x_i, y_i)_s$ at time $t + 1$ with the property that, $\forall (x_i, y_i)$, the potential $\omega(x_i, y_i)$ of the edges we consider at time $t + 1$ is smaller that the potential $\omega(x, y)$ of the edge $(x, y)_s$ we considered at time t, or x and y will be in the same heap.*

Proof. Let $(x, y)_s$ be a s-edge connecting H_i and H_j, i.e. $x \in H_i$, $y \in H_j$. Then according to our algorithm the following actions might be executed.

- x is the head of H_i and $y > x$ then $y = p(x)$ and x sends a *pred-request* message to y, resulting in a merge of H_i and H_j.
- x is the head of H_i and $x > y$ and y is a new id, then x sends a *scanack* to y with its own id and the edge $(y, x)_s$ is created connecting H_i and H_j. Then $\omega(y, x) = 2ord(x) + 2ord(y) + 0 < 2ord(x) + 2ord(y) + 1 = \omega(x, y)$.
- x forwards y to $p(x)$ by a *forward-head* message, such that $y \in S(p(x))$ and H_i and H_j are connected by $(p(x), y)_s$. Then $\omega(p(x), y) = 2ord(p(x)) + 2ord(y) + K(p(x), y) < 2ord(x) + 2ord(y) + K(x, y) = \omega(x, y)$.
- x receives a new id $z \in S(x)$ with $z = \max \{v \in N(x)\}$, such that $z > y$ and $z > x$. Then x sends a *scanack* containing z to y and the s-edge $(x, y)_s$ is substituted by s-edges $(x, z)_s$ and $(y, z)_s$. And H_i and H_j are connected via s-edges. Note

that since $p(x) > x$ and $z > x, y$, $ord(p(x)) < ord(x), ord(z) < ord(x)$ and $ord(z) < ord(y)$. The potential of the new edges is: $\omega(p(x), z) = 2ord(p(x)) + 2ord(z) + K(p(x), z) < 2ord(x) + 2ord(y) + K(x, y) = \omega_t(x, y)$. $\omega(y, z) = 2ord(y) + 2ord(z) + 0 < 2ord(x) + 2ord(y) + K(x, y) = \omega_t(x, y)$.

- x knows an id $z \in H_k$ with $z = \max\{v \in N(x)\}$, $z > y$ and $z \notin S(x)$. Then one of the following cases hold:

    1. $(x, z) \in E^0$, then according to Lemma 5 a node $u > x$ with $u \in H_i$ has scanned z resulting in the s-edge $(z, u)_s$ s-connecting H_i and H_k.

    2. x has received z by a *forward-from-predecessor* message. Then a node $u > x$ with $u \in H_i$ has scanned z resulting in the s-edge $(z, u)_s$ s-connecting H_i and H_k.

    3. z was in $S(x)$ in a previous round, then the edge $(x, z)_s$ existed s-connecting H_i and H_k.

    4. x has received z by a *forward-from-successor* message. Then there is a node $v \leq x$ in the sub heap rooted at x such that $(v, z) \in E^0$. Then according to Lemma 5 a node $w \in H_i$ with $w > v$ has scanned z and the s-edge $(z, w)_s$ existed s-connecting H_i and H_k. If $w > x$, H_i and H_k are s-connected by s-edges $(x_i, y_i)_s$ with $\forall(x_i, y_i) : (x < w < x_i \wedge x < w < y_i \wedge z \leq x_i \wedge z \leq y_i) \vee (x < w \leq x_i \wedge x < w \leq y_i \wedge z < x_i \wedge z < y_i)$. If $w < x$ then at least as many rounds have passed since w has scanned z as there are nodes on the path from w to x, because z has to be forwarded as many times. Then the edge $(z, w)_s$ has been forwarded or substituted t times or H_i and H_k have merged. Then H_i and H_k are s-connected by s-edges $(x_i, y_i)_s$ with $\forall(x_i, y_i) : (x < w < x_i \wedge x < w < y_i \wedge z \leq x_i \wedge z \leq y_i) \vee (x < w \leq x_i \wedge x < w \leq y_i \wedge z < x_i \wedge z < y_i)$.

   In each case x sends a *scanack* containing z to y and the s-edge $(y, z)_s$ is created. And H_i and H_j are s-connected over s-edges and in all cases the potential shrinks, since for each new s-edge it holds that at least one node is greater and the other node not smaller than the nodes in the edge they replace.

- x is the head of H_i and $x < y$, then H_i and H_j merge to one heap.

- x is the head of H_i and $x > y$ and y was in $N(x)$ in a previous round, then H_i and H_j are already s-connected by s-edges $(x_i, y_i)_s$ with greater ids by the same arguments as in the case before. Since the ids are greater, the potential shrinks also here.

Lemma 7. *If E_t is an s-connectivity set at round t, there exists an s-connectivity set E_{t+1} at round $t + 1$ such that $\Lambda(E_{t+1}) < \Lambda(E_t)$.*

Theorem 5. *After at most 4n+1 rounds, all heaps have been merged into one.*

3.4 Phase 3: Sorted List and Clique

Theorem 6. *If all nodes form one heap, it takes $\mathcal{O}(n)$ time until the network reaches a legal state.*

Combining Theorem 3, Theorem 4, Theorem 5 and Theorem 6 our main theorem Theorem 2 holds.

4   Message Complexity and Single Join and Leave Event

In this section we give an upper bound for the work spent by each node.

According to Theorem 2 it takes $\mathcal{O}(n)$ rounds to reach a legal state. In each round each active node sends a message to its predecessor and its successor (*forward-from-successor, forward-from-predecessor*) and receives a message from them (*forward-from-successor, forward-from-predecessor*). Also, a node sends at most one *activate/-deactivate* message to its successor at each round. This gives a resulting work of $\mathcal{O}(n)$ for each node or $\mathcal{O}(n^2)$ in total. By the following lemma we show that the additional messages sent and received during the linearization are at most $\mathcal{O}(n)$ for each node and $\mathcal{O}(1)$ as soon as a stable state is reached.

Theorem 7. *Each node sends and receives at most $\mathcal{O}(n)$ messages during the linearization phase (stabilization work) and at most $\mathcal{O}(1)$ messages in a legal state (maintenance work).*

4.1   Single Join and Leave Event

The case of arbitrary churn is hard to analyze formally. Thus, we will show that the clique can efficiently recover considering a single join or leave event in a legal state.

Theorem 8. *In a legal state it takes $\mathcal{O}(n)$ rounds and messages to recover and stabilize after a new node joins the network. It takes $\mathcal{O}(1)$ rounds and messages to recover the clique after a node leaves the network.*

5   Conclusion

In this paper we introduced a local self-stabilizing time-and work-efficient algorithm that forms a clique out of any weakly connected graph. By forming a clique our algorithm also solves the resource discovery problem, as each node is aware of any other node in the network. Our algorithm is the first algorithm that solves resource discovery in optimal message complexity. Furthermore our algorithm is self-stabilizing and thus can handle deletions of edges and joining or leaving nodes.

References

1. Aspnes, J., Shah, G.: Skip graphs. In: SODA, pp. 384–393 (2003)
2. Awerbuch, B., Scheideler, C.: The hyperring: a low-congestion deterministic data structure for distributed environments. In: SODA 2004, pp. 318–327 (2004)
3. Berns, A., Ghosh, S., Pemmaraju, S.V.: Brief announcement: a framework for building self-stabilizing overlay networks. In: PODC 2010, pp. 398–399 (2010)
4. Blin, L., Dolev, S., Potop-Butucaru, M.G., Rovedakis, S.: Fast self-stabilizing minimum spanning tree construction. In: Lynch, N.A., Shvartsman, A.A. (eds.) DISC 2010. LNCS, vol. 6343, pp. 480–494. Springer, Heidelberg (2010)
5. Blin, L., Maria, G.P.-B., Rovedakis, S.: Self-stabilizing minimum degree spanning tree within one from the optimal degree. J. Parallel Distrib. Comput. 71(3), 438–449 (2011)

6. Chor, B., Goldwasser, S., Micali, S., Awerbuch, B.: Verifiable secret sharing and achieving simultaneity in the presence of faults (extended abstract). In: FOCS, pp. 383–395 (1985)
7. Curt Cramer and Thomas Fuhrmann. Self-stabilizing ring networks on connected graphs. Technical Report 2005-5, University of Karlsruhe, TH (2005)
8. Edsger, W.D.: Self-stabilizing systems in spite of distributed control. Commun. ACM 17, 643–644 (1974)
9. Dolev, S., Tzachar, N.: Empire of colonies: Self-stabilizing and self-organizing distributed algorithm. Theor. Comput. Sci. 410(6-7), 514–532 (2009)
10. Hérault, T., Lemarinier, P., Peres, O., Pilard, L., Beauquier, J.: A model for large scale self-stabilization. In: IPDPS, pp. 1–10 (2007)
11. Jacob, R., Richa, A.W., Scheideler, C., Schmid, S., Täubig, H.: A distributed polylogarithmic time algorithm for self-stabilizing skip graphs. In: PODC, pp. 131–140 (2009)
12. Jacob, R., Ritscher, S., Scheideler, C., Schmid, S.: A self-stabilizing and local delaunay graph construction. In: Dong, Y., Du, D.-Z., Ibarra, O. (eds.) ISAAC 2009. LNCS, vol. 5878, pp. 771–780. Springer, Heidelberg (2009)
13. Malkhi, D., Naor, M., Ratajczak, D.: Viceroy: a scalable and dynamic emulation of the butterfly. In: PODC 2002, pp. 183–192 (2002)
14. Myasnikov, A.G., Shpilrain, V., Ushakov, A.: Group-based cryptography. In: Advanced Courses in Mathematics, Birkhäuser Verlag, CRM Barcelona (2008)
15. Naor, M., Wieder, U.: Novel architectures for p2p applications: The continuous-discrete approach. ACM Transactions on Algorithms 3(3) (2007)
16. Onus, M., Richa, A.W., Scheideler, C.: Linearization: Locally self-stabilizing sorting in graphs. In: ALENEX (2007)
17. Pease, M.C., Shostak, R.E., Lamport, L.: Reaching agreement in the presence of faults. J. ACM 27(2), 228–234 (1980)
18. Ratnasamy, S., Francis, P., Handley, M., Karp, R., Shenker, S.: A scalable content-addressable network. In: SIGCOMM 2001, pp. 161–172 (2001)
19. Rowstron, A., Druschel, P.: Pastry: Scalable, decentralized object location, and routing for large-scale peer-to-peer systems. In: Guerraoui, R. (ed.) Middleware 2001. LNCS, vol. 2218, pp. 329–350. Springer, Heidelberg (2001)
20. Shaker, A., Reeves, D.S.: Self-stabilizing structured ring topology p2p systems. In: Peer-to-Peer Computing, pp. 39–46 (2005)
21. Haeupler, B., Pandurangan, G., Peleg, D., Rajaraman, R., Sun, Z.: Discovery through Gossip. In: SPAA (2011)
22. Stoica, I., Morris, R., Liben-nowell, D., Karger, D., Frans, M., Dabek, K.F., Balakrishnan, H.: Chord: A scalable peer-to-peer lookup service for internet applications. In: SIGCOMM, pp. 149–160 (2001)
23. Harchol-Balter, M., Leighton, T., Lewin, D.: Resource discovery in distributed networks. In: PODC 1999, pp. 229–237 (1999)
24. Kutten, S., Peleg, D., Vishkin Deterministic, U.: resource discovery in distributed networks. In: SPAA 2001, pp. 77–83 (2001)
25. Konwar, K.M., Kowalski, D., Shvartsman, A.A.: Node discovery in networks In: J. Parallel Distrib. Comput. 69(4), 337–348 (2009)
26. Kniesburge, S., Koutsopoulos, A., Scheideler, C.: A Deterministic Worst-Case Message Complexity Optimal Solution for Resource Discovery (Pre-Print) In: arXiv:1306.1692 (2013)

Maintaining Balanced Trees
for Structured Distributed Streaming Systems[*]

Frédéric Giroire, Remigiusz Modrzejewski,
Nicolas Nisse, and Stéphane Pérennes

COATI, Joint Project I3S (CNRS & UNS) and INRIA, Sophia Antipolis, France
{frederic.giroire,remigiusz.modrzejewski,
stephane.perennes,nicolas.nisse}@inria.fr

Abstract. In this paper, we propose and analyze a simple localized algorithm to balance a tree. The motivation comes from live distributed streaming systems in which a source diffuses a content to peers via a tree, a node forwarding the data to its children. Such systems are subject to a high churn, peers frequently joining and leaving the system. It is thus crucial to be able to repair the diffusion tree to allow an efficient data distribution. In particular, due to bandwidth limitations, an efficient diffusion tree must ensure that node degrees are bounded. Moreover, to minimize the delay of the streaming, the depth of the diffusion tree must also be controlled. We propose here a simple distributed repair algorithm in which each node carries out local operations based on its degree and on the subtree sizes of its children. In a synchronous setting, we first prove that starting from any n-node tree our process converges to a balanced tree in $O(n^2)$ turns. We then describe a more restrictive model, adding a small extra information to each node, under which we adopt our algorithm to converge in $\Theta(n \log n)$ turns. We then exhibit by simulation that the convergence is much faster (logarithmic number of turns in average) for a random tree.

Keywords: Distributed algorithms, tree balancing, live streaming, peer-to-peer.

1 Introduction

Trees are inherent structures for data dissemination in general and particularly in peer-to-peer live streaming networks. Fundamentally, from the perspective of a peer, each atomic piece of content has to be received from some source and forwarded towards some receivers. Moreover, most of the actual streaming mechanisms ensure that a piece of information is not transmitted again to a peer that already possesses it. Therefore, this implies that dissemination of a single fragment defines a tree structure. Even in *unstructured* networks, whose main characteristic

---

[*] The research leading to these results has received funding from the European Project FP7 EULER, ANR CEDRE, ANR AGAPE, Associated Team AlDyNet, project ECOS-Sud Chile and région PACA.

T. Moscibroda and A.A. Rescigno (Eds.): SIROCCO 2013, LNCS 8179, pp. 177–188, 2013.

is lack of defined structure, many systems look into perpetuating such underlying trees, e.g. the second incarnation of Coolstreaming [7] or PRIME [9].

Unsurprisingly, early efforts into designing peer-to-peer video streaming concentrated on defining tree-based structures for data dissemination. These have been quickly deemed inadequate, due to fragility and unused bandwidth at the leaves of the tree. One possible fix to these weaknesses was introduced in SplitStream [3]. The proposed system maintains multiple concurrent trees to tolerate failures, and internal nodes in a tree are leaf nodes in all other trees to optimize bandwidth. The construction of intertwined trees can be simplified by a randomized process, as proposed in Chunkyspread [11], leading to a streaming algorithm performing better over a range of scenarios.

As found in [7], node churn is the main difficulty for live streaming networks, especially those trying to preserve structure. On the other hand, in [12] authors embrace change. Their stochastic optimization approach relies on constant random creating and breaking of relationships. To ensure network connectivity, nodes are said to keep open connections with hundreds of potential neighbours. Another approach, displayed in [8], is churn-resiliency by maintaining redundancy within the network structure. Although concentrating on a different field, authors of [10] face a similar to our own problem of maintaining balanced trees, needed for connecting wireless sensors. However, their solution is periodical rebuilding the whole tree from scratch. Our solution aims at minimizing the disturbance of nodes, whose ancestors were not affected by recent failures, as well as minimizing the redundancy in the network.

The analysis of these systems focus on the feasibility, construction time and properties of the established overlay network, see for example [3,11] and [4] for a theoretical analysis. But these works usually abstract over the issue of tree maintenance. Generally, in these works, when some elements (nodes or links) of the networks fail, the nodes disconnected from the root execute the same procedure as for initial connection. To the best of our knowledge, there are no theoretical analysis on the efficiency of tree maintenance in streaming systems, reliability is estimated by simulations or experiments as in [3].

In this paper, we tackle this issue by designing an efficient maintenance scheme for trees. Our distributed algorithm ensures that the tree recovers fast to a "good shape" after one or multiple failures occur. We give analytic upper bounds of the convergence time. To the best of our knowledge, this is the first theoretical analysis of a repair process for live streaming systems. While the $O(n^2)$ worst case bound seems high, simulations shown in Section 5 suggest that the average case is closer to $O(\log n)$, which is lower than the conceivable time of rebuilding a tree from scratch.

The problem setting is as follows. A single source provides live media to some nodes in the network. This source is the single reliable node of the network, all other peers may be subject to failure. Each node may relay the content to further nodes. Due to limited bandwidth, both source and any other node can provide media to a limited number $k \geq 2$ of nodes. The network is organized into a logical tree, rooted at the source of media. If node x forwards the stream towards node y, then x is the parent of y in the logical tree. Note that the delay between broadcasting a piece of media by the source and receiving by a peer is given by its distance from the root in the logical tree. Hence our goal is to minimize the tree depth, while following degree constraints.

As shown in [7], networks of this kind experience high rate of node joins and leaves. Leaves can be both graceful, where a node informs about imminent departure and network rearranges itself before it stops providing to the children, or abrupt (e.g. due to connection or hardware failure). In this work, we assume a *reconnection process*: when a node leaves, its children reattach to its parent. This can be done locally if each node stores the address of its grandfather in the tree. Note that this process is performed independently of the bandwidth constraint, hence after multiple failures, a node may become the parent of many nodes. The case of concurrent failures of father and grandfather can be handled by reattaching to the root of the tree. Other more sophisticated reconnection processes have been proposed, see for example [6].

This process can leave the tree in a state where either the bandwidth constraints are violated (the degree of a node is larger than k) or the tree depth is not optimal. Thus, we propose a distributed *balancing process*, where based on information about its degree and the subtree sizes of its children, a node may perform a local operation at each turn. We show that this balancing process, starting from any tree, converges to a balanced tree and we evaluate the convergence time.

Related Work. Construction of spanning trees has been studied in the context of self-stabilizing algorithms. Herault et al. propose in [6] a new analytic model for large scale systems. They assume that any pair of processes can communicate directly, under condition of knowing receiver's identifier, what is the case in Internet Protocol. They additionally assume a discovery service and a failure detection service. Under this model they propose and prove correctness of an algorithm constructing a spanning tree over a set of processes. Similar assumptions have been used by Caron et al. in [2] to construct a distributed prefix tree and by Bosilca et al. in [1] to construct a binomial graph (Chord-like) overlay.

In this paper we assume the results of these earlier works: nodes can reliably communicate, form connections and detect failures. We do not analyze these operations at message level. Furthermore, we analyze the overlay assuming it is already a spanning tree. However, it may have an arbitrary shape, e.g. be a path or a star (all nodes connected directly to the root). This can be regarded as maintaining the tree after connection or failure of an arbitrary number of nodes.

Our Results. In Section 2, we provide a formal definition of the problem and propose a distributed algorithm for the balancing process. The process works in a synchronous setting. At each turn, all nodes are sequentially scheduled by an adversary and must execute the process. In Section 3, we show that the balancing process always succeeds in $O(n^2)$ turns. Then, in Section 4, we study a restricted version of the algorithm in which a node performs an operation only when the subtrees of its children are balanced. In this case, we succeeded in obtaining a tight bound of $\Theta(n \log n)$ on the number of turns for the worst tree. Finally, we show that the convergence is in fact a lot faster in average for a random tree and takes a logarithmic number of turns.

Due to space limitations, only intuitions of some proofs are presented here. The full proofs can be found in [5].

2 Problem and Balancing Process

In this section, we present the main definitions and settings used throughout the paper, then we present our algorithm and prove some simple properties of it.

2.1 Notations

This section is devoted to some basic notations.

Let $n \in \mathbb{N}^*$. Let $T = (V, E)$ be a n-node tree rooted in $r \in V$. Let $v \in V$ be any node. The *subtree* T_v *rooted at* v is the subtree consisting of v and all its descendants. In other words, if $v = r$, then $T_v = T$ and, otherwise, let e be the edge between v and its parent, T_v is the subtree of $T \setminus e = (V, E \setminus \{e\})$ containing v. Let $n_v = |V(T_v)|$.

Let $k \geq 2$ be an integer. A node $v \in V(T)$ is *underloaded* if it has at most $k - 1$ children and at least one of these children is not a leaf. v is said *overloaded* if it has at least $k + 1$ children. Finally, a node v with k children is *imbalanced* if there are two children x and y of v such that $|n_x - n_y| > 1$. A node is *balanced* if it is neither underloaded, nor overloaded nor imbalanced. Note that a leaf is always balanced.

A tree is a *k-ary tree* if it has no nodes that are underloaded or overloaded, i.e., all nodes have at most k children and a node with $< k$ children has only leaf-children. A rooted k-ary tree T is *k-balanced* if, for each node $v \in V(T)$, the sizes of the subtrees rooted in the children of v differ by at most one. In other words, a rooted tree is k-balanced if and only if all its nodes are balanced.

As formalized by the next claim, k-balanced trees are good for our live streaming purpose since such overlay networks (k being small compared with n) ensure a low dissemination delay while preserving bandwidth constraints.

Claim 1. *Let T be a n-node rooted tree. If T is k-balanced, then each node of T is at distance at most $\lfloor \log_k n \rfloor$ from r.*

2.2 Distributed Model and Problem

Nodes are autonomous entities running the same algorithm. Each node v has a local memory where it stores the size n_v of its subtree, the size of the subtrees of its children and the size of the subtrees of its grand-children, i.e., for any child x of v and for any child y of x, v knows n_x and n_y.

Computations performed by the nodes are based only on the local knowledge, i.e., the information present in the local memory and that concerns only nodes

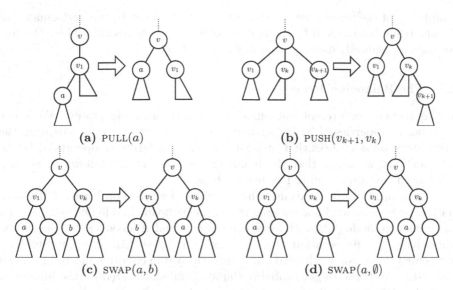

(a) PULL(a) (b) PUSH(v_{k+1}, v_k)

(c) SWAP(a, b) (d) SWAP($a, \emptyset$)

Fig. 1. Operations performed by node v in the balancing process

at distance at most 2. We consider a synchronous setting. That is, the time is slotted in turns. At each turn, any node may run the algorithm based on its knowledge and, depending on the computation, may do one of the following *operations*. In the algorithm we present, each operation done by a node v consists of rewiring at most two edges at distance at most 2 from v. More precisely, let v_1, v_k and v_{k+1} be children of v, a be a child of v_1 and b be a child of v_k (if any). The node v may

- replace the edge $\{v_1, a\}$ by the edge $\{v, a\}$. A grand-child a of v then becomes a child of v. This operation is denoted by PULL(a) and illustrated in Figure 1a;
- replace the edge $\{v, v_{k+1}\}$ by the edge $\{v_k, v_{k+1}\}$. A child v_{k+1} of v then becomes a child of another child v_k of v. This operation is denoted by PUSH (v_{k+1}, v_k), see Figure 1b;
- replace the edges $\{v_1, a\}$ and $\{v_k, b\}$ by the edges $\{v_1, b\}$ and $\{v_k, a\}$. The children v_1 and v_k of v exchange two of their own children a and b. This operation is denoted by SWAP(a,b) and an example is given in Figure 1c. Here, a or b may not exist, in which case, one of v_1 and v_k "wins" a new child while the other one "looses" a child. This case is illustrated in Figure 1d.

In all cases, the local memory of the at most $k^2 + 1$, including the parent of v, nodes that are concerned are updated. Note that each of these operations may be done using a constant number of messages of size $O(\log n)$.

In this setting, at every turn, all nodes sequentially run the algorithm. In order to consider the worst case scenario, the order in which all nodes are scheduled during one turn is given by an adversary. The algorithm must ensure that after a finite number of turns, the resulting tree is k-balanced. We are interested in time complexity of the worst case scenario of the repair. That is, the performance of the algorithm is measured by the maximum number of turns after which the tree becomes k-balanced, starting from any n-node tree.

2.3   The Balancing Process

In this section, we present our algorithm, called *balancing process*. We prove some basic properties of it. In particular, while the tree is not k-balanced, the balancing process ensures that at least one node performs an operation. In the next sections, we prove that the balancing process actually allows to reach a k-balanced tree after a finite number of steps.

At each turn, a node v executes the algorithm described on Figure 2. To summarize, an underloaded node does a PULL, an overloaded node does a PUSH and an imbalanced node (whose children are not overloaded) does a SWAP operation. Note that a SWAP operation may exchange a subtree with an empty subtree, but cannot create an overloaded node. Intuitively, the children affected by PUSH and PULL are chosen to get probably the least imbalance (reduce the biggest or merge the two small). It is important to emphasise that the balancing process requires no memory of the past operations.

Algorithm executed by a node v in a tree T. If v is not a leaf, let $(v_1, v_2, \cdots, v_d)$ be the $d \geq 1$ children of v ordered by subtree-size, i.e., $n_{v_1} \geq n_{v_2} \geq \cdots \geq n_{v_d}$.

1. **If** v is underloaded (then $d < k$), let a be a child of v_1 with biggest subtree size. **Then** node v executes PULL(a).   // *That is, a becomes a child of v.*
2. **Else if** v is overloaded (then $d > k \geq 2$), **then** node v executes PUSH(v_{k+1}, v_k).
   // *That is, v_{k+1} becomes a child of v_k.*
3. **Else if** v is imbalanced (then $d = k$) **and if** v_1 and v_k are not overloaded, let a and b be two children of v_1 and v_k respectively such that $|n_{v_1} - n_a + n_b - (n_{v_k} - n_b + n_a)|$ is minimum (a (resp. b) may be not defined, i.e., $n_a = 0$ (resp., $n_b = 0$), if v_1 (resp v_2) is underloaded).
   Then node v execute SWAP(a, b). // *That is, a and b exchange their parent.*

Fig. 2. Balancing Process

Note that if the tree if k-balanced, no operation are performed, and that, if the tree is not, at least one operation is performed.

Claim 2. *If T is not k-balanced, and all nodes execute the balancing process, then at least one node will do an operation.*

In the next section, we prove that, starting from any tree, the number of operations done by the nodes executing the balancing process is bounded. Together with the previous claim, it allows to prove

Theorem 1. *Starting from any tree T where each node executes the balancing process, after a finite number of steps, T eventually becomes k-balanced.*

Before proving the above result in next Section, we give a simple lower bound on the number of turns required by the Balancing Process. A *star* is a rooted tree where any non root-node is a leaf.

Lemma 1. *If the initial tree is a n-node star, then at least $\Omega(n)$ turns are needed before the resulting tree is k-balanced.*

3   Worst Case Analysis

In this Section we obtain an upper bound of $O(n^2)$ turns needed to balance the tree. We prove it using a potential function, whose initial value is bounded, integral and positive, may rise in a bounded number of turns and, otherwise, strictly decreases. For clarity of presentation we assume we want to obtain a 2-balanced tree. The proofs can be extended to larger k. Due to lack of space, most of them are only sketched here and can be found in [5].

Lemma 2. *Starting from any n-node rooted tree T, after having executed the Balancing Process during $O(n)$ turns, no node will do a PUSH operation anymore.*

This lemma is proved by tracking a potential function $\Phi(T) = \sum_{v \in V(T)} \max \{0, d_v - 3\}$, where d_v is the number of children of node v. Note that any node who started a turn with degree at least three, will perform a PUSH and receive at least one new child, thus finishing the turn with degree not greater than in the beginning. Thus, no operation can increase Φ. In each turn, either Φ decreases, or a node with no overloaded ancestors performs its last PUSH. As the value of Φ is bounded by the number of nodes, the lemma holds.

Let Q be the sum over all nodes $u \in T$ of the distance between u and the root.

Lemma 3. *Starting from any n-node rooted tree T, there are at most $O(n^2)$ distinct (not necessarily consecutive) turns with a PULL operation. More precisely, the sum of the sizes of the subtrees that are pulled during the whole process does not exceed n^2.*

Proof. First, by Lemma 2, there are no PUSH operations after $O(n)$ turns. Note that a SWAP operation does not change Q. Moreover, a PULL operation of a subtree T_v makes Q decrease by n_v. Since $Q = \sum_{u \in V(T)} d(u, r) \leq n^2$, the sum of the sizes of the subtrees that are pulled during the whole process does not exceed n^2. $\qquad\square$

Potential Function. To prove the main result of this section, we define a potential function and show that: (1) the initial value of the potential function is bounded; (2) its value may raise due to PULL operations, but in a limited number of turns and by a bounded amount; (3) a SWAP operation may not increase its value; (4) if no PUSH nor PULL operation are done, there exists at least one node doing a SWAP operation, strictly decreasing the potential function.

We tried simple potential functions first. However, they led either to an unbounded number of turns with non-decreasing value, or to a larger upper bound. For example, it would be natural to define the potential of a node as the difference between its subtree sizes. For this potential function, (1) (2) and (3) are true, but, unfortunately, for some trees the potential function does not decrease during a turn. This function can be patched so that each operation makes the potential decrease: multiplying the potential of a node by its distance to the root. However, the potential in this case can reach $O(n^3)$.

The potential function giving the $O(n^2)$ bound is defined as follows. Recall that we consider a n-node tree T rooted in r such that all nodes have at most two children. Let $E_0 = n$ and, for any $0 \leq i \leq \lceil \log(n+1) \rceil$, let $E_i = 2E_{i+1} + 1$. Note that $(E_i)_{i \leq \lceil \log(n+1) \rceil}$ is strictly decreasing, and $0 < E_{\lceil \log(n+1) \rceil} \leq 1$. Intuitively, E_i is the mean-size of a subtree rooted in a node at distance i from the root in a balanced tree with n nodes.

Let K_i be the set of nodes of T at distance exactly $i \geq 0$ from the root and $|K_i| = k_i$, and, for any $0 \leq i \leq \lceil \log(n+1) \rceil$, let $m_i = 2^i - k_i$. Intuitively, m_i represents the number of nodes, at distance i from the root, missing compared to a complete binary tree.

For any $v \in V(T)$ at distance $0 \leq i \leq \lceil \log(n+1) \rceil$ from the root, the *default* of v, denoted by $\mu(v)$, equals $n_v - \lceil E_i \rceil$ if $n_v > E_i$ and $\lfloor E_i \rfloor - n_v$ otherwise. Note that $\mu(v) \geq 0$ since n_v is an integer.

Let the *potential at distance i from r*, $0 \leq i \leq \lceil \log(n+1) \rceil$, be $P_i = m_i \cdot \lfloor E_i \rfloor + \sum_{u \in K_i} \mu(u)$. Finally, let us define the *potential* $\mathcal{P} = \sum_{0 \leq i \leq \lceil \log(n+1) \rceil} P_i$. Since $\mu(u) \leq n$ for any $u \in V(T)$, and $\sum_{0 \leq i \leq \lceil \log(n+1) \rceil} m_i + k_i \leq 2n$, then $\mathcal{P}(T) = O(n^2)$.

Lemma 4. *For any n-node rooted tree T, a* PULL *operation of a subtree T_v may increase the potential $\mathcal{P}$ by at most $2n_v$.*

This lemma is proved by case analysis. Let u be the node performing the operation, x its unique child and v the node being pulled. We show that the default increases by at most n_v for x, $\lfloor E_{j-1} \rfloor - \lfloor E_j \rfloor$ for nodes below it whose distance j from the root is $j \leq \lceil \log(n+1) \rceil$ and by at most n_w for every node whose distance from root is $\lceil \log(n+1) \rceil + 1$. Calculating the new potential, using all those inequations, the lemma holds.

Let v be a node at distance $\lceil \log(n+1) \rceil > i \geq 0$ from the root r of T. v is called i-*median* if it has one or two children a and b and $n_a > E_{i+1} > n_b$ (possibly v has exactly one child and $n_b = 0$).

Lemma 5. *For any n-node rooted tree T, a* SWAP *operation executed by any node v does not increase the potential $\mathcal{P}$. Moreover, if v is $(i-1)$-median then $\mathcal{P}$ strictly decreases by at least one.*

This lemma is proved by calculating the new potential, in all the possible cases of relative sizes of the children and E_i before and after the operation.

Let v be a node at distance $0 \leq i < \lceil \log(n+1) \rceil - 1$ from the root r of T. v is called i-*switchable* if it has one or two children a and b and $n_a > E_{i+1} > n_b$

(possibly v has only exactly child, and $n_b = 0$), $n_a - n_b \geq 2$ and none of its ancestors can execute a SWAP operation. Note that, if a node is i-*switchable*, then it is i-median.

Lemma 6. *Let T be a tree where no PUSH nor PULL operation is possible. If a node v is i-switchable, then either v can do a SWAP operation, or $0 \leq i < \lceil \log(n+1) \rceil - 2$ and it has a $(i+1)$-switchable child.*

To prove this lemma we first take care of nodes at distance $\lceil \log(n+1) \rceil$ from r, showing that in all the possible cases of its children sizes a SWAP can be performed. Then, for nodes at smaller distances to r, if an i-switchable node can not perform a SWAP, then in all possible cases one of its children is $(i+1)$-switchable.

Lemma 7. *At each turn when no PULL nor PUSH operations are done, if the tree is not balanced, then there is a i-switchable node, $0 \leq i < \lceil \log(n+1) \rceil - 1$.*

To prove this lemma, we define a $\mathcal{S}_i$-*situation*: for any $j < i$, all nodes at distance j from the root cannot do a SWAP operation, and for any $j \leq i$, $k_j = 2^j$ and, f or any node v at distance i from the root, $n_v \in \{\lceil E_i \rceil, \lfloor E_i \rfloor\}$. If the tree is in a $\mathcal{S}_{\lceil \log(n+1) \rceil - 1}$-situation, then it is balanced. Let j be the smallest integer such that T is not in a $\mathcal{S}_j$-*situation*. Then there is a node at distance $j - 1$ from the root, which in all possible cases is $(j-1)$-switchable.

Theorem 2. *Starting from any n-node rooted tree, the balancing process reaches a 2-balanced tree in $O(n^2)$ turns.*

Proof. By Lemma 2, after $O(n)$ turns, no PUSH operations are executed anymore and all nodes have at most two children. From then, there may have only pull or SWAP operations. Moreover, by Claim 2, there is at least one operation per turn while T is not balanced. From Lemma 3, there are at most $O(n^2)$ turns with a PULL operation. Once no PUSH operations are executed anymore, from Lemmata 3, 4 and 5, potential $\mathcal{P}$ can increase by at most $O(n^2)$ in total (over all turns). Moreover, by Lemma 5, if a i-median node executes a SWAP operation, the potential $\mathcal{P}$ strictly decreases by at least one.

By Lemma 7, at each turn when no pull nor PUSH operations are done, there is an i-switchable node, $0 \leq i < \lceil \log(n+1) \rceil - 1$. Thus, by Lemma 6, at each such turn, there is an i-switchable that can execute a SWAP operation. Since a i-switchable node is i-median $(0 \leq i < \lceil \log(n+1) \rceil - 1)$, by Lemma 5, the potential $\mathcal{P}$ strictly decreases by at least one.

The result then follows from the fact that $\mathcal{P} \leq n^2$.   □

4   Adding an Extra Global Knowledge to the Nodes

In this section, we assume an extra global knowledge: each node knows whether it has a descendant that is not balanced. This extra information is updated after each operation. Then, our algorithm is modified by adding the condition that any node v executing the balancing process can do a PULL or SWAP operation only if all its descendants are balanced. Adding this property allows to prove better upper bounds on the number of steps, by avoiding conflict between an operation performed by a node and an operation performed by one of its not balanced descendant. We moreover prove that this upper bound for our algorithm is asymptotically tight, reached when input tree is a path. The approach presented in this section is specific for $k = 2$. I.e., the objective of the Balancing Process is to reach a 2-balanced tree.

First, we define a function f used to bound the number of turns needed to balance a tree consisting of two balanced subtrees and a common ancestor. Let $f : \mathbb{N} \times \mathbb{N} \to \mathbb{N}$ be the function defined recursively as follows.

$$
\begin{aligned}
\forall a \geq 0, \qquad & f(a, a) = 0 \\
\forall a \geq 1, \qquad & f(a, a - 1) = 0 \\
\forall a \geq 2, \qquad & f(a, 0) = 1 + f(\lfloor \tfrac{a-1}{2} \rfloor, 0) \\
\forall a > 2, \forall 1 \leq b < a - 1, \; & f(a, b) = 1 + \max\left(f(\lceil \tfrac{a-1}{2} \rceil, \lfloor \tfrac{b-1}{2} \rfloor), f(\lfloor \tfrac{a-1}{2} \rfloor, \lceil \tfrac{b-1}{2} \rceil) \right)
\end{aligned}
$$

Lemma 8. *For any $a \geq 0$, $a \geq b \geq 0$, $f(a, b) \leq \max\{0, \log_2 a\}$.*

This lemma is proved by a simple induction on a. Now, we give a function bounding the number of turns needed to balance any tree of a given size. Let $g : \mathbb{N} \to \mathbb{N}$ be the function defined recursively as follows.

$$
\begin{aligned}
\forall n \in \{0, 1\}, \; g(n) &= 0 \\
\forall n > 1, \qquad g(n) &= \max_{a \geq b \geq 0, a+b=n-1}(\max\{g(a), g(b)\} + f(a, b))
\end{aligned}
$$

Using a simple induction on n, we obtain that:

Lemma 9. *For any $n \geq 0$, $g(n) \leq \max\{0, n \log_2 n\}$.*

We now state our main results:

Theorem 3. *Starting from any n-node rooted tree, the balancing process with global knowledge reaches a 2-balanced tree in $O(n \log n)$ turns.*

Note first that Lemma 2 still holds with the new balancing process, that is: no node is overloaded after $O(n)$ turns. Let now x be a node with all descendants balanced. Let y and z be the children of x. We show by induction on n_y that x becomes balanced in at most $f(n_y, n_z)$ turns. Then, by induction on n, we show that T can be balanced in at most $g(n)$ turns.

Next theorem shows that there are trees starting from which the balancing process actually uses a number of turns of the order of the above upper bound.

Theorem 4. *Starting from an n-node path rooted in one of its ends, the balancing process with global knowledge reaches a 2-balanced tree in $\Omega(n \log n)$ turns.*

The proof is done by an induction on the tree size.

5   Simulations

In the previous sections we obtained upper and lower bounds for the maximum number of turns needed to balance a tree of a given size. A significant gap between those bounds raises the question: which bound is closer to what happens for random instances? We investigate the performance of the algorithm running an implementation under a discrete event simulation. Scheduling of nodes within a turn is given by a simple adversary algorithm. First, it detects which nodes can perform no operation. It schedules them to move first, to ensure that they do not perform operations enabled by operations of other nodes. Then, it schedules the remaining nodes in a random order.

The process starts in a random tree. It is obtained by assigning random weights to a complete graph and building a minimum weight spanning tree over it. Figure 3 displays the number of turns it took to balance trees of progressing sizes. For each size the numbers are aggregated over 10000 different starting trees. The solid line marks the average, dotted lines the minimum and maximum numbers of turns and error bars show the standard deviation.

What can be seen from this figure, is that the number of turns spent to balance a random tree progresses logarithmically in regard to the tree size. This holds true both for average and the worst cases encountered. This is significantly less even than the lower bound on maximum time. This is because that comes from the particular case of star as the starting tree, which is randomly obtained with probability $\frac{1}{n!}$ and did not occur in our experiments for bigger values of n.

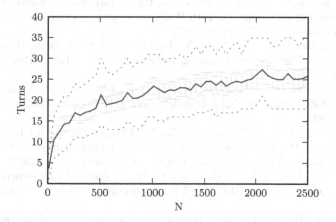

Fig. 3. Balancing a random tree

6   Conclusions and Future Research

We have proposed a distributed tree balancing algorithm and shown following properties. The algorithm does stop only when the tree is balanced. After at most $\Omega(n)$ turns there are no overloaded nodes in the tree, what corresponds to a broadcast tree where every node receives content. This bound is reached when the starting tree is a star. Balancing process after there are no overloaded nodes lasts at most $O(n^2)$ turns. With the additional restriction that a node acts only if all of its descendants are balanced, the number of turns to balance any tree is $O(n \log n)$. This bound is reached when the starting tree is a path.

An obvious, but probably hard, open problem is closing the gap between the $O(n^2)$ upper bound and the $\Omega(n)$ lower bound on balancing time. Another possibility is examination of the algorithm's average behaviour, which as hinted by simulations should yield $O(\log n)$ bound on balancing time.

The algorithm itself can be extended to handle well the case of trees that are not regular. Furthermore, in order to approach a practical system, moving to multiple trees would be highly beneficial. Allowing the algorithm to stop with more imbalance, where children are allowed to differ by a given threshold instead of one, could lead to a faster convergence.

References

1. Bosilca, G., Coti, C., Herault, T., Lemarinier, P., Dongarra, J.: Constructing resiliant communication infrastructure for runtime environments. In: International Conference in Parallel Computing (2009)
2. Caron, E., Datta, A., Petit, F., Tedeschi, C.: Self-stabilization in tree-structured peer-to-peer service discovery systems. In: IEEE Symposium on Reliable Distributed Systems, pp. 207–216 (2008)
3. Castro, M., Druschel, P., Kermarrec, A., Nandi, A., Rowstron, A., Singh, A.: Split-Stream: high-bandwidth multicast in cooperative environments. In: Proceedings of the Nineteenth ACM Symposium on Operating Systems Principles, p. 313 (2003)
4. Dan, G., Fodor, V., Chatzidrossos, I.: On the performance of multiple-tree-based peer-to-peer live streaming. In: 26th IEEE International Conference on Computer Communications, pp. 2556–2560 (2007)
5. Giroire, F., Remigiusz, M., Nisse, N., Pérennes, S.: Maintaining Balanced Trees for Structured Distributed Streaming Systems. Research Report RR-8309, INRIA (May 2013)
6. Herault, T., Lemarinier, P., Peres, O., Pilard, L., Beauquier, J.: A model for large scale self-stabilization. In: IEEE Parallel and Distributed Processing Symposium, pp. 1–10 (2007)
7. Li, B., Qu, Y., Keung, Y., Xie, S., Lin, C., Liu, J., Zhang, X.: Inside the new coolstreaming: Principles, measurements and performance implications. In: 27th IEEE International Conference on Computer Communications (2008)
8. Li, Z., Xie, G., Hwang, K., Li, Z.: Churn-resilient protocol for massive data dissemination in p2p networks. IEEE Parallel and Distributed Systems 22(8), 1342–1349 (2011)
9. Magharei, N., Rejaie, R.: Prime: Peer-to-peer receiver-driven mesh-based streaming. IEEE/ACM Transactions on Networking 17(4), 1052–1065 (2009)
10. Pan, M.-S., Tsai, C.-H., Tseng, Y.-C.: The orphan problem in zigbee wireless networks. IEEE Transactions on Mobile Computing 8(11), 1573–1584 (2009)
11. Venkataraman, V., Yoshida, K., Francis, P.: Chunkyspread: Heterogeneous unstructured tree-based peer-to-peer multicast. In: 14th IEEE International Conference on Network Protocols, pp. 2–11 (2006)
12. Zhang, S., Shao, Z., Chen, M.: Optimal distributed p2p streaming under node degree bounds. In: 18th IEEE International Conference on Network Protocols, pp. 253–262 (2010)

Rendezvous of Two Robots
with Constant Memory

Paola Flocchini[1], Nicola Santoro[2],
Giovanni Viglietta[2], and Masafumi Yamashita[3]

[1] EECS, University of Ottawa, Ottawa, Canada
flocchin@site.uottawa.ca
[2] SCS, Carleton University, Ottawa, Canada
{santoro,viglietta}@scs.carleton.ca
[3] Kyushu University, Fukuoka, Japan
mak@csce.kyushu-u.ac.jp

Abstract. We study the impact that persistent memory has on the classical *rendezvous* problem of two mobile computational entities, called robots, in the plane. It is well known that, without additional assumptions, rendezvous is impossible if the entities have no persistent memory, even if the system is semi-synchronous and movements are rigid. It has been recently shown that if each entity is endowed with $O(1)$ bits of persistent visible memory (called lights), they can rendezvous even if the system is asynchronous.

In this paper we investigate the rendezvous problem in two weaker settings in systems of robots endowed with visible lights: in FSTATE, a robot can only see its own light, while in FCOMM a robot can only see the other robot's light. Among other things, we prove that, with rigid movements, finite-state robots can rendezvous in semi-synchronous settings, and finite-communication robots are able to rendezvous even in asynchronous ones. All proofs are constructive: in each setting, we present a protocol that allows the two robots to rendezvous in finite time.

1 Introduction

1.1 Framework and Background

Rendezvous is the process of two computational mobile entities, initially dispersed in a spatial universe, meeting within finite time at a location, non known *a priori*. When there are more than two entities, this task is known as *Gathering*. These two problems are core problems in distributed computing by mobile entities. They have been intensively and extensively studied when the universe is a connected region of $\mathbb{R}^2$ in which the entities, usually called *robots*, can freely move; see, for example, [1,3,4,8,9,11,14,15,16,17,18].

Each entity is modeled as a point, it has its own local coordinate system of which it perceives itself as the centre, and has its own unit distance. Each entity operates in cycles of LOOK, COMPUTE, MOVE activities. In each cycle, an entity observes the position of the other entities expressed in its local coordinate system (LOOK); using that observation as input, it executes a protocol (the same

T. Moscibroda and A.A. Rescigno (Eds.): SIROCCO 2013, LNCS 8179, pp. 189–200, 2013.
© Springer International Publishing Switzerland 2013

for all robots) and computes a destination point (COMPUTE); it then moves to the computed destination point (MOVE). Depending on the activation schedule and the synchronization level, three basic types of systems are identified in the literature: a *fully synchronous* system (FSYNCH) is equivalent to a system where there is a common clock and at each clock tick all entities are activated simultaneously, and COMPUTE and MOVE are instantaneous; a *semi-synchronous* system (SSYNCH) is like a fully synchronous one except that, at each clock tick, only some entities will be activated (the choice is made by a fair scheduler); in a *fully asynchronous* system (ASYNCH), there is no common notion of time, each COMPUTE and MOVE of each robot can take an unpredictable (but finite) amount of time, and the interval of time between successive activities is finite but unpredictable. The focus of almost all algorithmic investigations in the continuous setting has been on *oblivious* robots, that is when the memory of the robots is erased at the end of each cycle, in other words the robots have no persistent memory (e.g., for an overview see [10]).

The importance of *Rendezvous* in the continuous setting derives in part from the fact that it separates FSYNCH from SSYNCH for oblivious robots. Indeed, *Rendezvous* is trivially solvable in a fully synchronous system, without any additional assumption. However, without additional assumptions, *Rendezvous* is *impossible* for oblivious robots if the system is semi-synchronous [19]. Interestingly, from a computational point of view, *Rendezvous* is very different from the *Gathering* problem of having $k \geqslant 3$ robots meet in the same point; in fact, *Gathering* of oblivious robots is always *possible* for any $k \geqslant 3$ even in ASYNCH without any additional assumption other than multiplicity detection [3]. Furthermore, in SSYNCH, $k \geqslant 3$ robots can gather even in spite of a certain number of faults [1,2,7], and converge in spite of inaccurate measurements [5]; see also [12]. The *Rendezvous* problem also shows the impact of certain factors. For example, the problem has a trivial solution if the robots are endowed with *consistent compasses* even if the system is fully asynchronous. The problem is solvable in ASYNCH even if the local compasses have some degree of inconsistency of an appropriate angle [13]; the solution is no longer trivial, but does exist.

In this paper, we are interested in determining what type and how much persistent memory would allow the robots to rendezvous. What is known in this regard is very little. On the one hand, it is well known that, in absence of additional assumptions, without persistent memory rendezvous is impossible even in SSYNCH [19]. On the other hand, a recent result shows that rendezvous is possible even in ASYNCH if each robot has $O(1)$ bits of persistent memory *and* can transmit $O(1)$ bits in each cycle *and* can remember (i.e., can persistently store) the last received transmission [6] (see also [20] for size-optimal solutions). The conditions of this result are overly powerful. The natural question is whether the simultaneous presence of these conditions is truly necessary for rendezvous.

1.2   Main Contributions

In this paper we address this question by weakening the setting in two different ways, and investigate the *Rendezvous* problem in these weaker settings. Even though its use is very different, in both settings, the amount of persistent memory of a robot is constant.

We first examine the setting where the two robots have $O(1)$ bits of *internal* persistent memory but cannot communicate; this corresponds to the *finite-state* (FSTATE) robots model. Among other contributions, we prove that FSTATE robots with rigid movements can rendezvous in SSYNCH, and that this can be done using only six internal states. The proof is constructive: we present a protocol that allows the two robots to rendezvous in finite time under the stated conditions.

We then study the *finite-communication* (FCOMM) setting, where a robot can transmit $O(1)$ bits in each cycle and remembers the last received transmission, but it is otherwise oblivious: it has no other persistent memory of its previous observations, computations and transmissions. We prove that two FCOMM robots with rigid movements are able to rendezvous even in ASYNCH; this is doable even if the different messages that can be sent are just 12. We also prove that only three different messages suffice in SSYNCH. Also for this model all the proofs are constructive.

Finally, we consider the situation when the movement of the robots is not rigid, that is it can be interrupted by an adversary. The only constraint on the adversary is that a robot moves at least a distance $\delta > 0$ (otherwise, rendezvous is clearly impossible). We show that, with knowledge of δ, three internal states are sufficient to solve *Rendezvous* by FSTATE robots in SSYNCH, and three possible messages are sufficient for FCOMM robots in ASYNCH.

These results are obtained modeling both settings as a system of robots endowed with a constant number of *visible lights*: a FSTATE robot can see only its own light, while a FCOMM robot can see only the other robot's light. Our results seem to indicate that "it is better to communicate than to remember". In addition to the specific results on the *Rendezvous* problem, an important contribution of this paper is the extension of the classical model of oblivious silent robots into two directions: adding finite memory, and enabling finite communication.

Due to space limitations, several details and proofs are omitted; they can be found in http://arxiv.org/abs/1306.1956.

2   Model and Terminology

The general model we employ is the standard one, described in [10]. The two robots are autonomous computational entities modeled as points moving in $\mathbb{R}^2$. Each robot has its own coordinate system and its own unit distance, which may differ from each other, and it always perceives itself as lying at the origin of its own local coordinate system. Each robot operates in cycles that consist of three phases: LOOK, COMPUTE, and MOVE. In the LOOK phase it gets the position (in its local coordinate system) of the other robot; in the COMPUTE phase, it computes a destination point; in the MOVE phase it moves to the computed destination point, along a straight line. Without loss of generality, the LOOK phase is assumed to be instantaneous. The robots are anonymous and oblivious, meaning that they do not have distinct identities, they execute the same algorithm in each COMPUTE phase, and the input to such algorithm is the snapshot coming from the previous LOOK phase.

We study two settings; both can be described as restrictions of the model of visibile lights introduced in [6]. In that model, each robot carries a persistent memory of constant size, called *light*; the value of the light is called *color* or *state*, and it is set by the robot during each COMPUTE phase. Other than their own light, the robots have no other memory of past snapshots and computations.

In the first setting, that of silent finite-state (FSTATE) robots, the light of a robot is visible only to the robot itself; i.e., the colored light merely encodes an *internal state*. In the second setting, of oblivious finite-communication (FCOMM) robots, the light of a robot is visible only to the other robot; i.e., they can communicate with the other robot through their colored light, but by their next cycle they forget even the color of their own light (since they do not see it). The color a robot sees is used as input during the computation.

In the *asynchronous* (ASYNCH) model, the robots are activated independently, and the duration of each COMPUTE, MOVE and inactivity is finite but unpredictable. As a consequence, the robots do not have a common notion of time, they can be seen while moving, and computations can be made based on obsolete observations. In the *semi-synchronous* (SSYNCH) model the activationsof robots can be logically divided into global rounds; in each round, one or both robots are activated, obtain the same snapshot, compute, and perform their move. It is assumed that the activation schedule is fair, i.e., each robot is activated infinitely often.

Depending on whether or not the adversary can stop a robot before it reaches its computed destination, the movements are called *non-rigid* and *rigid*, respectively. In the case of non-rigid movements, there exists a constant $\delta > 0$ such that if the destination point's distance is smaller than δ, the robot will reach it; otherwise, it will move towards it by at least δ. Note that, without this assumption, an adversary could make it impossible for any robot to ever reach its destination, following a classical Zenonian argument.

The two robots solve the *Rendezvous* problem if, within finite time, they move to the same point (not determined *a priori*) and do not move from there. A rendezvous algorithm for SSYNCH (resp., ASYNCH) is a protocol that allows the robots to solve the *Rendezvous* problem under any possible schedule in SSYNCH (resp., ASYNCH). A particular class of algorithms, denoted by $\mathcal{L}$, is that where each robot may only compute a destination point of the form $\lambda \cdot other.position$, for some $\lambda \in \mathbb{R}$ obtained as a function only of the light of which the robot is aware (i.e., its internal state in the FSTATE model, or the other robot's color in the FCOMM model). The algorithms of this class are of interest because they operate also when the coordinate system of a robot is not self-consistent (i.e., it can unpredictably rotate, change its scale or undergo a reflection).

3   Finite-State Robots

We fist consider FSTATE robots and we start by identifying a simple impossibility result for algorithms in $\mathcal{L}$.

Theorem 1. *In* SSYNCH, *Rendezvous of two* FSTATE *robots is unsolvable by algorithms in* $\mathcal{L}$, *regardless of the amount of their internal memory.*

Thus the computation of the destination must take into account more than just the lights (or states) of which the robot is aware.

The approach we use to circumvent this impossibility result is to have each robot use its own unit of distance as a computational tool; recall that the two robots might have different units, and they are not known to each other. We propose Algortihm 1 for *Rendezvous* in SSYNCH. Each robot has six internal states, namely S_{start}, S_1, S_2^{left}, S_2^{right}, S_3, and S_{finish}. Both robots are assumed to begin their execution in S_{start}. Each robot lies in the origin of its own local coordinate system and the two robots have no agreement on axes orientations or unit distance. Intuitively, the robots try to reach a configuration in which they both observe the other robot at distance not lower than 1 (their own unit). From this configuration, they attempt to meet in the midpoint. If they never meet because they arc never activated simultaneously, at some point one of them notices that its observed distance is lower than 1. This implies a breakdown of symmetry that enables the robots to finally gather. In order to reach the desired configuration in which they both observe a distance not lower than 1, the two robots first try to move farther away from each other if they are too close. If they are far enough, they memorize the side on which they see each other (left or right), and try to switch positions. If only one of them is activated, they gather; otherwise they detect a side switch and they can finally apply the above protocol. This is complicated by the fact that the robots may disagree on the distances they observe. To overcome this difficulty, they use their ability to detect a side switch to understand which distance their partner observed. If the desired configuration is not reached because of a disagreement, a breakdown of symmetry occurs, which is immediately exploited to gather anyway. As soon as the two robots coincide at the end of a cycle, they never move again, and *Rendezvous* is solved.

Theorem 2. *In* SSYNCH, *Rendezvous of two* FSTATE *robots is solvable with six internal states. This result holds even without unit distance agreement.*

4   Finite-Communication Robots

4.1   Asynchronous

It is not difficult to see that algorithms in $\mathcal{L}$ are not sufficient to solve the problem.

Theorem 3. *In* ASYNCH, *Rendezvous of two* FCOMM *robots is unsolvable by algorithms in* $\mathcal{L}$, *regardless of the amount of colors employed.*

Algorithm 1. Rendezvous for rigid SSYNCH with no unit distance agreement and six internal states

---

1: $dist \leftarrow \|other.position\|$
2: **if** $dist = 0$ **then**
3:     terminate
4: **if** $other.position.x > 0$ **then**
5:     $dir \leftarrow$ right
6: **else if** $other.position.x < 0$ **then**
7:     $dir \leftarrow$ left
8: **else if** $other.position.y > 0$ **then**   ▷ $other.position.x = 0$
9:     $dir \leftarrow$ right
10: **else**
11:     $dir \leftarrow$ left
12: **if** $me.state = S_{\text{start}}$ **then**
13:     **if** $dist < 1$ **then**
14:         $me.state \leftarrow S_1$
15:         $me.destination \leftarrow other.position \cdot (1 - 1/dist)$
16:     **else**
17:         $me.state \leftarrow S_2^{dir}$
18:         $me.destination \leftarrow other.position$
19: **else if** $me.state = S_1$ **then**
20:     **if** $dist \leqslant 1$ **then**
21:         $me.state \leftarrow S_{\text{finish}}$
22:         $me.destination \leftarrow (0,0)$
23:     **else**
24:         $me.state \leftarrow S_2^{dir}$
25:         $me.destination \leftarrow other.position$
26: **else if** $me.state = S_2^d$ **then**
27:     **if** $dir = d$ **then**
28:         $me.state \leftarrow S_{\text{finish}}$
29:         $me.destination \leftarrow other.position$
30:     **else if** $dist < 1/2$ **then**   ▷ side switch detected
31:         $me.state \leftarrow S_{\text{finish}}$
32:         $me.destination \leftarrow (0,0)$
33:     **else**
34:         $me.destination \leftarrow other.position/2$
35:         **if** $dist < 1$ **then**
36:             $me.state \leftarrow S_3$
37: **else if** $me.state = S_3$ **then**
38:     $me.state \leftarrow S_{\text{finish}}$
39:     **if** $dist < 1/4$ **then**
40:         $me.destination \leftarrow (0,0)$
41:     **else**   ▷ $1/4 \leqslant d < 1/2$
42:         $me.destination \leftarrow other.position$
43: **else**   ▷ $me.state = S_{\text{finish}}$
44:     **if** $dist \leqslant 1$ **then**
45:         $me.destination \leftarrow (0,0)$
46:     **else**
47:         $me.destination \leftarrow other.position$

We now describe an algorithm (which is not in $\mathcal{L}$) that solves the problem. Also this algorithm uses the local unit distance as a computational tool, but in a rather different way since a robot cannot remember and has to infer information by observing the other robot's light.

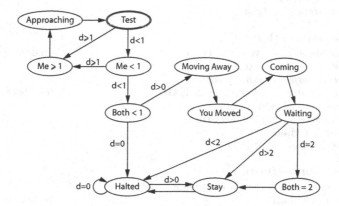

Fig. 1. State transitions in Algorithm 2

Intuitively, the two robots try to reach a configuration in which both robots see each other at distance lower than 1. To do so, they first communicate to the other whether or not the distance they observe is smaller than 1 (recall that they may disagree, because their unit distances may differ). If one robot acknowledges that its partner has observed a distance not smaller than 1, it reduces the distance by moving toward the midpoint.

The process goes on until both robots observe a distance smaller than 1. At this point, if they have not gathered yet, they try to compare their distance functions, in order to break symmetry. They move away from each other in such a way that their final distance is the sum of their respective unit distances. Before proceeding, they attempt to switch positions. If, due to asynchrony, they failed to be in the same state at any time before this step, they end up gathering. Instead, if their execution has been synchronous up to this point, they finally switch positions. Now, if the robots have not gathered yet, they know that their distance is actually the sum of their unit distances. Because each robot knows its own unit, they can tell if one of them is larger. If a robot has a smaller unit, it moves toward its partner, which waits.

Otherwise, if their units are equal, they apply a simple protocol: as soon as a robot wakes up, it moves toward the midpoint and orders its partner to stay still. If both robots do so, they gather in the middle. If one robot is delayed due to asynchrony, it acknowledges the order to stay still and tells the other robot to come.

Theorem 4. *In* ASYNCH, *Rendezvous of two* FCOMM *robots is solvable with 12 colors. This result holds even without unit distance agreement.*

Algorithm 2. Rendezvous for rigid ASYNCH with no unit distance agreement and 12 externally visible states

---

1: $dist \leftarrow \|other.position\|$
2: **if** $other.state = (\text{TEST})$ **then**   ▷ testing distances
3:     **if** $dist \geqslant 1$ **then**
4:         $me.state \leftarrow (\text{ME} \geqslant 1)$
5:     **else**
6:         $me.state \leftarrow (\text{ME} < 1)$
7: **else if** $other.state = (\text{ME} \geqslant 1)$ **then**   ▷ reducing distances
8:     $me.state \leftarrow (\text{APPROACHING})$
9:     $me.destination \leftarrow other.position/2$
10: **else if** $other.state = (\text{APPROACHING})$ **then**   ▷ test distances again
11:     $me.state \leftarrow (\text{TEST})$
12: **else if** $other.state = (\text{ME} < 1)$ **then**
13:     **if** $dist \geqslant 1$ **then**
14:         $me.state \leftarrow (\text{ME} \geqslant 1)$
15:     **else**
16:         $me.state \leftarrow (\text{BOTH} < 1)$
17: **else if** $other.state = (\text{BOTH} < 1)$ **then**
18:     **if** $dist = 0$ **then**   ▷ we have gathered
19:         $me.state \leftarrow (\text{HALTED})$
20:     **else**
21:         $me.state \leftarrow (\text{MOVING AWAY})$
22:         **if** $dist < 1$ **then**   ▷ moving away by $1 - dist/2$
23:             $me.destination \leftarrow other.position \cdot (^1\!/_2 - 1/dist)$
24: **else if** $other.state = (\text{MOVING AWAY})$ **then**
25:     $me.state \leftarrow (\text{YOU MOVED})$
26: **else if** $other.state = (\text{YOU MOVED})$ **then**
27:     $me.state \leftarrow (\text{COMING})$
28:     $me.destination \leftarrow other.position$
29: **else if** $other.state = (\text{COMING})$ **then**
30:     $me.state \leftarrow (\text{WAITING})$
31: **else if** $other.state = (\text{WAITING})$ **then**
32:     **if** $dist > 2$ **then**   ▷ my unit is smaller
33:         $me.state \leftarrow (\text{STAY})$
34:         $me.destination \leftarrow other.position$
35:     **else if** $dist = 2$ **then**   ▷ our units are equal
36:         $me.state \leftarrow (\text{BOTH} = 2)$
37:     **else**   ▷ my unit is bigger or we have gathered
38:         $me.state \leftarrow (\text{HALTED})$
39: **else if** $other.state = (\text{BOTH} = 2)$ **then**
40:     $me.state \leftarrow (\text{STAY})$
41:     **if** $dist = 2$ **then**   ▷ moving to the midpoint
42:         $me.destination \leftarrow other.position/2$
43: **else if** $other.state = (\text{STAY})$ **then**
44:     $me.state \leftarrow (\text{HALTED})$
45: **else**   ▷ $other.state = (\text{HALTED})$
46:     **if** $dist = 0$ **then**   ▷ we have gathered
47:         $me.state \leftarrow (\text{HALTED})$
48:         terminate
49:     **else**   ▷ maintain position while I come
50:         $me.state \leftarrow (\text{STAY})$
51:         $me.destination \leftarrow other.position$

Proof. We show that Algorithm 2, also depicted in Figure 1, correctly solves *Rendezvous*. Both robots start in state (TEST), and then update their state to (ME $\geqslant 1$) or (ME < 1), depending if they see each other at distance greater or lower than 1 (they may disagree, because their distance functions may be different). If robot r sees robot s set to (ME $\geqslant 1$), it starts approaching it by moving to the midpoint, in order to reduce the distance. No matter if r approaches s several times before s is activated, or both robots approach each other at different times, one of them eventually sees the other set to (APPROACHING). When this happens, their distance has reduced by at least a half, and at least one robot turns (TEST) again, thus repeating the test on the distances. At some point, both robots see each other at distance lower than 1 during a test, and at least one of them turns (BOTH < 1). If they have not gathered yet, they attempt to break symmetry by comparing their distance functions. To do so, when a robot sees the other set to (BOTH < 1), it turns (MOVING AWAY) and moves away by its own unit distance minus half their current distance. This move will be performed at most once by each robot, because if one robot sees the other robot still set to (BOTH < 1), but it observes a distance not lower than 1, then it knows that it has already moved away, and has to wait. When a robot sees its partner set to (MOVING AWAY), it shares this information by turning (YOU MOVED). If only one robot turns (YOU MOVED), while the other is still set to (MOVING AWAY), then the second robot turns (COMING) and reaches the other robot, which just turns (WAITING) and stays still until they gather. Otherwise, if both robots see each other set to (YOU MOVED), they both turn (COMING) and switch positions. At least one of them then turns (WAITING). Now, if a robot sees its partner set to (WAITING) and they have not gathered yet, it knows that their current distance is the sum of their unit distances. If such distance is greater than 2, then the robot knows that its partner's unit distance is bigger, and it moves toward it, while ordering it to stay still. Vice versa, if the distance observed is smaller than 2, the observing robot stays still and orders its partner to come. Finally, if the distance observed is exactly 2, the observing robot knows that the two distance functions are equal, and turns (BOTH $= 2$). In this case, a simple protocol allows them to meet. If a robot sees the other set to (BOTH $= 2$) at distance 2, it turns (STAY) and moves to the midpoint. If both robots do so, they eventually gather. Indeed, even if the first robot reaches the midpoint while the other is still set to (BOTH $= 2$), it now sees its partner at distance 1, and knows that it has to wait. On the other hand, whenever a robot sees its partner set to (STAY), it turns (HALTED), which tells its partner to reach it. This guarantees gathering even if only one robot attempts to move to thee midpoint.

4.2 Semi-Synchronous

In SSYNCH the situation is radically different from the ASYNCH case. In fact, it is possible to find a simple solution in $\mathcal{L}$ that uses the minimum number of colors possible, and operates correctly without unit distance agreement, starting from any arbitrary color configuration, and with interruptable movements (see Algorithm 3).

Algorithm 3. Rendezvous for non-rigid SSYNCH with three externally visible states

1: **if** *other.state* = A **then**
2:      *me.state* $\leftarrow$ B
3:      *me.destination* $\leftarrow$ *other.position*/2
4: **else if** *other.state* = B **then**
5:      *me.state* $\leftarrow$ C
6: **else**   ▷ *other.state* = C
7:      *me.state* $\leftarrow$ A
8:      *me.destination* $\leftarrow$ *other.position*

Theorem 5. *In* SSYNCH, *Rendezvous of two* FCOMM *robots is solvable by an algorithm in* $\mathcal{L}$ *with only three distinct colors. This result holds even if starting from an arbitrary color configuration, without unit distance agreement, and with non-rigid movements.*

Note that the number of colors used by the algorithm is optimal. This follows as a corollary of the impossibility result when lights are visible to both robots:

Lemma 1. *[20] In* SSYNCH, *Rendezvous of two robots with persistent memory visible by both of them is unsolvable by algorithms in* $\mathcal{L}$ *that use only two colors.*

5   Movements: Knowledge vs. Rigidity

In this section, we consider the *Rendezvous* problem when the movement of the robots can be interrupted by an adversary; previously, unless otherwise stated, we have considered rigid movements, i.e., in each cycle a robot reaches its computed destination point. Now, the only constraint on the adversary is that a robot, if interrupted before reaching its destination, moves by at least $\delta > 0$ (otherwise, rendezvous is clearly impossible). We prove that, for rendezvous with lights, knowledge of δ has the same power as rigidity of the movements. Note that knowing δ implies also that the robots can agree on a unit distance.

5.1   FState Robots

Both robots start in state A. If a robot sees its partner at distance lower than $\delta/2$, it moves in the opposite direction, to the point at distance $\delta/2$ from its partner. On the other hand, if the distance observed is not lower than δ, it moves toward the point located $\delta/4$ before the midpoint.

It is easy to see that after sufficiently many turns, the robots find themselves at a distance in the interval $[\delta/2, \delta)$, and both in state A. From now on, all their movements are rigid.

Theorem 6. *In non-rigid* SSYNCH, *Rendezvous of two* FSTATE *robots with knowledge of* δ *is solvable with three colors.*

5.2 FComm Robots

The idea of the Algorithm is simple. Both robots begin their execution in state START, and attempt to position themselves at a distance in the interval $(\delta, 2\delta]$. To do so, they adjust their position by moving by $\delta/2$ at each step. When a robot sees its partner at the desired distance, it turns READY and stops. It is easy to show that, even if its partner is still moving, it will end its move at a distance in the interval $(\delta, 2\delta]$. When a robot sees its partner set to READY, it turns COME and moves to the midpoint; the midpoint is eventually reached, because the distance traveled is not greater than δ.

We can conclude that:

Theorem 7. *In non-rigid* ASYNCH, *Rendezvous of two* FCOMM *robots with knowledge of δ is solvable with three colors.*

6 Open Problems

Our results, showing that rendezvous is possible in SSYNCH for FSTATE robots and in ASYNCH for FCOMM robots, seem to indicate that "it is better to communicate than to remember". However, determining the precise computational relationship between FSTATE and FCOMM is an open problem. To settle it, it must be determined whether or not it is possible for FSTATE robots to rendezvous in ASYNCH.

Although minimizing the amount of constant memory was not the primary focus of this paper, the number of states employed by our algorithms is rather small. An interesting research question is to determine the smallest amount of memory necessary for the robots to rendezvous when rendezvous is possible, and devise optimal solution protocols.

The knowledge of δ in non-rigid scenarios is quite powerful and allows for simple solutions. It is an open problem to study the *Rendezvous* problem for FSTATE and FCOMM robots when δ is unknown or not known precisely.

This paper has extended the classical models of oblivious silent robots into two directions: adding finite memory, and enabling finite communication. It thus opens the investigation in the FSTATE and FCOMM models of other classical robots problems (e.g., *Pattern Formation, Flocking,* etc.); an exception is *Gathering* because, as mentioned in the introduction, it is already solvable without persistent memory and without communication [3].

Acknowledgments. This work has been supported in part by NSERC, and by Prof. Flocchini's URC.

References

1. Agmon, N., Peleg, D.: Fault-tolerant gathering algorithms for autonomous mobile robots. SIAM Journal on Computing 36, 56–82 (2006)
2. Bouzid, Z., Das, S., Tixeuil, S.: Gathering of mobile robots tolerating multiple crash faults. In: Proceedings of 33rd IEEE International Conference on Distributed Computing Systems, ICDCS (2013)

3. Cieliebak, M., Flocchini, P., Prencipe, G., Santoro, N.: Distributed computing by mobile robots: Gathering. SIAM Journal on Computing 41(4), 829–879 (2012)
4. Cohen, R., Peleg, D.: Convergence properties of the gravitational algorithm in asynchronous robot systems. SIAM Journal on Computing 34, 1516–1528 (2005)
5. Cohen, R., Peleg, D.: Convergence of autonomous mobile robots with inaccurate sensors and movements. In: Durand, B., Thomas, W. (eds.) STACS 2006. LNCS, vol. 3884, pp. 549–560. Springer, Heidelberg (2006)
6. Das, S., Flocchini, P., Prencipe, G., Santoro, N., Yamashita, M.: The power of lights: synchronizing asynchronous robots using visible bits. In: Proceedings of the 32nd International Conference on Distributed Computing Systems (ICDCS), pp. 506–515 (2012)
7. Défago, X., Gradinariu, M., Messika, S., Raipin-Parvédy, P.: Fault-tolerant and self-stabilizing mobile robots gathering. In: Dolev, S. (ed.) DISC 2006. LNCS, vol. 4167, pp. 46–60. Springer, Heidelberg (2006)
8. Degener, B., Kempkes, B., Langner, T., Meyer auf der Heide, F., Pietrzyk, P., Wattenhofer, R.: A tight runtime bound for synchronous gathering of autonomous robots with limited visibility. In: Proceedings of 23rd ACM Symposium on Parallelism in Algorithms and Architectures (SPAA), pp. 139–148 (2011)
9. Dieudonné, Y., Petit, F.: Self-stabilizing gathering with strong multiplicity detection. Theoretical Computer Science 428(13) (2012)
10. Flocchini, P., Prencipe, G., Santoro, N.: Distributed Computing by Oblivious Mobile Robots. Morgan & Claypool (2012)
11. Flocchini, P., Prencipe, G., Santoro, N., Widmayer, P.: Gathering of asynchronous robots with limited visibility. Theo. Comp. Sci. 337(1-3), 147–168 (2005)
12. Izumi, T., Bouzid, Z., Tixeuil, S., Wada, K.: Brief Announcement: The BG-simulation for Byzantine mobile robots. In: Peleg, D. (ed.) Distributed Computing. LNCS, vol. 6950, pp. 330–331. Springer, Heidelberg (2011)
13. Izumi, T., Souissi, S., Katayama, Y., Inuzuka, N., Defago, X., Wada, K., Yamashita, M.: The gathering problem for two oblivious robots with unreliable compasses. SIAM Journal on Computing 41(1), 26–46 (2012)
14. Kamei, S., Lamani, A., Ooshita, F., Tixeuil, S.: Asynchronous mobile robot gathering from symmetric configurations without global multiplicity detection. In: Kosowski, A., Yamashita, M. (eds.) SIROCCO 2011. LNCS, vol. 6796, pp. 150–161. Springer, Heidelberg (2011)
15. Lin, J., Morse, A.S., Anderson, B.D.O.: The multi-agent rendezvous problem. parts 1 and 2. SIAM Journal on Control and Optimization 46(6), 2096–2147 (2007)
16. Pagli, L., Prencipe, G., Viglietta, G.: Getting close without touching. In: Even, G., Halldórsson, M.M. (eds.) SIROCCO 2012. LNCS, vol. 7355, pp. 315–326. Springer, Heidelberg (2012)
17. Prencipe, G.: Impossibility of gathering by a set of autonomous mobile robots. Theoretical Computer Science 384(2-3), 222–231 (2007)
18. Souissi, S., Défago, X., Yamashita, M.: Using eventually consistent compasses to gather memory-less mobile robots with limited visibility. ACM Transactions on Autonomous and Adaptive Systems 4(1), 1–27 (2009)
19. Suzuki, I., Yamashita, M.: Distributed anonymous mobile robots: Formation of geometric patterns. SIAM Journal on Computing 28, 1347–1363 (1999)
20. Viglietta, G.: Rendezvous of two robots with visible bits. Technical Report arXiv:1211.6039 (2012)

Pattern Formation by Mobile Robots with Limited Visibility*

Yukiko Yamauchi and Masafumi Yamashita

Graduate School of Information Science and Electrical Engineering,
Kyushu University, Japan
{yamauchi,mak}@inf.kyushu-u.ac.jp

Abstract. We investigate the pattern formation problem by mobile robots with *limited* visibility that can observe the positions of robots within limited distance. For robots with *unlimited* visibility, Fujinaga et al. (DISC 2012) showed that asynchronous oblivious robots have the same formation power as fully-synchronous non-oblivious robots, that is, starting from any initial configuration I, target pattern F is formable if and only if $\rho(I)$ divides $\rho(F)$ where $\rho(\cdot)$ is the geometric symmetricity. We first show that fully-synchronous oblivious robots with limited visibility cannot form F even when $\rho(I)$ divides $\rho(F)$. Hence, limited visibility substantially weakens the formation power of oblivious robots. Secondly, we show that despite limited visibility, semi-synchronous robots with rigid moves, and fully-synchronous robots with non-rigid moves have the same formation power as robots with unlimited visibility. Consequently, local memory is necessary and sufficient for these robots.

1 Introduction

A mobile robot system consists of a set of autonomous mobile robots each of which observes the locations of other robots (Look phase), computes its next location (Compute phase) and moves to the next location (Move phase). Each robot repeats "Look-Compute-Move" cycles locally without explicitly exchanging messages. One of the most important problems in self-organization of a mobile robot system is the *pattern formation problem*, that is, starting from an initial deployment of robots, form a given target pattern.

In this paper, we focus on pattern formation by mobile robots with *limited* visibility. While unlimited visibility allows a robot to observe the positions of all other robots irrespective of distance, limited visibility provides observation of the positions of robots within limited distance V. Existing pattern formation algorithms [5,7,8] show that unlimited visibility provides rich solutions to a mobile robot system. However, the formation power of robots with limited visibility has not been discussed yet.

* This work is supported in part by JSPS Grant-in-Aid for Scientific Research on Innovative Areas "Molecular Robotics" (No. 24104003, and No. 24104519), and JSPS KAKENHI (No. 22300004, No. 24650008, and No. 23700019).

T. Moscibroda and A.A. Rescigno (Eds.): SIROCCO 2013, LNCS 8179, pp. 201–212, 2013.
© Springer International Publishing Switzerland 2013

We model these robots by a set of points on a Euclidean space. The robots have very weak capabilities. Each robot does not have the access to the global coordinate system and uses its own *local coordinate system*. These robots are *anonymous* in the sense that they have no ID, and they execute the same algorithm. These robots are *oblivious* if the Compute phase depends only on the observation (i.e., Look phase) of the current cycle, otherwise *non-oblivious*. The executions of the Look-Compute-Move cycles are neither instantaneous nor synchronized. We call this synchronization model the *asynchronous* (ASYNC) model. Even in the asynchronous phase, the Look phase is assumed to be instantaneous, in the sense that it returns the locations of all robots at a time. Another stronger synchronization models are the *semisynchronous* (SSYNC) model and the *fully-synchronous* (FSYNC) model. In the SSYNC model, Look-Computer-Move cycles are instantaneous, and in the FSYNC model, all robots execute the i-th instantaneous cycle simultaneously.

Suzuki and Yamashita first investigated the pattern formation problem in SSYNC model and FSYNC model [7,8]. They characterized the class of patterns formable by mobile robots and showed the effect of local memory and synchronization. They showed that in the SSYNC model, the gathering problem for two oblivious robots is unsolvable, despite that it is trivially solvable for non-oblivious robots, which differentiates non-oblivious from oblivious robots. FSYNC robots have stronger formation power than SSYNC by definition, however, all patterns formable by non-oblivious FSYNC robots are also formable by oblivious SSYNC robots, except gathering of two robots. Later, Flocchini et al. [3] introduced the ASYNC model. Fujinaga, Yamauchi, Kijima, and Yamashita showed that oblivious ASYNC robots have the same formation power as non-oblivious FSYNC robots, except the gathering of two robots [5].

Let P be a set of distinct points. We assume that any point does not have the multiplicity, i.e., no two robots are located on the same point [1]. The *symmetricity* $\rho(P)$ of P is defined to be 1 if there is a point at the center of the smallest enclosing circle $c(P)$ of P. Otherwise, $\rho(P)$ is the number of angles θ between $(0, 2\pi]$ such that rotating P by θ results in P. Then, [7,8] showed that a target pattern F is formable if and only if $\rho(I)$ divides $\rho(F)$.

All these existing pattern formation algorithms [5,7,8] depend on unlimited visibility of each mobile robot. Based on the smallest enclosing circle of robots, [7,8] constructs a global coordinate system, and [5] embeds the target pattern and assign a robot to each point in the embedded target pattern by using the clock-wise matching [4]. However the pattern formation power of robots with limited visibility has not been discussed yet.

Several papers discussed gathering and convergence of robots with limited visibility as a first step towards self-organization of these robots. Ando, Oasa, Suzuki, and Yamashita investigated the point convergence problem of oblivious SSYNC robots with limited visibility [1]. Flocchini, Prencipe, Santoro, and Widmayer investigated the gathering problem in ASYNC robots with common

---

[1] In this paper, we assume that robots do not have the *multiplicity test capability*, in other words, they cannot count the number of robots at a point.

compass [2]. Pagli, Prencipe, and Viglietta proposed a gathering algorithm for the ASYNC model, that avoids collisions of robots under the assumption of a common compass [6].

Our main focus is the difference in formation power caused by limited visibility. We first show that oblivious FSYNC robots with limited visibility cannot form pattern F even when the symmetricity $\rho(I)$ of the initial configuration I divides $\rho(F)$. SSYNC robots and ASYNC robots have weaker formation power. Hence, limited visibility substantially weakens the pattern formation power of oblivious robots irrespective of asynchrony. Second, we show the formation power of non-oblivious robots, each of which can record the history of local views and outputs during execution. We first consider the robots with *rigid* moves, i.e., in the Move phase, robots move according to the algorithm without stopping on the way to the destination. We present a pattern formation algorithm that first gathers robots until they observe each other by using [1], and starts existing pattern formation algorithm for unlimited visibility [5,7,8]. Because the convergence phase may increase the symmtericity, we put a symmtericity control phase between the two phases, that decreases the symmtericity below $\rho(F)$. Rigid moves guarantees that robots can reconstruct their initial positions, hence, they can regain $\rho(I)$. Finally, we present a symmetricity control algorithm for non-oblivious FSYNC robots with *non-rigid* moves, that allows robots stop on the way to the destination. We will show that even when moves are non-rigid, the FSYNC robots can regain $\rho(I)$ with their entire local history during the convergence phase. Consequently, non-oblivious FSYNC robots with limited visibility have the same formation power as robots with unlimited visibility.

2 Robot Model and Pattern Formation

Robot System: Let $R = \{r_1, r_2, \ldots, r_n\}$ $(n \geq 3)$ be a set of anonymous robots in a Euclidean space. Each robot r_i does not have an identifier, and we use r_i just for description.

We consider discrete time $0, 1, 2, \ldots$. Let $p_i(t)$ (in the global coordinate system Z_0) be the position of r_i at time t $(r_i \in R)$. $P(t) = \{p_1(t), p_2(t), \ldots, p_n(t)\}$ is a configuration of robots at time t. The robots initially occupy distinct locations, i.e., $|P(0)| = n$. We denote the distance (in Z_0) between two points p, q by $dist(p, q)$.

The robots do not agree on the coordinate system, and each robot r_i has its *local coordinate system* $Z_i(t)$ such that the origin of $Z_i(t)$ is the position of r_i at time t (i.e., $\mathbf{0} = Z_i(t)[p_i(t)]$). However, we assume all local coordinate systems are right-handed. We denote by $Z_i(t)[P(t)]$ the set of points $P(t)$ observed in $Z_i(t)$. We assume that during a move phase, the origin of $Z_i(t)$ is fixed to the point of r_i in the look phase, and does not change.

Each robot can observe the positions of robots in distance V (in Z_0). Let $R_i(t)$ be the set of robots visible from r_i at time t, and $P_i(t) = \{p_j | r_j \in R_i(t)\}$. In a look phase, r_i obtains $S_i(t) = Z_i(t)[P_i(t)]$ at some time t in the look phase. We call $S_i(t)$ the *local view* of r_i at time t. Because the visibility range V is common to all robots, $r_i \in S_j(t)$ means $r_j \in S_i(t)$. We define an undirected graph $G(t) = (R, E(t))$ for $P(t)$, called *mutual visibility graph* at time t where $(r_i, r_j) \in E(t)$ if and only if $r_i \in S_j(t)$ [2]. We assume that $G(0)$ is connected. In the following, we use $P(t)$ and $G(t)$ ($S_i(t)$ and the subgraph of $G(t)$ induced by $R_i(t)$, respectively) interchangeably.

An algorithm is a function, say ψ, that returns a curve to the next location in the Euclidean space, and each robot moves along the curve. The move phase is *rigid* if r_i moves along ψ and to the endpoint of ψ. On the other hand, a *non-rigid* move phase may finish while r_i is still on the way to the next location. However, we assume that each robot moves at least δ (in Z_0), or reaches the next location if the length of the curve is shorter than δ. We consider non-rigid moves without any explicit explanation when it is clear from the context.

An execution is a sequence of configurations, $P(0), P(1), P(2), \ldots$. The execution is not uniquely determined even when it starts from a fixed initial configuration I. Rather, there are many possible executions depending on the distance that the robots move, the activation schedule of robots in SSYNC and ASYNC model, and the length of the move phase in ASYNC model. We denote the transition from $P(t)$ to $P(t+1)$ by $P(t) \to P(t+1)$.

Pattern Formation: A target pattern F is given to every robot r_i as a set of points $Z_0[F] = \{Z_0[p] : p \in F\}$. (Remember that r_i does not have access to the global coordinate system Z_0.) We assume that $|F| = n$. In the following, as long as it is clear from the context, we identify p with $Z_0[p]$, and write, e.g., "F is given to r_i", instead of "$Z_0[F]$ is given to r_i".

Let $\mathbb{T}$ be the set of all coordinate systems, which can be identified with the set of all transformations consisting of transformations, rotations, and uniform scalings. Let $\mathcal{P}_n$ be the set of all patterns of n points. For any $P, P' \in \mathcal{P}_n$, P is *similar* to P', if there exists $Z \in \mathbb{T}$ such that $Z[P] = P'$, denoted by $P \simeq P'$.

We say that algorithm ψ forms pattern $F \in \mathcal{P}_n$ from initial configuration I, if for any execution $P(0)(= I), P(1), \ldots$, there exists a time instant t such that $P(t') \simeq F$ for all $t' > t$.

For any $P \in \mathcal{P}_n$, let $C(P)$ be the smallest enclosing circle of P, and $c(P)$ be the center of $C(P)$. Formally, the *symmetricity* $\rho(P)$ of P is defined by

$$\rho(P) = \begin{cases} 1 & \text{if } c(P) \in P, \\ |\{Z \in \mathbb{T} : P = Z[P]\}| & \text{otherwise.} \end{cases}$$

We can also define $\rho(P)$ in the following way [7]: P can be divided into regular k-gons with co-center $c(P)$, and $\rho(P)$ is the maximum of such k. Here, any point is a regular 1-gon with an arbitrary center, and any pair of points $\{p, q\}$ is a regular 2-gon with its center $(p + q)/2$.

A point on the circumference of $C(P)$ is said to be "on circle $C(P)$". The radius of $C(P)$ is denoted by $r(P)$.

---

[2] Because V is common to all robots, $G(t)$ is a unit disk graph.

3   Pattern Formation by Oblivious Robots

Clearly, we have the following lemma even for the robots with limited visibility in the same way as robots with unlimited visibility [7].

Lemma 1. *Let* $F, I \in \mathcal{P}_n$ *for any* $n \geq 3$. *Then,* F *is not formable from any initial configuration* I *by oblivious FSYNC robots with limited visibility, if* $\rho(I) > \rho(F)$.

In the following, we will show Theorem 1.

Theorem 1. *Let* $F, I \in \mathcal{P}_n$ *for any* $n \geq 3$. *There exist infinitely many* F *such that for any pattern formation algorithm* ψ *for oblivious FSYNC robots with limited visibility, there exists an initial configuration* I_ψ *such that* F *is not formable from* I_ψ *even when* $\rho(I_\psi)$ *divides* $\rho(F)$.

Proof. We consider an adversary with arbitrarily small δ. Hence, each robot can be stopped almost every point on the output of compute phase.

Let ψ be an arbitrary pattern formation algorithm for oblivious FSYNC robots with limited visibility, that forms any target pattern F from an initial configuration I when $\rho(I)$ divides $\rho(F)$. Algorithm ψ at robot r_i at t is a function from F and the local view of r_i, say $S_i(t)$, to a curve, denoted by $\psi(F, S_i(t))$.

Fig. 1 (a) show an initial configuration I with $\rho(I) = 2$ and Fig. 1 (c) show a target pattern F with $\rho(F) = 2$ that have an execution in which the symmetricity becomes 4, and F is no more formable. The basic structure of I is the 16 black robots and 4 white robots connected by edges of length V, and they form a pattern with symmetricity 4. Once the mutual visibility graph is disconnected, there exists an execution where robots never form F. Hence, the black robots cannot move in I because any movement disconnects the mutual visibility graph unless their coordinate systems are agreed. On the other hand, the white robots can move. We add four white robots with two different local views S_A and S_B so that (i) $\rho(I) = 2$, (ii) they are not seen by the 16 robots outside their local view, and (iii) they move symmetric position for $c(I)$. From (ii), the 4 white robots move symmtrically without knowing the difference between S_A and S_B. Together with (ii), (iii) increases the symmetricity of the 24 robots to $4 > \rho(F)$. We will show the existence of such local views (S_A and S_B), and outputs of ψ ($\psi(F, S_A)$ and $\psi(F, S_B)$), that satisfies (iii).

Consider an initial configuration I' and a target pattern F shown in Fig. 1 (b) and (c). Algorithm ψ forms F from I' because $\rho(I') = \rho(F) = 2$. I' consists of 24 identical robots connected by edges of length V. Hence, algorithm ψ cannot move the black robots in I' because any movement can disconnect the mutual visibility graph. On the other hand, ψ should move at least one of the white robots, otherwise the configuration never changes. Hence, ψ outputs some curve when given a local view S_r in Fig. 1 (d). Let this output be $\psi(F, S_r)$.

Given local view of S_r, we consider a polar coordinate system whose origin is the other robot, say r_O as shown in Fig. 1 (d). Let the endpoint of the output curve of $\psi(F, S_r)$ be point $p = (x, \theta)$. Then, we have the following two cases.

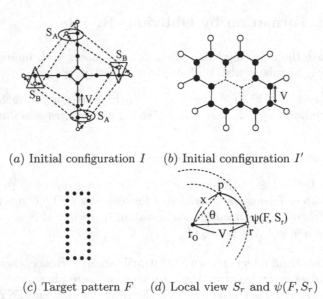

(a) Initial configuration I (b) Initial configuration I'

(c) Target pattern F (d) Local view S_r and $\psi(F, S_r)$

Fig. 1. Initial configurations and non-empty output of ψ

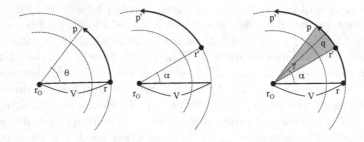

(a) Case 1   (b) Local view $S_{r'}$ (c) Common radial coordinate

Fig. 2. Case 1

Case 1: Curve rp is an arc of the circle centered at r_O (Fig. 2 (a)). Consider r' in Fig. 2 (b). The local view of r' is identical to S_r. Choose α so that curve rp and curve $r'p'$ has a common point q as shown in Fig. 2 (c). Now we use S_r (curve rq) as S_A, and $S_{r'}$ (curve $r'q$) as S_B. We can choose small enough α and β as shown in Fig. 3 (a) and (b) so that the white robots in S_A and S_B are not seen by other robots outside the local views.

Case 2: Otherwise. Consider local view $S_{r'}$ shown in Fig. 4 (a) where $x \leq y < V$. Let $r'p'$ be the output curve of $\psi(F, S_{r'})$. We have the following two cases. (Note that the following discussion holds for $\theta = 0$.)

Case 2(a): $\psi(F, S_{r'})$ is non-empty (Fig. 4 (a)). There exists at lease one point $q' = (z, \alpha)$ on curve $r'p'$ that satisfies $x \leq z < V$. The curve rp also contains a point q whose radial coordinate is z as shown in Fig. 4 (b). Hence, we use curve rq as $\psi(F, S_A)$ and curve $r'q'$ as $\psi(F, S_B)$.

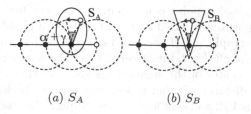

(a) S_A (b) S_B

Fig. 3. Two local views for Case 1

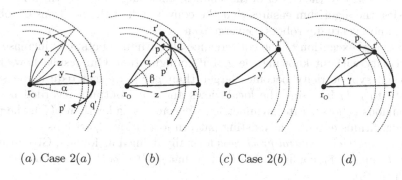

(a) Case 2(a)   (b)   (c) Case 2(b)   (d)

Fig. 4. Case 2

Case 2(b): $\psi(F, S_{r'})$ is empty (Fig. 4 (c)). In this case, we use the curve rr' as $\psi(F, S_A)$ and the empty curve as $\psi(F, S_B)$ Fig. 4 (d).

We can find at least one point r' that satisfies Case 2(a) with small enough α and β, or Case 2(b) with small enough γ so that the white robots in S_A and S_B are not seen by other robots in I. Consequently, we have the theorem for FSYNC model. We note that there exists infinitely many I' and F such that we can construct an initial configuration I from which F is not formable even when $\rho(I)$ divides $\rho(F)$.   □

4   Pattern Formation by Non-oblivious Robots

In this section, we show the following theorems.

Theorem 2. *Let $F, I \in \mathcal{P}_n$ for any $n \geq 3$. Then F is formable from any initial configuration I by non-oblivious FSYNC/SSYNC robots with limited visibility and rigid moves, if and only if $\rho(I)$ divides $\rho(F)$.*

Theorem 3. *Let $F, I \in \mathcal{P}_n$ for any $n \geq 3$. Then F is formable from any initial configuration I by non-oblivious FSYNC robots with limited visibility, if and only if $\rho(I)$ divides $\rho(F)$.*

We present pattern formation algorithms that consist of two phases. First, the algorithm make the robots converge until each robot observes all other robots. Because each robot knows F ($|F| = n$) and V, the robot can check the termination locally. After that, the algorithm starts pattern formation in the same way as the robots with unlimited visibility, for example, [3,7,5].

Ando et al. proposed a point convergence algorithm for oblivious ASYNC robots [1]. Their algorithm is based on the following simple idea: (i) robots in a connected visibility graph "get closer", and (ii) robots that are mutually visible at time t remain so for all $t' > t$. Intuitively, (i) is achieved by making each robot r_i move toward the center of the smallest enclosing circle of its local view. To predict the moving distance of other robots, each robot checks a common point of circles with radius $V/2$ and centered at the midpoint between r_i and r_j for all r_j in its local view. Then, (ii) is achieved by stopping r_i at the furthest common point on the way to the center of the smallest enclosing circle. By the above two properties, the algorithm ensures that the convex hull of the positions of robots shrinks, and eventually robots converge to a point.

During any execution of the convergence algorithm, robots may increase the symmetricity without knowing the global configuration. Our result shows that even when symmetricity becomes larger than $\rho(F)$, we can reduce the symmetricity to $\rho(I)$ by using the local views and local outputs recorded by the non-oblivious robots[3]. After symmetricity becomes smaller than $\rho(F)$, the proposed algorithms execute the existing pattern formation algorithms.

The *symmetricity control problem* is formally defined as follows: Given target pattern F and initial configuration P such that $\rho(F) < \rho(P)$, form some pattern F' such that $\rho(F') \leq \rho(F)$.

In the proposed symmetricity control algorithm, we use the sequence of local views recorded at each robot. For any execution $P(0), P(1), \ldots$, *local history* of robot r_i at time t is a sequence $H_i(t) = S_i(0), S_i(1), \ldots, S_i(t')$ where $S_i(j)$ is the local view at j-th activation of r_i ($0 \leq j \leq t'$). Note that in the FSYNC model $t' = t$ at each robot, while in the SSYNC or ASYNC model $t' \leq t$. Let $\varrho(P(t))$ be the symmetricity of configuration $P(t)$ considering the local history of robots. Let $H(t) = \{(p_i(t), Z_i(t)[h_i(t)]) : r_i \in R\}$.

$$\varrho(P(t)) = \begin{cases} 1 & \text{if } c(P(t)) \in P(t), \\ |\{Z \in \mathbb{T} : H(t) = Z[H(t)]\}| & \text{otherwise.} \end{cases}$$

Clearly, $\varrho(P(t)) \leq \rho(P(t))$. We can also define the partition of $P(t)$ into regular $\varrho(P(t))$-gons co-centered at $c(P(t))$ in the same way as (geometric) symmetricity.

We modify Ando's convergence algorithm and obtain the following convergence algorithm ψ_A.

1. Each robot r_i records $S_i(t) = Z_i(t)[P_i(t)]$ at each time t, and
2. Each robot r_i stops execution of Ando's algorithm when all robots can observe each other, i.e., $|S_i(t)| = n$ and the radius of the smallest enclosing circle of n robots is smaller than $V/2$.

Clearly, the FSYNC/SSYNC robots can agree on the termination of ψ_A.

Remember that adversary cannot change the local coordinate system of robots after execution starts. Because the output of ψ_A is a line segment, any movement

---

[3] We also note that the convergence algorithm [1] allows two or more robots occupy the same location. The proposed symmetricity control algorithm resolves multiplicity, however, we omit the detail for simplicity.

of r_i just translates the local coordinate system of r_i without any rotation. Hence, each robot can recognize the rotation between two consecutive outputs of ψ_A. We confirm the property of ψ_A as follows.

Property 1. Algorithm ψ_A has the following properties:

1. The output of ψ_A at robot r_i at time t depends only on $S_i(t)$.
2. The output of ψ_A at robot r_i is always a line segment.
3. Robot r_i knows the rotation angle of $\psi(S_i(t))$ from $\psi(S_i(t+1))$ for all $t > 0$.

4.1   Symmetricity Control with Rigid Moves

We start with FSYNC robots with rigid moves. From Property1 and rigid moves, after the termination of ψ_A, each robot can recognize its initial position by using its local history.

Consider an execution $P(0), P(1), P(2), \ldots$ of ψ_A where ψ_A terminates at time t. In $P = P(t)$, each robot can observe all other robots. By definition, we can partition P into regular $\rho(P)$-gons $P_1, P_2, \ldots, P_{n/\rho(P)}$ co-centered at $c(P)$. An important observation in [8] is that, all robots that observe P even in its local coordinate system, agree on an order of P_i's such that the distance of the points in P_i from $c(P)$ is no greater than that of P_{i+1}, and that each robot is conscious of the group P_i it belongs to. Note that there may exist some i, j such that $C(P_i) = C(P_j)$[4].

Now, each robot starts the execution of the symmetricity control algorithm ψ_B described in the following. Algorithm ψ_B first make $r(P_i) \neq r(P_j)$ for all $1 \leq i, j \leq n/\rho(P(t))$ by repeating the following procedure: If $C(P_i) = C(P_j)$ for some $1 \leq i < j \leq n/\rho(P(t))$, then the robots in $C(P_i)$ move toward $c(P)$ for distance $(r(P_i) - r(P_k))/2$ where P_k is the largest k in $C(P_i)$. If $C(P_i)$ contains more than three $\rho(P)$-gons, then the robots in $\rho(P)$-gon with the smallest index move inside. Let $P(t')$ be the first discrete time after t where $r(P_i) \neq r(P_j)$ for all $1 \leq i, j \leq \rho(P(t))$. All robots can recognize the termination of this phase. For simplicity, in the following, we assume that a terminal configuration of ψ_A satisfies the above condition.

At time t', each robot r_i translates its initial position $Z_i(t')[p_i(0)]$ in a polar coordinate system with origin $c(P(t'))$, unit distance $dist(c(P(t)), p_i)$, and polar axis $c(P(t'))p_i$ shown in Fig. 5 (a). We denote this polar coordinate system by $Z_i(c(P(t')), p_i(t'))$. Let $(\pi_i, \ell_i) = Z_i(c(P(t')), p_i(t'))[Z_i(t')[p_i(0)]]$. We have the following lemma for the translated initial positions of robots.

Lemma 2. *Let $P(0), P(1), \ldots$ be an execution of ψ_A where ψ_A terminates at $P(t) = P$. Assume $\rho(P) > \rho(P(0))$. Let $P_1, P_2, \ldots, P_{n/\rho(P)}$ be the partition of P into regular $\rho(P)$-gons co-centered at $c(P)$. Then, there exists at least one regular $\rho(P)$-gon P_i, where (π_j, ℓ_j) of all $r_j \in P_i$ are not identical.*

---

[4] Note that $C(P_i)$ is a circle centered at $c(P)$ and containing all robots in P_i, and $r(P_i)$ is the radius of $C(P_i)$.

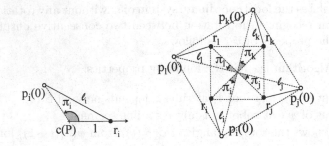

(a) Polar coordinate at r_i (b) Identical initial positions of P_i

Fig. 5. Initial position in polar coordinate system: black point is a robot, and white point connected by a solid line is its initial position

Proof. Assume that for every P_i ($1 \le i \le n/\rho(P)$), (π_j, ℓ_j) of all $r_j \in P_i$ are identical. Hence, the initial positions of robots in P_i are regular $\rho(P)$-gons co-centered at $c(P)$ as shown in Fig. 5 (b). Consequently, the initial positions of all robots are regular $\rho(P)$-gons co-centered at $c(P)$ and we have $\rho(P) = \rho(P(0))$, which is a contradiction. $\square$

Algorithm ψ_B make each robot $r_i \in P_j$ circulate on $C(P_j)$ into the clockwise direction to show π_i and ℓ_i. First, if the smallest enclosing circle $C(P_{n/\rho(P)})$ consists of only two robots, ψ_B moves the robots in $C(P_{n/\rho(P)-1})$ to the smallest enclosing circle to form a regular tetragon. After that, in the first time step, ψ_B circulates r_i in the clockwise direction for distance $(\pi r(P_j))/(|P_j|\ell_i)$, and in the next time step, ψ_B circulates r_i in the clockwise direction for distance $(\pi r(P_j))/(|P_j|\pi_i)$. (Note that $(\pi r(P_j))/|P_j|$ is the half length of the arc between neighboring robots in P_j.) From Property 2, in these two moves, one of the regular $\rho(P)$-gons becomes non-regular. Algorithm ψ_B repeats the above procedure until the symmetricity becomes smaller than $\rho(F)$.

We can directly apply ψ_B to SSYNC robots, because asynchrony just give chance to reduce symmetricity. Consequently, we have Theorem 2.

4.2   Symmetricity Control with Non-rigid Moves

A non-rigid move stops a robot on the way to the endpoint of the output curve of the Compute phase. Hence, the robot cannot obtain its local position from its local history. We will show that, even when moves are non-rigid, the symmetricity considering local histories is always smaller than $\rho(I)$.

Lemma 3. *For any execution $P(0)(= I), P(1), P(2), \dots$ of ψ_A of non-oblivious FSYNC robots with limited visibility, for any $t > 0$, $\varrho(P(t)) \le \varrho(P(0)) = \rho(I)$.*

Proof. Assume that there exists an execution $P(0), P(1), P(2), \dots$ that does not satisfy the lemma. Hence, there exists t such that $\forall t' > t$, $\varrho(C_{t'}) > \rho(C_0)$. Let t be the smallest such value. Hence, $\varrho(P(t)) \le \rho(I) < \varrho(P(t+1))$. Let

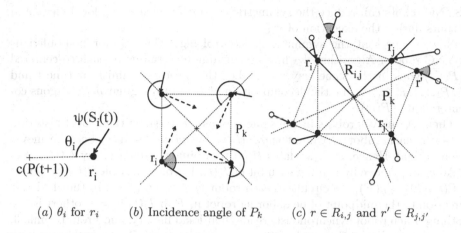

(a) θ_i for r_i (b) Incidence angle of P_k (c) $r \in R_{i,j}$ and $r' \in R_{j,j'}$

Fig. 6. Incidence angle and moving distance to form P_k

$c = c(P(t + 1))$ and $P_1, P_2, \ldots, P_{n/\varrho(P(t+1))}$ be the partition of $P(t + 1)$ into regular $\varrho(P(t + 1))$-gons co-centered at c. From definition, after t, the robots in P_i $(1 \le i \le n/\varrho(P(t + 1)))$ move globally symmetrically against c.

In $P(t + 1)$, for robot r_i, define the incidence angle θ_i of its move in $P(t) \to P(t + 1)$ as the clockwise rotation angle that overlaps r_i's movement to line $c\, p_i(t+1)$ in $P(t) \to P(t+1)$ (Fig. 6 (a)). Then, for each P_k, the incidence angle of $r_i, r_j \in P_k$ are identical. Otherwise, the rotation angle of $\psi_A(S_i(t + 1))$ from $\psi_A(S_i(t))$ is different from the rotation angle of $\psi_A(S_j(t + 1))$ from $\psi_A(S_j(t))$ (Fig. 6 (b)).

There exists at least one $\varrho(P(t + 1))$-gon P_k such that there exist $r_i, r_j \in P_k$ whose move distances in $P(t) \to P(t + 1)$ are different. Otherwise, all P_k's form regular $\varrho(P(t + 1))$-gon in $P(t)$ and $\varrho(P(t)) = \varrho(P(t + 1))$.

Once robots move closer enough to see each other, the robots in the same $\rho(P(t + 1))$-gon P_k $(0 \le k \le n/\rho(P(t + 1)))$ obtain a symmteric coordinate system as in Fig. 5 (a). In the following, we consider that local histories of these robots are observed by this coordinate system. Without loss of generality, we asseume that r_i and r_j occupy neighboring corners of P_k. Let $r_{j'}$ be the other neighbor of r_j on P_k. Let $R_{i,j}$ ($R_{i,j'}$, respectively) be the set of robots that are in the sector defined by r_i and r_j (r_j and $r_{j'}$, respectively) in $P(t)$ (Fig. 6 (c)). $R_{i,j}$ is not empty and the robots in $R_{i,j}$ connect r_i to r_j, otherwise, the mutual visibility graph is disconnected. This also holds for $R_{i,j'}$. Each robot $r \in R_{i,j}$ has a corresponding robot r' in $R_{i,j'}$ that forms the same $\rho(P(t + 1))$-gon in $P(t+1)$. $R_{i,j}$ also defines r''s incidence angle. The incidence angles of robots in $R_{i,j'}$ are completely determined by $R_{i,j}$. However, it is impossible to construct the same local views for all robots in $R_{i,j'}$ that coincide with incidence angles determined by $R_{i,j}$, because the moving distance of r_i is different from that of r_j. Hence, there exists no such transition $P(t) \to P(t + 1)$. $\qquad\square$

Let $P(0), P(1), P(2), \ldots$ be an execution of ψ_A of FSYNC robots where the convergence terminates in $P(t')$. From Lemma 3, when $\rho(P(t)) > \rho(F)$, the

FSYNC robots can reduce the symmetricity to $\rho(I)$ by showing local views and outputs during the execution of ψ_A.

Now, we present the symmetricity control algorithm ψ_C for non-oblivious FSYNC robots. Algorithm ψ_C first assigns different radius to each co-centered $\rho(P(t'))$-gons in the same way as ψ_B. Let this procedure finish at time t and $P_1, P_2, \ldots, P_{n/\rho(P(t))}$ be the decomposition of $P(t)$ into regular $\rho(P(t))$-gons co-centered at $c(P(t))$.

Then, ψ_C moves robots in the same way as ψ_B to show their local histories. However, each robot r_i cannot show $H_i(t)$ by move distance, because moves are non-rigid. Hence, ψ_C translates $H_i(t)$ to a binary sequence, say $T_i(t)[\cdot]$, and make r_i stop/move by reading each bit of $T_i(t)$. In configuration $P(t+k)$, unless $\rho(P(t+k)) \leq \rho(F)$, ψ_C circulates each robot $r_i \in P_j$ on $C(P_j)$ in the clockwise direction to the midpoint of neighboring robot in P_j if $T_i(t)[k] = 1$, otherwise do nothing. Each robot synchronously reads their local history, and From Lemma 3, it is guaranteed that they find a difference in finite time. Consequently, we have Theorem 3.

5   Concluding Remark

We investigated the pattern formation problem by mobile robots with limited visibility. The results show that limited visibility weakens the formation power of oblivious robots, however, does nothing for non-oblivious FSYNC robots. Non-oblivious SSYNC robots with rigid moves still have the same power as unlimited visibility robots.

References

1. Ando, H., Oasa, Y., Suzuki, I., Yamashita, M.: A distributed memoryless point convergence algorithm for mobile robots with limited visibility. IEEE Trans. on Robotics and Automata 15(5), 818–828 (1999)
2. Flocchini, P., Prencipe, G., Santoro, N., Widmayer, P.: Gathering of asynchronous robots with limited visibility. Theor. Comput. Sci. 337, 147–168 (2005)
3. Flocchini, P., Prencipe, G., Santoro, N., Widmayer, P.: Arbitrary pattern formation by asynchronous, anonymous, oblivious robots. Theor. Comput. Sci. 407, 412–447 (2008)
4. Fujinaga, N., Ono, H., Kijima, S., Yamashita, M.: Pattern formation through optimum matching by oblivious CORDA robots. In: Lu, C., Masuzawa, T., Mosbah, M. (eds.) OPODIS 2010. LNCS, vol. 6490, pp. 1–15. Springer, Heidelberg (2010)
5. Fujinaga, N., Yamauchi, Y., Kijima, S., Yamashita, M.: Asynchronous pattern formation by anonymous oblivious mobile robots. In: Aguilera, M.K. (ed.) DISC 2012. LNCS, vol. 7611, pp. 312–325. Springer, Heidelberg (2012)
6. Pagli, L., Prencipe, G., Viglietta, G.: Getting close without touching. In: Even, G., Halldórsson, M.M. (eds.) SIROCCO 2012. LNCS, vol. 7355, pp. 315–326. Springer, Heidelberg (2012)
7. Suzuki, I., Yamashita, M.: Distributed anonymous mobile robots: Formation of geometric patterns. SIAM J. on Comput. 28(4), 1347–1363 (1999)
8. Yamashita, M., Suzuki, I.: Characterizing geometric patterns formable by oblivious anonymous mobile robots. Theor. Comput. Sci. 411, 2433–2453 (2010)

Optimal Gathering of Oblivious Robots in Anonymous Graphs*

Gabriele Di Stefano[1] and Alfredo Navarra[2]

[1] Dipartimento di Ingegneria e Scienze dell'Informazione e Matematica,
Università degli Studi dell'Aquila, Italy
gabriele.distefano@univaq.it
[2] Dipartimento di Matematica e Informatica,
Università degli Studi di Perugia, Italy
alfredo.navarra@unipg.it

Abstract. The paper presents general results about the gathering problem on graphs. A team of robots placed at the vertices of a graph, have to meet at some vertex and remain there. Robots operate in Look-Compute-Move cycles; in one cycle, a robot perceives the current configuration in terms of robots disposal (Look), decides whether to move towards one of its neighbors (Compute), and in the positive case makes the computed move (Move). Cycles are performed asynchronously for each robot.

So far, the goal has been to provide feasible resolution algorithms with respect to different assumptions about the capabilities of the robots as well as the topology of the underlying graph. In this paper, we are interested in studying the quality of the resolution algorithms in terms of the minimum number of asynchronous moves performed by the robots.

We provide results for general graphs that suggest resolution techniques and provide feasibility properties. Then, we apply the obtained theory on specific topologies like trees and rings. The resulting algorithms for trees and rings are then compared with the existing ones, hence showing how the old solutions can be far apart from the optimum.

1 Introduction

The gathering task in robot based computing systems represents one of the most fundamental problems widely considered in the literature. The basic requirement of the problem is to devise a distributed algorithm that allows a team of robots to meet at some common place. Different assumptions on the capabilities of the robots as well as on the environment where they move, lead to very different scenarios (see [7,16] for a survey).

In this paper, we are interested in robots placed on the vertices of a graph. Robots are equipped with visibility sensors and motion actuators, and operate in *Look-Compute-Move* cycles (see, e.g. [10]). The Look-Compute-Move model

---

* Work supported by the Research Grant 2010N5K7EB 'PRIN 2010' ARS Techno-Media (Algoritmica per le Reti Sociali Tecno-mediate)' from the Italian Ministry of University and Research.

T. Moscibroda and A.A. Rescigno (Eds.): SIROCCO 2013, LNCS 8179, pp. 213–224, 2013.

assumes that in each cycle a robot takes a snapshot of the current global configuration (Look), then, based on the perceived configuration, takes a decision to
stay idle or to move to one of its adjacent vertices (Compute), and in the latter
case it moves to this neighbor (Move). Cycles are performed asynchronously, i.e.,
the time between Look, Compute, and Move operations is finite but unbounded,
and it is decided by an adversary for each robot. Hence, robots may move based
on outdated perceptions. In fact, due to asynchrony, by the time a robot takes
a snapshot of the configuration, this might have drastically changed. The scheduler determining the Look-Compute-Move cycles timing is assumed to be fair,
that is, each robot performs its cycle within finite time and infinitely often.

So far, the problem has been focused on the feasibility of resolution algorithms
for various initial configurations.A main distinction has been considered for the
graph topology, by letting robots move on rings [6,11,14], grids [4,8], or trees [7].
Another crucial property concerns robots' capabilities. Robots are assumed to
be oblivious (without memory of the past), uniform (running the same deterministic algorithm), autonomous (without a common coordinate system, identities
or chirality), asynchronous (without central coordination), without the capability to communicate. Neither vertices nor edges are labeled (i.e., the graph is
anonymous) and no local memory is available on vertices.

An important capability associated to robots concerns the so called *multiplicity detection* (see, e.g. [17]). During the Look phase, a robot may perceive
whether a vertex is occupied by more than one robot in different ways. Here
we assume the so called *global strong* multiplicity detection [2], meaning that
robots perceive the actual number of robots among all the vertices. The *global
weak* form considers robots able to detect only whether a vertex is occupied by
more than one robot, but not the exact number. The *local* versions instead of
global refer to the corresponding ability of a robot in perceiving the information
about multiplicities only concerning the vertex where it currently resides.

In this paper, the aim is to provide a general theory, valid for any input
graph and any configuration. Based on this theory, we want to devise resolution
algorithms that also minimize the number of asynchronous moves performed by
the robots in order to accomplish the gathering task, when possible. In particular,
our algorithms will be based on the concept of *Weber-point* [15] on weighted
graphs [9]. A Weber-point for a discrete set of sample points in the Euclidean
space is the point minimizing the sum of distances to the sample points. On
graphs with robots, there might occur more than one Weber-point. These are
all the vertices of the graph that minimize the sum of the shortest paths from
each robot toward each of such vertices.

If an algorithm is able to assure the gathering on a Weber-point by letting
move robots along the shortest paths towards such a vertex, we talk about *exact*
algorithm. This is clearly optimal with respect to the number of asynchronous
moves performed by the robots. However, sometimes it is just not possible to
gather the robots on a Weber-point, but still the algorithm might be optimal.

If from the one hand our assumption on the multiplicity detection refers to
the global strong version, that is the most powerful option, on the other hand

we do not make any assumption concerning the initial disposal of the robots on the graph (see, e.g. [11,13]), that is, initial configurations may also contain multiplicities. Moreover, it is possible to show configurations where exact gathering cannot be accomplished without the global strong assumption.

2   Definitions

In this section we provide all the basic definitions and notation necessary for the understanding of the proposed results.

A simple undirected graph $G = (V, E)$, with vertex set V and edge set E, will represent the topology where robots are placed on. A function $\ell : V \longrightarrow \mathbb{N}$, represents the number of robots on each vertex of G, and we call (G, ℓ) a *configuration* whenever $\sum_{v \in V} \ell(v)$ is bounded and greater than zero. A configuration is *final* if all the robots are on a single vertex u (i.e., $\ell(u) > 0$ and $\ell(v) = 0$, $\forall v \in V \setminus \{u\}$). The distance $d(u, v)$ between two vertices u, v in V is the number of edges of a shortest path connecting u to v.

Two graphs $G = (V_G, E_G)$ and $H = (V_H, E_H)$ are *isomorphic* if there is a bijection φ from V_G to V_H such that $uv \in E_G$ if and only if $\varphi(u)\varphi(v) \in E_H$.

An *automorphism* on a graph G is an isomorphism from G to itself, that is, a permutation of the vertices of G that maps edges to edges and non-edges to non-edges. The set of all automorphisms of G forms a group called *automorphism group* of G and denoted by $\mathrm{Aut}(G)$.

We extend the concept of isomorphism to configurations in a natural way: two configurations (G, ℓ) and (G', ℓ') are isomorphic if G and G' are isomorphic via a bijection φ and for each vertex v in G, $\ell(v) = \ell'(\varphi(v))$. An *automorphism* on a configuration (G, ℓ) is an isomorphism from (G, ℓ) to itself and the set of all automorphisms of (G, ℓ) forms a group that we call *automorphism group* of (G, ℓ), denoted by $\mathrm{Aut}((G, \ell))$.

Given an isomorphism $\varphi \in \mathrm{Aut}((G, \ell))$, the *cyclic subgroup* of order k generated by φ is given by $\{\varphi^0, \varphi^1 = \varphi, \varphi^2 = \varphi \circ \varphi, \ldots, \varphi^k\}$ where φ^0 is the identity.

If H is a subgroup of $\mathrm{Aut}((G, \ell))$, the *orbit* of a vertex v of G is $Hv = \{\gamma(v) \mid \gamma \in H\}$.

If $|\mathrm{Aut}(G)| = 1$, that is, G admits only the identity automorphism, then G is said *asymmetric*, otherwise it is said *symmetric*. Analogously, if $|\mathrm{Aut}((G, \ell))| = 1$, we say that (G, ℓ) is *asymmetric*, otherwise it is *symmetric*.

Definition 1. *Given a configuration (G, ℓ), with $G = (V, E)$, the centrality of each $v \in V$, is $c_{G, \ell}(v) = \sum_{u \in V} d(u, v) \cdot \ell(u)$.*

A vertex $v \in V$ is a Weber-point if it has the minimal centrality, that is, $c_{G, \ell}(v) = \min\{c_{G, \ell}(u) \mid u \in V\}$.

Whenever clear by the context, we refer to the centrality of a vertex v by $c_G(v)$, $c_\ell(v)$, or simply $c(v)$. By definition, a Weber-point is a vertex that has the overall minimal distance from all the robots in the configuration. Then, an algorithm that gathers all the robots on a Weber-point via shortest paths is optimum w.r.t. the total number of moves. More formally, a gathering algorithm must define the sequence of moves for each robot, leading to a final configuration. A move is the change of the position of a single robot from a vertex u to an adjacent vertex v. This equals to change the configuration from, say (G, ℓ) to (G, ℓ'), where $\ell(w) = \ell'(w) \; \forall w \in V \setminus \{u, v\}$, $\ell'(u) = \ell(u) - 1$ and $\ell'(v) = \ell(v) + 1$. A robot perceives its position on the graph G if (G, ℓ) is asymmetric. Whereas, if G admits a non-identity isomorphism φ, a robot cannot distinguish its position at u from $\varphi(u)$. As a consequence, two robots (e.g., one on u and one on $\varphi(u)$) can decide to move simultaneously, as any algorithm is unable to distinguish between them. This fact greatly increases the difficulty to devise a gathering algorithm for symmetric configurations.

We say that an algorithm *assures* the gathering if it achieves the gathering regardless any possible sequence of the moves it allows, and possible simultaneous moves. We propose to measure the efficiency of a gathering algorithm by counting the number of moves that it requires to gather all the robots from an arbitrary initial configuration to a single vertex. We say that an algorithm is *optimal* if it requires the minimum possible number of moves. We say that an algorithm is *exact* if it achieves the gathering with a number of moves equal to the centrality of a Weber-point in the initial configuration. Of course, this is a lower bound for each algorithm. As we will see, not all the optimal gathering algorithms can be exact. We say that an algorithm is r-approximate if r is the ratio between the moves it requires and the moves required by an optimal algorithm.

3   General Graphs

In this section, we provide the core of the paper by means of general results that allow to define optimal and exact gathering algorithms, or to recognize when this is not possible. Actually, the gathering problem is so characterized:

Proposition 1. *Gathering is achieved on a configuration $((V, E), \ell)$ if and only if there exists a $v \in V$ such that $c(v) = 0$.*

Proof. If gathering is achieved, all the robots are on a vertex v. Then $c(v) = 0$ as $\ell(v') = 0$ for all the vertices $v' \neq v$, and $d(v, v) = 0$ when $\ell(v) \neq 0$. On the other hand, by definition of $c(v)$, $d(u, v) \cdot \ell(u)$ must be equal to zero for all the vertices $u \neq v$. This implies that $\ell(u) = 0$ for all the vertices $u \neq v$, and hence all the robots are on v.   □

Along the text, we say that a robot on a vertex u *moves towards a vertex* v if it moves to a vertex adjacent to u along a shortest path between u and v.

Theorem 1. *Given a configuration $((V, E), \ell)$ with Weber-points in $X \subseteq V$, a move of a robot towards a Weber-point x gives rise to a configuration $((V, E), \ell')$ with Weber-points in $X' \subseteq V$ such that:*

   1. $c_{\ell'}(v) = c_\ell(v) - 1$ for each $v \in X'$;
   2. $x \in X'$;
   3. $X' \subseteq X$.

Proof. If a robot moves from w to w', then, for each $v \in V$, $c_{\ell'}(v) = \sum_{u \in V \setminus \{w,w'\}}$ $d(u,v) \cdot \ell(u)+ \ d(w,v) \cdot \ell'(w)+ \ d(w',v) \cdot \ell'(w') \ = \ \sum_{u \in V \setminus \{w,w'\}} d(u,v) \cdot \ell(u)+$ $d(w,v) \cdot (\ell(w) - 1)+ d(w',v) \cdot (\ell(w') + 1) = c_\ell(v)- \ d(w,v)+ \ d(w',v)$.

Being w' adjacent to w, for each $v \in V$, $c_\ell(v) - 1 \leq c_{\ell'}(v) \leq c_\ell(v) + 1$ and $c_{\ell'}(v) = c_\ell(v) - 1$ if w' is on the shortest path from w to v (that is the robot on w moves towards v). This immediately proves points 1 and 2. For point 3, it is sufficient to note that no vertex can became a Weber-point if it was not as such before the move, since the variation of its centrality is of at most one unit. $\square$

When the configuration admits a unique Weber-point, the above theorem suggests an exact gathering algorithm that also exploits concurrency among robots. In fact, regardless other robots, each one can move towards the only Weber-point via the shortest path, until finalizing the gathering.

Corollary 1. *If a configuration admits only one Weber-point then the gathering can be achieved by an exact algorithm.*

Other situations where the optimal gathering can be achieved are stated by the next corollary.

Corollary 2. *Given a configuration $((V, E), \ell)$, if there exists a real function $f : V \ \rightarrow \mathbb{R}^+$ and f admits only one minimum on the set of Weber-points, then gathering can be achieved by an exact algorithm.*

On the negative side, the next theorem provides a sufficient condition for a configuration to be not gatherable, but we first need the following definition:

Definition 2. *Let $C = ((V, E), \ell)$ be a configuration. An isomorphism $\varphi \in Aut(C)$ is called* partitive *on $V' \subseteq V$ if the cyclic subgroup $H = \{\varphi^0, \varphi^1 = \varphi, \varphi^2 = \varphi \circ \varphi, \ldots, \varphi^k\}$ generated by φ has order $k > 1$ and is such that $|Hu| = k$ for each $u \in V'$.*

Note that, in the above definition, the orbits Hu, for each $u \in V'$ form a partition of V'. The associated equivalence relation is defined by saying that x and y are equivalent if and only if there exists a $\gamma \in H$ with $\gamma(x) = y$. The orbits are then the equivalence classes under this relation; two elements x and y are equivalent if and only if their orbits are the same; i.e., $Hx = Hy$. Moreover, note that $\ell(u) = \ell(v)$ whenever u and v are equivalent.

Theorem 2. *Let $C = ((V, E), \ell)$ be a non-final configuration. If there exists $\varphi \in Aut(C)$ partitive on V then C is not gatherable.*

Proof. Assume there exists an algorithm $\mathcal{A}$ that starting from configuration C reaches a final configuration F, where all robots lie on vertex x.

If a robot on a vertex v moves to w by following algorithm $\mathcal{A}$, a new configuration $C' = ((V, E), \ell')$ is reached. There are two cases depending on whether v and w are equivalent (i.e., $Hv = Hw$) or not.

Assuming that v and w are equivalent, then there exists $i \leq k$ such that $\varphi^i(v) = w$. Let $\gamma = \varphi^i$. Note that, for each h, $\gamma^h(v)\gamma^{h+1}(w) \in E$. Then, there

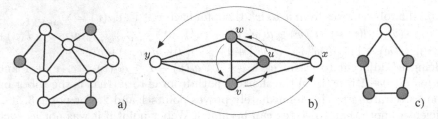

Fig. 1. A gray vertex indicates the presence of one robot. a) Configuration admitting a partitive isomorphism: the sets of the partition are the three central vertices, the vertices with robots, and the three remaining vertices. b) Configuration admitting a non-partitive isomorphism that maps v in u, u in w, w in v, x in y and y in x. c) Configuration admitting a non-partitive isomorphism with two sets of the partition with size two, and one with size one.

exists a cycle $(v = \gamma^0(v), w = \gamma^1(v), \gamma^2(v), \ldots, \gamma^j(v))$ of equivalent vertices. Now, if the algorithm moves a robot from v to w, it cannot avoid that a robot on $\gamma^h(v)$ moves to $\gamma^{h+1}(v)$, for each $1 \le h \le j$. This makes unchanged $\ell(u)$ for each $u \in Hu$. It follows that $\varphi \in \text{Aut}(C')$, that is, it is still a partitive isomorphism in the new configuration.

On the other hand, if $Hv \neq Hw$, the algorithm cannot avoid that one robot on $\varphi^h(v)$ moves to $\varphi^h(w)$, for each $0 \le h \le |Hv|$. As $|Hv| = |Hw|$, all the vertices $\varphi^h(w)$, $0 \le h \le |Hv|$, are distinct, and then $\ell'(u) = \ell(u) - 1$ for each $u \in Hv$, and $\ell'(u) = \ell(u) + 1$ for each $u \in Hw$. This implies that φ is still a partitive isomorphism in $\text{Aut}(C')$.

Hence, for any move decided by $\mathcal{A}$, a new configuration can arise such that φ is still an isomorphism for it. But this contradicts $\varphi \notin \text{Aut}(F)$. In fact, $|Hx| = 1$ since vertex x (where the gathering is finalized) cannot be mapped to a different vertex, whereas $|Hx|$ should be greater than one as φ is partitive. $\square$

In Figure 1a, it is shown a partitive configuration where each vertex belongs to an orbit of size three. By the above theorem we deduce that the gathering cannot be assured, since each move allowed by an algorithm can be executed synchronously by all the three robots due to an adversary. This would always produce a new partitive configuration. Figure 1b, shows a configuration admitting an isomorphism which is not partitive. In this case the gathering is possible even though not the exact one. In fact, each of the three occupied vertices are Weber-points, but moving from one to another may produce the same configuration if the three robots move concurrently in the same direction. Hence, a gathering algorithm can move the three robots towards the two empty vertices. Once all the three robots have moved, a multiplicity is created. The multiplicity either contains all the robots or just two. In the first case the gathering has been accomplished. In the second case, the gathering is finalized by letting move the single robot towards the multiplicity. Finally, Figure 1c shows a configuration admitting a non-partitive isomorphism but the gathering cannot be assured as shown in [12]. It follows that, there exist configurations not admitting partitive isomorphisms but still not gatherable.

It is worth noting how most of the configurations proved to be not gatherable for the rings [13], trees [7], and grids [4] fall into the hypothesis of Theorem 2. Considering the ring case [13], for instance, *periodic* configurations (i.e., invariant with respect to not full rotations), or configurations admitting an *edge-edge symmetry* (i.e., invariant to reflection on an even ring) are not gatherable.

The next theorem suggests the gathering point in some circumstances.

Theorem 3. *Given a configuration $C = ((V, E), \ell)$, if there exists an automorphism $\varphi \in Aut(C)$ that is partitive on $V \setminus \{v\}$, with $l(v) = 0$, then, any gathering algorithm can not assure the gathering on a vertex $u \neq v$.*

The above theorem implies that some configurations can be gathered only at some predetermined vertices, regardless they are Weber-points or not. Hence, in such cases the optimality of the provided solutions cannot be measured with respect to the minimum distances of the robots towards Weber-points.

In the next sections, we show how the obtained results can be applied and extended with respect to specific topologies like trees and rings.

4   Trees

In this section, we characterize the gathering on tree topologies. We provide a general algorithm that always achieves the exact gathering starting from configurations not falling into the hypothesis of Theorem 2. To this aim, we exploit interesting properties resulting from the tree topology.

Let (T, ℓ) be a configuration for a tree T, and a and b two of its vertices. We denote by P_{ab} the path between a and b of length $|P_{ab}|$. Tree T can be decomposed into three subtrees: The one containing a when removing from T the edge incident to a in P_{ab}, and denoted by T_a; The one containing b when removing the edge incident to b in P_{ab}, and denoted by T_b; And the third one obtained from T by removing both T_a and T_b, and denoted by T_{ab}. Let $L_a = \sum_{v \in T_a} \ell(v)$ and $L_b = \sum_{v \in T_b} \ell(v)$, that is the number of robots in T_a and T_b, respectively.

Theorem 4. *Let (T, ℓ) be a configuration for a tree T. Given two distinct Weber-points a and b, T_{ab} does not contain any robots.*

Corollary 3. *Let (T, ℓ) be a configuration for a tree T. Given two distinct Weber-points a and b, $L_a = L_b$.*

Theorem 5. *Let (T, ℓ) be a configuration for a tree T. The Weber-points form a path.*

Proof. If (T, ℓ) admits only one or two adjacent Weber-points, then the claim holds. Otherwise, let a and b be two non-adjacent Weber-points. Let α be a vertex between a and b in P_{ab}, and p_α be its distance from a. By Corollary 3, it follows that $c_T(\alpha) = c_{T_a}(a) + L_a p_\alpha + c_{T_b}(b) + L_b(|P_{ab}| - p_\alpha) = c_{T_a}(a) + c_{T_b}(b) + L_b|P_{ab}| = c_T(a)$, and then all the vertices in P_{ab} are Weber-points.

To show that all the Weber-points lie on a path, we prove that any subgraph $K_{1,3}$ (that is a star graph of four vertices) of T cannot consist of all Weber-points. By contradiction let us assume that such a subgraph exists. Let a, b, c be the three vertices connected to the center of the star. By Theorem 4, there are no robots on T_{ab}, T_{bc}, and T_{ac}, respectively. This would imply that T is empty. $\square$

Theorem 6. *Given a configuration (T, ℓ), if the number of robots is odd, then there exists only one Weber-point.*

The above theorem, along with Corollary 1, implies the existence of a simple exact gathering algorithm when the number of robots is odd. A complete characterization about the existence of exact gathering algorithms on trees is given by the next theorem. It shows that an exact algorithm exists unless there is an automorphism that maps each vertex to a different one. This is a lighter condition with respect to the case given in Theorem 2.

Theorem 7. *Let $C = (T, \ell)$ be a configuration for a tree $T = (V, E)$. There exists an exact gathering algorithm for C if and only if for each $\varphi \in Aut(C)$ there exists $v \in V$ such that $\varphi(v) = v$.*

Proof. ($\Rightarrow$) Assume there is a $\varphi \in Aut(C)$ such that $\varphi(v) \neq v$ for each $v \in V$. We show that in this case there is no gathering algorithm for C.

A center is a vertex of a graph having a minimal eccentricity, that is the greatest distance to any other vertex. A tree can have only one or two centers. In the latter case the two centers are joined by an edge (see, e.g., [18]).

If T has one center x, any automorphism must map x to itself, a contradiction. Then T has two centers x and y such that $\varphi(x) = y$ and $\varphi(y) = x$. Moreover, if we consider the two subtrees T_x and T_y rooted in x and y, respectively, and obtained from T by removing edge xy, each child of x (y resp.) must be mapped to a child of y (x, resp.) and then, recursively, each vertex in T_x (T_y resp.) must be mapped to a vertex of T_y (T_x, resp.). Hence the two trees are isomorphic. The isomorphism that maps each vertex in T_x to a vertex in T_y and viceversa gives rise to orbits each of size two, then it is partitive. Hence, by Theorem 2, no gathering algorithm exists for C.

($\Leftarrow$) From Theorem 5, let P_{ab} be the path of Weber-points and let us suppose that T has one center x. There must be a vertex v in P_{ab} nearest to x (possibly, x coincides with v). By Corollary 2, it follows that all the robots can move towards v via the shortest paths, and eventually finalizing the exact gathering.

Let us suppose now that T has two centers joined by edge xy. If xy is in T_a (T_b, respectively) then, by Corollary 2, all the robots can move towards a (b, resp.), the gathering point. If xy is in T_{ab} but either x or y is not in P_{ab}, there must be a vertex w in P_{ab} nearest to xy (possibly, w coincides either with x or with y). Again by by Corollary 2, all the robots can move towards w and gather there. Finally, let us assume that xy is an edge of P_{ab}. As, by hypothesis, for each $\varphi \in Aut(C)$ there exists $v \in V$ such that $\varphi(v) = v$, then the two subtrees T_x and T_y obtained by removing xy cannot be isomorphic. Then it is always possible to determine which tree between T_x and T_y is less than the other with

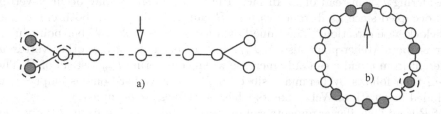

Fig. 2. A gray vertex indicates the presence of one robot; Dashed circled vertices are Weber-points. Dashed line stands for an undefined sequence of empty vertices. Vertices pointed by an arrow represent the gathering vertices with respect to algorithms in [7] for a), and in [13] for b).

respect to a natural ordering on labeled trees (see, e.g. [1,3]), where the label of a vertex is given by function ℓ, and each tree is represented by the string obtained by reading the labels from the root downwards to the leaves. Without loss of generality, assuming T_x greater than T_y, all the robots in T_x can move towards x. In this way, after each move, T_x remains always greater than T_y. Once all the robots in T_x are at x, they move to y. As soon as one robot moves from x to y, the path of Weber-points will be P_{yb}, xy is not in P_{yb} and we can proceed as before: all the robots can move towards y and gather there.   □

The gathering algorithm provided by the above theorem exploits similar properties of that presented in [7]. However, this new version accomplishes the exact gathering while the old one always gather robots on a center of the underlying tree. Considering Figure 2a, it is easy to provide configurations where our algorithm performs the gathering in two moves, while the old algorithm requires n moves, i.e., it is a $\frac{n}{2}$-approximate algorithm.

5   Rings

In this section, we characterize the gathering on ring topologies. Before providing a gathering algorithm for achieving exact gathering starting from the same settings of [13], we provide some useful properties concerning the disposal of Weber-points on rings.

Lemma 1. *Given a configuration (P, ℓ) where P is a path graph, the set of Weber-points is constituted either by one occupied vertex, or by one subpath whose extremes are occupied.*

Theorem 8. *Given a configuration (R, ℓ) on a ring R, if an empty vertex u is a Weber-point then also its neighbors are Weber-points.*

Proof. Let v be a neighbor of u. If R is even, let xy be the antipodal edge of uv, with x being closer to u than to v. Let P_{xy} be the path obtained from R by removing edge xy. All the distances from the robots to v do not change when considering P_{xy} instead of R. In fact, the only ambiguity may occur if vertex x is occupied since its distance to v in R can be evaluated in both clock- and anticlock-wise directions. The same holds for u. Since v is a Weber-point in R, it must be a Weber-point also in P_{xy}, as the centrality of any other vertex w different from u and v can only increase when considering P_{xy} instead of R. The claim then follows by Lemma 1 since on P_{xy}, if a Weber-point is empty, it is included in a path of Weber-points whose extremes are occupied.

If R is odd, similar arguments can be applied by removing from R the vertex x antipodal to uv. This can be done since $d(x, u) = d(x, v)$ and hence the contribution of x to $c(u)$ and $c(v)$ is the same.   □

By the above theorem, as for the path case, if there exist a sequence of vertices that are Weber-points, then the extremes of such a sequence are Weber-points occupied by robots. It is worth noting that on rings there might occur more than one of such sequences.

As further application of our theory, we now provide an exact algorithm for asymmetric configurations on rings[1], in the same settings as in [13], that is, starting from configurations without multiplicities, and assuming the global weak multiplicity detection.

Theorem 9. *Given an asymmetric configuration (R, ℓ) without multiplicities on a ring R of n vertices, it is always possible to accomplish the exact gathering even with the global weak multiplicity detection.*

Proof. In order to prove the theorem we need to specify what a robot perceives during the Look phase. Depending on the direction it looks at vertices of the ring, two possible sequences may arise: $(\ell(v_0), \ell(v_1), \ldots, \ell(v_{n-1}))$ and $(\ell(v_0), \ell(v_{n-1}), \ldots, \ell(v_1))$, with v_0 being the vertex where it resides.

From Theorem 8, there exists in R at least one Weber-point occupied by a robot. We consider two different cases: either there are only isolated Weber-points (i.e., no two Weber-points are adjacent) or not.

In the first case, as the configuration is asymmetric, among all the possible Weber-points which are occupied by robots, there must be one whose view represents the lexicographical maximum among all the views. Let r be the robot occupying such a Weber-point, and without loss of generality, let $(\ell(v_0), \ell(v_1), \ldots, \ell(v_{n-1}))$ be its maximum view.

Let r' be a robot on v_i, where $i > 0$ is the smallest index such that $\ell(v_i) = 1$. The gathering algorithm makes move r' towards r. By Theorem 1, v_0 remains a Weber-point. Moreover the view of r remains maximum as it has been increased. This is repeated until a multiplicity is created on v_0.

In the second case, the algorithm considers the robots on Weber-points with views such that v_1 or v_{n-1} is a Weber-point. If both v_1 and v_{n-1} are Weber-points, the robot chooses the maximum view, otherwise it chooses the view in the direction of the adjacent Weber-point. Let r be the robot with the maximal view w.r.t. these constraints, and r' be the first robot seen by r according to its view. By Theorem 8, r and r' determine a path P of Weber-points.

---

[1] Asymmetric configurations are referred as *rigid* configurations in [13].

Consider the two views of r and r', respectively, in the opposite direction with respect to P. The algorithm moves r (r', resp.) towards r' (r resp.) if its view is lexicographically bigger than that of r' (r, resp.). In doing so, the path P is shortened and will be again selected in the subsequent steps as r has increased its maximum view. As above, this is repeated until a multiplicity is created.

In both cases, the algorithm finalizes the exact gathering by moving, at turn, the closest robot to the multiplicity towards it along a shortest path.   □

The proposed algorithm differs from that in [13] simply in the choice of the vertex where a multiplicity is created. Once this is done, the two algorithms proceed in the same way. The algorithm proposed in [13] considers the longest interval I of empty vertices. Among the two intervals of empty vertices neighboring to I, the shortest one was reduced by moving the robot limiting I. Ties were broken by the asymmetry of the configuration. The described move was repeated until creating a multiplicity.

In Figure 2b, it is shown a configuration where our algorithm requires 25 moves while the algorithm in [13] takes 35 moves. It is easy to provide worsen instances where I is far apart from Weber-points, hence resulting in a much larger difference with our algorithm in terms of computed moves.

Concerning the symmetric configurations studied in [5,6], is it worth noting that the gathering vertices selected by those algorithms turn out to be the right choice according to Theorem 3. However, such algorithms are not optimal as the performed moves are not always along the shortest paths towards the gathering vertex. The design of optimal algorithms for symmetric configurations with either global strong or other types of multiplicity detection remains an open problem.

6 Conclusion

We have studied the gathering problem under the Look-Compute-Move cycle model with the global strong multiplicity detection assumption. A new theory holding for general graphs has been devised, and a characterization for pursuing optimal gathering in terms of computed moves has been addressed. We have also compared the obtained results with existing gathering algorithms on tree and ring topologies. This study is a first step towards the optimization of the gathering task in terms of computed moves. It opens a wide research area for reconsidering previous strategies with respect to the new twofold objective function that requires not only to accomplish the gathering task but also the minimum number of moves.

References

1. Aho, A., Hopcroft, J., Ullman, J.: Data Structures and Algorithms. Addison-Wesley (1983)
2. Bouzid, Z., Das, S., Tixeuil, S.: Gathering of Mobile Robots Tolerating Multiple Crash Faults. In: Proceedings of the The 33rd IEEE International Conference on Distributed Computing Systems, ICDCS (2013)

3. Buss, S.: Alogtime algorithms for tree isomorphism, comparison, and canonization. In: Gottlob, G., Leitsch, A., Mundici, D. (eds.) KGC 1997. LNCS, vol. 1289, pp. 18–33. Springer, Heidelberg (1997)
4. D'Angelo, G., Di Stefano, G., Klasing, R., Navarra, A.: Gathering of robots on anonymous grids without multiplicity detection. In: Even, G., Halldórsson, M.M. (eds.) SIROCCO 2012. LNCS, vol. 7355, pp. 327–338. Springer, Heidelberg (2012)
5. D'Angelo, G., Di Stefano, G., Navarra, A.: Gathering of six robots on anonymous symmetric rings. In: Kosowski, A., Yamashita, M. (eds.) SIROCCO 2011. LNCS, vol. 6796, pp. 174–185. Springer, Heidelberg (2011)
6. D'Angelo, G., Di Stefano, G., Navarra, A.: How to gather asynchronous oblivious robots on anonymous rings. In: Aguilera, M.K. (ed.) DISC 2012. LNCS, vol. 7611, pp. 326–340. Springer, Heidelberg (2012)
7. D'Angelo, G., Di Stefano, G., Navarra, A.: Gathering asynchronous and oblivious robots on basic graph topologies under the look-compute-move model. In: Alpern, S., Fokkink, R., Gasieniec, L., Lindelauf, R., Subrahmanian, V. (eds.) Search Theory: A Game Theoretic Perspective, pp. 197–222. Springer (2013)
8. Devismes, S., Lamani, A., Petit, F., Raymond, P., Tixeuil, S.: Optimal grid exploration by asynchronous oblivious robots. In: Richa, A.W., Scheideler, C. (eds.) SSS 2012. LNCS, vol. 7596, pp. 64–76. Springer, Heidelberg (2012)
9. Eiselt, H.A., Marianov, V. (eds.): Foundations of Location Analysis. International Series in Operations Research & Management Science, vol. 155. Springer (2011)
10. Flocchini, P., Prencipe, G., Santoro, N.: Distributed Computing by Oblivious Mobile Robots. In: Synthesis Lectures on Distributed Computing Theory. Morgan & Claypool Publishers (2012)
11. Kamei, S., Lamani, A., Ooshita, F., Tixeuil, S.: Asynchronous mobile robot gathering from symmetric configurations without global multiplicity detection. In: Kosowski, A., Yamashita, M. (eds.) SIROCCO 2011. LNCS, vol. 6796, pp. 150–161. Springer, Heidelberg (2011)
12. Klasing, R., Kosowski, A., Navarra, A.: Taking advantage of symmetries: Gathering of many asynchronous oblivious robots on a ring. Theoretical Computer Science 411, 3235–3246 (2010)
13. Klasing, R., Markou, E., Pelc, A.: Gathering asynchronous oblivious mobile robots in a ring. Theoretical Computer Science 390, 27–39 (2008)
14. Kranakis, E., Krizanc, D., Markou, E.: The Mobile Agent Rendezvous Problem in the Ring. Morgan & Claypool (2010)
15. Kupitz, Y., Martini, H.: Geometric aspects of the generalized Fermat-Torricelli problem. Number 6 in Intuitive Geometry. Bolyai Society Math Studies (1997)
16. Pelc, A.: Deterministic rendezvous in networks: A comprehensive survey. Networks 59(3), 331–347 (2012)
17. Prencipe, G.: Impossibility of gathering by a set of autonomous mobile robots. Theoretical Computer Science 384, 222–231 (2007)
18. Santoro, N.: Design and Analysis of Distributed Algorithms. John Wiley & Sons (2007)

Probabilistic Connectivity Threshold
for Directional Antenna Widths
(Extended Abstract)

Hadassa Daltrophe, Shlomi Dolev*, and Zvi Lotker

Ben-Gurion University, Beer-Sheva, 84105, Israel
{hd,dolev}@cs.bgu.ac.il,
zvilo@bgu.ac.il

Abstract. Consider the task of maintaining connectivity in a wireless network where the network nodes are equipped with directional antennas. Nodes correspond to points on the unit disk and each uses a directional antenna covering a sector of a given angle α, where the orientation of the sector is either random or not.

The width required for a connectivity problem is to find out the necessary and sufficient conditions of α that guarantee connectivity when an antenna's location is uniformly distributed and the direction is either random or not.

We prove basic and fundamental results about this (reformulated) problem. We show that when the number of network nodes is big enough, the required $\check{\alpha}$ approaches zero. Specifically, on the unit disk it holds with high probability that the threshold for connectivity is $\check{\alpha} = \Theta(\sqrt[4]{\frac{\log n}{n}})$. This is shown by the use of Poisson approximation and geometrical considerations. Moreover, when the model is relaxed to allow orientation towards the center of the area, we demonstrate that $\check{\alpha} = \Theta(\frac{\log n}{n})$ is a necessary and sufficient condition.

Keywords: Wireless networks; Directional antennas; Connectivity threshold.

1 Introduction

Communication among wireless devices is of great interest in the scope of current wireless technology, where devices are part of sensor networks, mobile ad-hoc networks and RFID devices that take part in the emerging ubiquitous computing, and even satellite networks. These communication networks are usually extremely dynamic, where devices frequently join and leave (or crash) and, therefore, require probabilistic techniques and analysis. Imagine, for example, sensor networks that use directed antennas (saving expensive energy and increasing communication capacity) among the sensors

---

* Partially supported by a Russian-Israeli grant from the Israeli Ministry of Science and Technology and the Russian Foundation for Basic Research, the Rita Altura Trust Chair in Computer Sciences, the Lynne and William Frankel Center for Computer Sciences, Israel Science Foundation (grant number 428/11), Cabarnit Cyber Security MAGNET Consortium, Grant from the Institute for Future Defense Technologies Research named for the Medvedi of the Technion, MAFAT, and Israeli Internet Association.

T. Moscibroda and A.A. Rescigno (Eds.): SIROCCO 2013, LNCS 8179, pp. 225–236, 2013.

that should be connected even though they are deployed by an airplane that drops them from the air (just as in a smart dust scenario). What is the density of those sensors needed to ensure their connectivity? Is there a way to renew connectivity after some portion of the sensors stops functioning- maybe by deploying only an additional fraction, uniformly distributed in the area with random orientation of the antennas? In this work, we try for the first time to suggest and analyze ways to ensure connectivity in such probabilistic scenarios. Namely, we have studied the problem of arranging randomly scattered wireless sensor antennas in a way that guarantees the connectivity of the induced communication graph. The main challenge here is to minimize energy consumption while preserving node connectivity.

In order to save power, increase transmission capacity and reduce interference [11], antennas should communicate along a wedge-shaped area, that is, an angular and practically infinite section of a certain angle α whose apex is the antenna.

The smaller the angle is, the better it is in terms of energy saving. When knowing nothing about the future positioning of the antennas, each antenna may be directed to a random direction that may stay fixed forever. Therefore, we wish to find the minimum $\alpha > 0$ so that no matter what finite set of locations the antennas are given, with high probability they can communicate with each other. Our goal is to specify necessary and sufficient conditions for the width of wireless antennas that enable one to build a connected communication network when antennas locations and directions are randomly and uniformly chosen.

Throughout this paper, we relate to an undirected graph, where two antennas are connected by an edge if and only if each lies in each other's wedge. However, our calculations hold for the directed case as well. Specifically, Theorem 1 hold for both cases, and the result proven by Theorem 2 also implies a connectivity threshold for the directed graph case.

Previous results that handle wireless directional networks [5,1] assume coordinated locations and orientations for the antennas. They show that a connected network can be built with antennas of width $\alpha = \pi/3$. The same model's assumptions were used by [2] to study graph connectivity in the presence of interference and in [10] to optimize the transmission range as well as the hop-stretch factor of the communication network. A different model of a *directed* graph of directional antennas of bounded transmission range was studied in [4,7].

In contrast to the above worst case approaches, to the best of our knowledge, we consider for the first time the connectivity problem in a probabilistic perspective. Namely, we are interested in the minimal communication angle that implies high probability for the graph to be connected as a function of the number of nodes. This approach significantly reduces the required communication angle and is more general in the sense that it also allows omission of the use of a directing procedure. A particular example of a system for which our results can be applied is a MIMO (multiple input multiple output) based system [15]. Wireless sensor networks using MIMO wireless links have recently emerged as one of the most significant technical breakthroughs in modern communications [3]. Our results imply that the use of *multiple* antenna elements (lobes) may improve the transmission width by a polynomial factor. The MIMO technique can ensure that a directed to the center antenna exists; hence, we reduce the square factor of the gap between Theorems 2 and Theorem 1, namely, the connectivity threshold is $\Theta\left(\frac{\log n}{n}\right)$ instead of $\Theta\left(\sqrt[4]{\frac{\log n}{n}}\right)$.

The probabilistic setting of the problem is related to other research in the field of continuum percolation theory [12]. The model for the points here is a Poisson point process, and the focus is on the existence of a connected component under different models of connections. For example, [16] studied the number of neighbors that implies connectivity. Works [14,8] are focused on the minimal number r such that two points are connected if and only if their metric distance is $\leq r$. In [9] the authors generalized the results in [14,8] and proved that for a fractal in $\mathbb{R}^d$, it holds with high probability that $r \approx (\frac{\log n}{n})^{1/d}$, where $\approx$ means that the quantity is bounded between two absolute constants.

Our main results (summarized in Theorems 1 and 2) handle two different models. The first is related to the case where all the antennas are directed to one reference point (specifically, we used the center of a disk). The second model generalizes the results by dealing with randomly chosen locations and directions with no prior knowledge. Assuming that the number of nodes is big enough, we show in both cases that with high probability, the threshold $\check{\alpha}$ approaches zero.

We obtain the following results that we believe are important for both their combinatorial and computational geometric perspectives and in their implications in the design of wireless networks.

Theorem 1. $\alpha = \Theta(\frac{\log n}{n})$ *is necessary and sufficient for the (asymptotical) connectivity of n nodes that choose their transmission direction to the center. Specifically, there are two constants $0 < c_1 < c_2$ such that:*

$$\lim_{n \to \infty} \Pr\left[G(n, c_1 \log n/n) \text{ is disconnected}\right] = 1 \text{ and}$$

$$\lim_{n \to \infty} \Pr\left[G(n, c_2 \log n/n) \text{ is connected}\right] = 1$$

Theorem 2. $\alpha = \Theta(\sqrt[4]{\frac{\log n}{n}})$ *is necessary and sufficient for the (asymptotical) connectivity of n nodes that choose their transmission direction uniformly. Specifically, there are two constants $0 < c_1 < c_2$ such that:*

$$\lim_{n \to \infty} \Pr\left[G\left(n, c_1 \sqrt[4]{\log n/n}\right) \text{ is disconnected}\right] = 1 \text{ and}$$

$$\lim_{n \to \infty} \Pr\left[G\left(n, c_2 \sqrt[4]{\log n/n}\right) \text{ is connected}\right] = 1$$

Remark 1. From the calculation in the proofs of Theorems 1 and 2, one can choose any $c_1 < 1/2$ and $c_2 > \pi^2$.

Remark 2. The connectivity threshold in the random direction settings (Theorem 2) holds for an undirected graph. We also prove that for a directed graph, $\alpha = \Theta(\sqrt[3]{\frac{\log n}{n}})$ is the necessary and sufficient threshold for graph connectivity, though *not* necessarily for graph strong connectivity.

Remark 3. Due to space restrictions some of the proofs are omitted from this extended abstract. The full proofs and can be found in [6].

2 Preliminaries

Let $P = \{p_1(x,y,\theta), ..., p_n(x,y,\theta)\}$ be a set of n *points*
(or *nodes*). The point location (x,y) is chosen independently from the uniform distribution over the unit
disk $\mathbb{D}$ in the plane (or over the unit disk boundary).
The antenna direction θ of each point is chosen independently and uniformly over $[0, \pi]$. Each point represents a communication station by fixing two opposite
wedges of angle α with direction θ at each node (see
Figure 1).

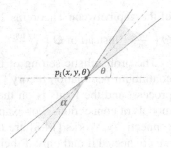

Fig. 1. Fixing at each point p_i two opposite wedges of angle α with direction θ

Definition 1 (The communication graph). *Given nodes u and v with communication
angle α, we say that u sees v if u lies in the wedge (on the intercepted arc) of v. The
communication graph $G(P,E,\alpha)$ is an undirected graph that consists of the node set
P, its communication angle α and the set of edges $E = \{(u,v) | u \text{ sees } v \text{ and } v \text{ sees } u\}$.
(Since E is straightly defined by $P(x,y,\theta)$ and α, we omit E in the sequel from the
graph notation, i.e., $G(P,E,\alpha) = G(P,\alpha)$).*

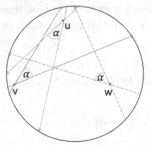 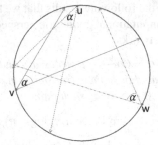

Fig. 2. The communication graph $G = (V,E) = (\{u,v,w\}, \{(u,v)\})$ on the disk (on the left) and
on the boundary (on the right). Note that G is not connected.

Definition 2 (Connectivity threshold problem). *Given a set P and its induced graph
G, our goal is to find the critical angle$\equiv \breve{\alpha}$, such that G is connected with probability
$1 - o(1)$ as n tends to infinity iff $\alpha \geq \breve{\alpha}$.*

Notations

- The *intercepted arc* of a point u with angle α is the part of a circle that lies between
  the two rays of α that intersect it.
- Let $arc(u)$ be the intercepted *arc* of u and let $|arc(u)|$ denote its length.
- Let $wedge(u)$ be the wedge *area* of u and let $|wedge(u)|$ denote its area.
- Let $*$ be an equivalence relation such that for a point $u \in P$, u^* is the antipodal point
  of u (i.e., u and u^* are opposite through the center).
- Throughout, we use the term w.h.p. as a shortcut for "with probability $1 - o(1)$ as
  n tends to infinity."
- Following we sometimes use the term "random point" instead of "random variable."

- Let $\mathbb{D}$ denote the unit disk (in $\mathbb{R}^2$) and let $\partial\mathbb{D}$ denote its boundary.
- Let $\partial\mathbb{D}^2$ be the space of pairs $\{u, u^*\}$ of antipodal points in the unit disk boundary $\partial\mathbb{D}$.
- Let $X = \{X_1, ..., X_n\}$ be a set of uniform random variables defined over $\mathbb{D}$.
- Let $Y = \{Y_1, ..., Y_n\}$ be a set of uniform random variables defined over $\partial\mathbb{D}$.
- Let $Z = \{Z_1, ..., Z_n\}$ be a set of uniform random variables defined over $\partial\mathbb{D}^2$.
- Let $(Y_1, R_1), ..., (Y_n, R_n)$ be pairs of uniform random variables defined over $\partial\mathbb{D}^2 \times [0, 1]$.
- Let $(Z_1, \chi_1), ..., (Z_n, \chi_n)$ be pairs of uniform random variables defined over $\partial\mathbb{D}^2 \times \{0, 1\}$.

Poisson Distribution and Approximation

Throughout this paper, the wireless network is modeled as nodes located randomly on the plane according to a Poisson point process. We use standard tools from continuum percolation and refer the reader to [12,13] for a general introduction of the topic. The discrete Poisson approximation is a random process that yields random points in $\mathbb{R}^d$ with density λ. See [13] Chapter 5 for a precise definition.

Definition 3. *The number of points in the set P of points in a unit disk is a random variable with distribution $Pr(|P| = n) = \frac{e^{-\lambda}\lambda^n}{n!}$.*

This follows immediately from the definition of a Poisson process.

The connectivity threshold problem can be translated into the mathematical framework of "balls and bins." We have n balls that are thrown independently and uniformly at random into m bins. The distribution of the number of balls in a given bin is approximately a Poisson variable with a density parameter $\lambda = n/m$. By the "coupon collector" principle, we get that the number of balls that need to be thrown until all bins have at least one ball w.h.p. is $m \log m$ (see Theorem 5.13 at [13]). In the sequel, we will use these results in a variety of settings.

With relation to the connectivity problem, the "balls" represent the set P of nodes distributed over the disk (the disk's boundary), and the "bins" are slices of the disk area (boundary) defined by the wedge area (by the intercepted arc) of the nodes (note that the bins in this setting are not disjoint, and we will refer to this later).

Let us call the scenario in which the number of balls in the bins are taken to be independent Poisson random variables with mean $\lambda = m/n$ the *Poisson case*, and the scenario where m balls are thrown into n bins independently and uniformly at random the *exact case*. We justify the use of Poisson approximation instead of calculating the exact case by using the following Theorem (which is given by Corollary 5.11 at [13]):

Theorem 3. *Let Λ be an event whose probability is either monotonically increasing or monotonically decreasing in the number of balls. If Λ has probability p in the Poisson case, then Λ has probability at most $2p$ in the exact case.*

Throughout this paper, we use Poisson distribution as well as uniform distribution since in all cases the probability is monotone (decreasing). N denotes the random Poisson variable with the parameter $\lambda = f(n)$, and n denotes the exact number of points in the uniform case.

Following, we also use the Chernoff bound for a sum of Poisson trials (a.k.a. Bernoulli trials) (Theorem 4.5 in [13]):

Theorem 4. *Let $X_1,...,X_n$ be independent Poisson trials such that $\Pr(X_i) = p_i$. Let $X = \sum_{i=1}^{n} X_i$ and $E(X) = \mu$. Then, for $0 < \delta < 1$:*

$$\Pr(X \le (1-\delta)\mu) \le e^{-\mu\delta^2/2}$$

Definition 4 (Covering problem). *We would like to find the minimal communication angle $\bar{\alpha}$, such that the wedge area (intercepted arc) induced by the set P defined over $\mathbb{D}$ (over $\partial\mathbb{D}$), covers the whole disk (boundary): $\bigcup_{u \in P} wedge(u) = \mathbb{D} \, (\bigcup_{u \in P} arc(u) = \partial\mathbb{D})$ with high probability.*

Note that the disk cover is a necessary but *not* a sufficient condition for the graph to be connected, as illustrated in Figure 3.

Using this notation, we rewrite the connectivity problem terms of the boundary setting:

Definition 5 (Connectivity problem). *We would like to find the minimal communication angle $\check{\alpha}$, such that the nodes' wedge area (intercepted arc) induces a connected graph with high probability.*

The relation between the covering and the connectivity problems is given by the following Lemma which is explicitly proven in Lemma 2.2 of [9].

Lemma 1. *Given $\bar{\alpha}$ which is the minimal angle that induces a cover, with high probability, $\check{\alpha} = 3\bar{\alpha}$ is the expected connectivity threshold.*

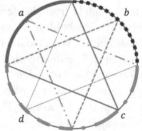

Fig. 3. The disk is covered by the nodes, however, the induced graph is not connected (the graph contain only two edges)

3   Centered Angles

In this section, we consider the case where the antennas' communication angle α is directed to the center o of the disk. We define three different models, prove their equivalence and use one of them to resolve the connectivity threshold.

The diagram at the right illustrates the way we have proven that the three models are equivalent up to $O(\cdot)$ notation:

The equivalence of these three models imply the following corollary:

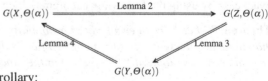

Corollary 1. *Let $\check{\alpha}_{G(Y,\alpha)}$ be the critical α of $G(Y,\alpha)$, then there exists a constant $c > 0$ such that the critical α of $G(X,\alpha)$ is $\check{\alpha}_{G(X,c\alpha)} = \Theta(\check{\alpha}_{G(Y,\alpha)})$.*

We prove each of the three equivalences separately, producing a new set of points from the given one, as follows:

Definition 6. *Let $\phi : \mathbb{D} \longrightarrow \partial \mathbb{D}^2$ be a function defined as follows. ϕ projects every point $u \in \mathbb{D}$ to the antipodal pair $\{\bar{u}, \bar{u}^*\} \in \partial \mathbb{D}^2$ located on the intersection of $\partial \mathbb{D}$ and the line goes through u and o (the center of the disk).*

Proposition 1. *Given the set $\{X_1, ..., X_n\}$, one can produce the set $\{Z_1, ..., Z_n\}$ by $Z_i \equiv X_i \circ \phi^{-1}$, which implies that Z_i is independently identically uniformly distributed over $\partial \mathbb{D}^2$.*

Proposition 2. *ϕ defines for every edge $(u,v) \in G(X,\alpha)$ a connected path: $P_{\bar{u},\bar{v}} = \{(\bar{u}, \bar{v}^*), (\bar{v}^*, \bar{v}), (\bar{v}, \bar{u}^*), (\bar{u}^*, \bar{u})\} \in G(Z, \alpha)$.*

Lemma 2. *Given the communication graphs $G(X, \alpha)$ and $G(Z, \alpha)$ such that $Z_i = X_i \circ \phi^{-1}$, then if $G(X, \alpha)$ is connected, it implies that $G(Z, \alpha)$ is connected.*

Definition 7. *Let $\varphi : \partial \mathbb{D}^2 \times \{0, 1\} \longrightarrow \partial \mathbb{D}$ be a function that gets the pair $\{u, u^*\} \in \partial \mathbb{D}^2$ and a bit b and returns one node $u' \in \partial \mathbb{D}$ from the pair, e.g., $\varphi(\{u, u^*, 0\}) = u$ and $\varphi(\{u, u^*, 1\}) = u^*$.*

Proposition 3. *Given the set $\{(Z_1, \chi_1), ..., (Z_n, \chi_n)\}$, one can produce the set $\{Y_1, ..., Y_n\}$ by $Y_i \equiv (Z_i, \chi_i) \circ \varphi^{-1}$, which implies that Y_i is independently identically uniformly distributed.*

Lemma 3. *Given the communication graphs $G(Z, \alpha)$ and $G(Y, 3\alpha)$ such that $Y_i = (Z_i, \chi_i) \circ \varphi^{-1}$. If $G(Z, \alpha)$ is connected, then w.h.p $G(Y, 3\alpha)$ is connected.*

Definition 8. *Let $\psi : \partial \mathbb{D} \times [0, 1] \longrightarrow \mathbb{D}$ be a function defined as follows. For every point $\bar{u} \in \partial \mathbb{D}$, ψ gets a radius r_u and returns the point $u \in \mathbb{D}$ located on the line going through $\bar{u}$ and o, such that $dist(u, \bar{u}) = \sqrt{r_u}$.*

Proposition 4. *Given the set $(Y_1, R_1), ..., (Y_n, R_n)$, one can produce the set $X_1, ..., X_n$ by $X_i \equiv (Y_i, R_i) \circ \psi^{-1}$, which implies that X_i is independently identically uniformly distributed over $\mathbb{D}$.*

Lemma 4. *Given the communication graphs $G(Y, \alpha/2)$ and $G(X, \alpha)$ such that $X_i = (Y_i, R_i) \circ \psi^{-1}$, and given that $G(Y, \alpha/2)$ is connected, then $G(X, \alpha)$ is connected.*

3.1   Finding the Connectivity Threshold

Given the set L of n uniformly distributed points on $\partial \mathbb{D}$, such that the angle α of every node is directed to the center, we show that the threshold for $G(Y, \alpha)$ is $\check{\alpha}(Y) = \Theta(\log n/n)$ as presented in the following two lemmas.

Lemma 5 (Sufficient condition for connectivity). *Given the set (Y, α), there exists a constant $c > 0$ such that when $\alpha \geq c \frac{\log n}{n}$, the communication graph $G(Y, \alpha)$ is connected w.h.p.*

Proof. The proof provides a cover of the disk's boundary, and then by Lemma 1 we can achieve the expected cover.

Given a node $u \in G(Y, \alpha)$ we divide $arc(u)$ to three equal length bins denoted (from left to right) by b_ℓ, b_{mid} and b_r. Since $|arc(u)| = 2\alpha$, the length of each bin is $2\alpha/3$. Let ℓ be a node at b_ℓ, and u_ℓ be a node that lies at $arc(\ell)$. The angles are centered; hence, the nodes u, ℓ and u_ℓ are connected. In the same way, we expand the cover to the left (with relation to $arc(u)$). The same considerations are valid for the right side of the arc and for u, r and u_r, respectively. Note that the existence of ℓ, r, u_ℓ and u_r is an outcome of the connectedness of $G(Y, \alpha)$.

By the coupon collector principle, we know that the number of balls that need to be thrown until all bins have at least one ball with high probability is $m \log m$, where m is the number of bins. Similarly, when placing n balls, $n/\log n$ bins will promise that in each bin, there will be at least one ball.

Dividing the circumference to $2\alpha/3$ cells, we have $\frac{2\pi r}{2\alpha/3} = \frac{3\pi}{\alpha}$ bins (note that $r = 1$). Assigning $\frac{3\pi}{\alpha} = \frac{n}{\log n}$ bins, we get that $\alpha = O(\log n/n)$ as expected.

Lemma 6 (Necessary condition for connectivity). *Given the set (Y, α), there exists a constant $c > 0$ such that if $\alpha < c\frac{\log n}{n}$, then w.h.p. the induced communication graph $G(Y, \alpha)$ is not connected.*

Proof. Let $A = \{A_i\}_{i=1}^{\frac{n}{c \log n}}$ be a set of $\frac{n}{c \log n}$ disjoint arc intervals induced by $\frac{n}{c \log n}$ nodes of $G(Y, \alpha)$. We show that there exists an arc at A that does not contain any node; hence, its antipodal arc is not covered, which yields by Lemma 1 that $G(Y, \alpha)$ is not connected. The existence of A is proven in Proposition 5 below.

Let X_i be a discrete Poisson random variable over the number of balls j in the bin i. Let χ_i be an indicator random variable that is 1 when the ith bin is empty and 0 otherwise. The density parameter of X is $\lambda = n/|A| = n/(n/c \log n) = c \log n$. Thus, the probability that a bin i is empty is $\Pr(\chi_i = 1) = \Pr(X_i = 0) = e^{-c \log n} = \left(\frac{1}{n}\right)^c$. By the union bound, we get that the probability that a bin i is not empty is $\Pr(\chi_i = 0) = 1 - \left(\frac{1}{n}\right)^{1/c}$. Using the independency property of the Poisson variables, the probability that in all the bins there is at least one ball is

$$Pr\left(\chi_1 = 0 \cap \chi_2 = 0 \cap ... \cap \chi_{|A|} = 0\right) = \left(1 - \left(\frac{1}{n}\right)^c\right)^{|A|} =$$

$$= \left(1 - \left(\frac{1}{n}\right)^c\right)^{\frac{n}{c \log n}} \leq \exp\left(-n^{\frac{1-c}{c \log n}}\right)$$

When setting $0 < c < 1$, we get that $\exp\left(-\frac{n^{1-c}}{c \log n}\right) \xrightarrow{n \to \infty} 0$, which implies that w.h.p. there exists an empty bin, i.e., an arc fragment that is *not* covered.

Proposition 5. *Let $I = \{I_i\}_{i=1}^n$ be a set of intervals of length $|I_i| \leq \frac{\log n}{n}$. There exists a positive constant c and natural n_0 such that for all $n \geq n_0$ and for all I, there exist subset $I' = \{I_{i_j}\}_{j=1}^{\frac{n}{c \log n}} \subseteq I$ of $\frac{n}{c \log n}$ disjoint intervals.*

4   Random Angle Direction

Given the set P of N Poisson distributed points (variables) on a disk, we now assume that the direction of the antenna is a random variable θ_i. Hence, each point can be represented by three parameters (x, y, θ) where x and y indicate the point location, distributed over $\mathbb{D}$, and θ distributed over $[0, \pi]$ is the direction of the antenna. Since the problem has three dimensions, it makes sense to use a three-dimensional object, such that the probability is representing by the volume of the object. Observe the set P lies over a torus $\mathbb{T}$ in $\mathbb{R}^3$, such that the unit disk is swept

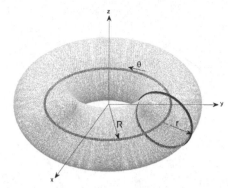

Fig. 4. The torus $\mathbb{T} \in \mathbb{R}^3$. The blue circle (with the "minor radius") is swept around the axis defining the red circle. The radius of the red circle R, (the "major radius") is the distance from the center of the tube to the center of the torus.

around an axis with length 2π (all the possible directions[1]). At this setting, $\mathbb{T}$'s interior volume is $V_{\mathbb{T}} = \pi r^2 2\pi R = \{R = r = 1\} = 2\pi^2$. To achieve a probability space, we normalize this number to be equal to one.

Our goal is to find the minimal angle that promises that the induced communication graph is connected; hence, we would like to find the set of points that induces the minimal communication area and ensures that these points induce a cover (which in turn yields a connected graph due to the relation between the covering and the connectivity problems, see Lemma 1). Observing that when the node is located on the boundary and the node's direction is close to the tangent direction, the communication area is minimal, we focus on the set B of points that their location is α–*closed* to the disk's boundary and their direction is α–*closed* to the tangent's direction. At the three-dimensional representation, B lies at the external ring (a.k.a. annulus) T_{ex} of $\mathbb{T}$ (see Figure 5), i.e., $B = \{(x_i, y_i, \theta_i) \in T_{ex} \cap P : \theta_i$ is the tangent direction$\}$, such that B induces a minimal communication volume.

Proposition 6. *For any constant $c \geq 4$ there exists natural n_0, such that for all $n \geq n_0$, tangent angle induce a wedge of size $\leq c\left(\alpha(n)\right)^3$.*

Proposition 7. *For any constant $c \geq 2$ there exists natural n_0 such that for all $n \geq n_0$, B lies over a volume $V_B \geq c\left(\alpha(n)\right)^2$.*

Each node $i = (x_i, y_i, \theta_i) \in B$ defines a ball (spherical cap) H_i of the set of nodes that can communicate with i: $H_i = \{(x_j, y_j, \theta_j) \in P \cap T_{ex} : \theta_j \in [\theta_i - \alpha, \theta_i + \alpha]\}$ (i.e., H_i is the $3D$ shape symmetric to the $2D$ sector defined by $wedge(i)$).

Proposition 8. *For any constant $c \geq 4$ there exists natural n_0 such that for all $n \geq n_0$, the volume of the ball H_i is $V_{H_i} \leq c\left(\alpha(n)\right)^4$.*

---

[1] Note that in this setting, θ is distributed over $[0, 2\pi]$ instead of $[0, \pi]$. However, these are equal in $O(\cdot)$ notation.

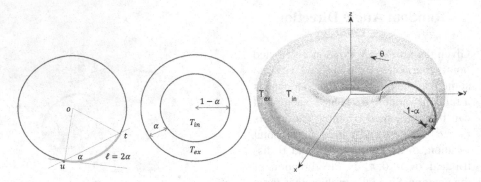

Fig. 5. When the node is located on the boundary and the node's direction is close to the tangent direction, the communication area is minimal. We are interested in the points that are α-close to $\mathbb{T}$'s crust. We calculate their volume by eliminating the interior torus with $r = 1 - \alpha$ from $V_{\mathbb{T}}$.

Proof. From proposition 6, we find that the area of $wedge(i)$ is $\leq c\alpha^3$ (for $c \geq 4$). We multiply this by α to insert the direction constraint (see Proposition 9) and achieve the expected volume.

Remark 4. For the directed communication settings, we can avoid multiplying by α since the nodes in the ball H_i do may not be directed to the nodes that reside in B. Hence, the volume V_{H_i} is $\leq \alpha^3$, and the connectivity threshold becomes $\check{\alpha} = \Theta(\sqrt[3]{\frac{\log n}{n}})$.

Proposition 9. *Given the nodes u, v each with an angle α, the possible directions that induce adjacency between v and u is α.*

Proposition 10. *There exist a constant $c > 0$ and natural n_0 such that for all $n \geq n_0$, there are at least $cn^{1/2}$ nodes in the set B.*

Proposition 11. *Given a volume $V = \sqrt[4]{\log n/n}$ at $\mathbb{R}^3$, throwing $\sqrt{n}$ balls with volume $V^4 = \log n/n$ to V, there exist a constant $c > 0$ and natural $n_0 > 0$ such that for all $n \geq n_0$, with high probability there is a subset of at least $c\sqrt{n}$ disjoint balls.*

Lemma 7 (Sufficient condition for connectivity). *Given n nodes that are uniformly distributed on the unit disk, there exists a constant $c > 0$ such that if $\alpha \geq c\sqrt[4]{\frac{\log n}{n}}$, then w.h.p. the induced communication graph is connected.*

Proof. The set B represents the nodes that induce the minimal communication volume (V_{H_i}). To achieve the space cover, we can divide it to disjoint cells of size V_{H_i} and use probabilistic consideration to promise that w.h.p. there exists at least one ball in every cell.

Proposition 8 implies that for a constant $c \geq 4$, $V_{H_i} \leq c\alpha^4$; hence, we have $\frac{1}{c\alpha^4}$ cells. By the coupon collector principle, when we set the number of nodes n to be $\frac{1}{c\alpha^4} \log \frac{1}{c\alpha^4}$, it yields that with high probability there is no empty cell. As a result, $\alpha = c\sqrt[4]{\frac{\log n}{n}}$.

Lemma 8 (Necessary condition for connectivity). *Given n nodes that are uniformly distributed on the unit disk, there exists a constant $c > 0$ such that if $\alpha < c\sqrt[4]{\frac{\log n}{n}}$, then w.h.p. the induced communication graph is not connected.*

Proof. We show that when $\alpha = c\sqrt[4]{\frac{\log n}{n}}$, the graph G is not connected since B induce a volume that is not covered w.h.p.

Let $\mathbb{H} = \underset{H_i \cap H_j = \emptyset}{\bigcup H_i}$ be a set of $n^{1/2}$ disjoint balls. The existence of this set is proven in Proposition 10 and 11. To complete the proof of Lemma 8, we show that w.h.p. there exists a ball $H_i \in \mathbb{H}$ that is empty, i.e., the node $i \in B$ is not connected.

Let χ_i denote an indicator random variable that is 1 when the ball H_i is empty and 0 otherwise. Let X_i be a discrete Poisson random variable over the number of nodes in H_i. The density parameter of X_i is $\lambda = n/V_{H_i} = n/\alpha^4 = \frac{n}{c\log n/n} = \frac{c}{\log n}$. Therefore, the probability that a given ball H_i is empty is

$$\Pr(\chi_i = 1) = \Pr(X_i = 0) = e^{-\lambda} = e^{-\frac{c}{\log n}} = (1/n)^{\frac{c}{\log^2 n}}$$

Using the independency property of the Poisson variable, we get that the probability that *all* of the balls in $\mathbb{H}$ are not empty is

$$\Pr\left(\chi_1 = 0 \cap \chi_2 = 0 \cap \ldots \cap \chi_{\sqrt{n}} = 0\right) = \Pr(\chi_1 = 0)\Pr(\chi_2 = 0)\ldots\Pr(\chi_{\sqrt{n}} = 0) =$$

$$= \left(1 - \left(\frac{1}{n}\right)^{\frac{c}{\log^2 n}}\right)^{\sqrt{n}} \leq \exp\left(-n^{-\frac{c}{\log^2 n}+\frac{1}{2}}\right)$$

Since $\exp\left(-n^{-\frac{c}{\log^2 n}+\frac{1}{2}}\right) \xrightarrow{n\to\infty} 0$, it implies that, with high probability, there exists an empty ball. Hence, the graph is not connected.

5 Concluding Remarks

In this paper, we have analyzed the connectively threshold for directional antennas. Our results show that if one can adjust the direction of the antennas, then in order to guarantee the network connectivity, the angle should be $\Theta\left(\frac{\log(n)}{n}\right)$. In contrast, if the direction of the antenna is a random variable, then the angle should be $\Theta\left(\sqrt[4]{\frac{\log(n)}{n}}\right)$. This gives a polynomial gap between the two models.

One of the simplest ways to increase the capacity of the network is by directional antennas. Our work defines theoretical bounds on how small the angle of the directional antennas can be in order to maintain connectivity. If we compare the classical results on connectivity in the unit disk graph model to our results, we find two main differences. The first difference is that in the unit disk graph model, the minimal graph that maintains the connectivity of the network is the Euclidean minimal spanning tree. In the directional antennas, the minimal graph is closer to the Hamiltonian cycle in nature. More accurately, if our points are located on the boundary of a disk, then it is the Hamiltonian cycle. Moreover, our analysis indicates that the network is totally different near the critical angle from a usual network graph implied by the unit disk; while in the unit disk graph model the communication moves over short distances, on the directional antenna model, the communication prefers long distances, i.e., forms a point p to the antipodal point p^*.

We believe that much more work is needed in order to understand the geometry of the network with directed antennas, and that this paper is a step in this direction.

Throughout the paper, we have assumed that the wireless antennas are scattered on a unit disk. We believe that the disk assumption can be relaxed to accommodate a more general and realistic setting, namely; our results hold for any *convex fat* object with curvature> 0, denoted by $\mathbb{S}$.

References

1. Ackerman, E., Gelander, T., Pinchasi, R.: Ice-creams and wedge graphs. In: Computational Geometry (2012)
2. Aschner, R., Katz, M.J., Morgenstern, G.: Do directional antennas facilitate in reducing interferences? In: Fomin, F.V., Kaski, P. (eds.) SWAT 2012. LNCS, vol. 7357, pp. 201–212. Springer, Heidelberg (2012)
3. Biglieri, E.: MIMO wireless communications. Cambridge University Press (2007)
4. Caragiannis, I., Kaklamanis, C., Kranakis, E., Krizanc, D., Wiese, A.: Communication in wireless networks with directional antennas. In: Proceedings of the Twentieth Annual Symposium on Parallelism in Algorithms and Architectures, pp. 344–351. ACM (2008)
5. Carmi, P., Katz, M.J., Lotker, Z., Rosn, A.: Connectivity guarantees for wireless networks with directional antennas. Computational Geometry 44(9), 477–485 (2011)
6. Daltrophe, H., Dolev, S., Lotker, Z.: Probabilistic connectivity threshold for directional antenna widths. The Lynne and William Frankel Center for Computer Science, Technical Reports 13-04 (2013)
7. Damian, M., Flatland, R.: Spanning properties of graphs induced by directional antennas. In: Electronic Proc. 20th Fall Workshop on Computational Geometry. Stony Brook University, Stony Brook (2010)
8. Gupta, P., Kumar, P.: Critical power for asymptotic connectivity. In: Proceedings of the 37th IEEE Conference on Decision and Control, pp. 1106–1110. IEEE (1998)
9. Kozma, G., Lotker, Z., Stupp, G.: On the connectivity threshold for general uniform metric spaces. Information Processing Letters 110(10), 356–359 (2010)
10. Kranakis, E., MacQuarrie, F., Morales-Ponce, O.: Stretch factor in wireless sensor networks with directional antennae. In: Lin, G. (ed.) COCOA 2012. LNCS, vol. 7402, pp. 25–36. Springer, Heidelberg (2012)
11. Kumar, U., Gupta, H., Das, S.: A topology control approach to using directional antennas in wireless mesh networks. In: IEEE International Conference on Communications, ICC 2006, vol. 9, pp. 4083–4088 (2006)
12. Meester, R., Roy, R.: Continuum percolation. Cambridge University Press (1996)
13. Mitzenmacher, M., Upfal, E.: Probability and Computing: Randomized Algorithms and Probabilistic Analysis. Cambridge University Press (2005)
14. Penrose, M.: A strong law for the longest edge of the minimal spanning tree. The Annals of Probability 27(1), 246–260 (1999)
15. Tse, D., Viswanath, P.: Fundamentals of wireless communication. Cambridge University Press (2005)
16. Xue, F., Kumar, P.: The number of neighbors needed for connectivity of wireless networks. Wireless Networks 10(2), 169–181 (2004)

Broadcasting in Ad Hoc
Multiple Access Channels*

Lakshmi Anantharamu and Bogdan S. Chlebus

Department of Computer Science and Engineering, University of Colorado Denver,
Denver, Colorado, USA

Abstract. We study dynamic broadcasting in multiple access channels
in adversarial settings. There is an unbounded supply of anonymous stations attached to the channel. There is an adversary who injects packets
into stations to be broadcast on the channel. The adversary is restricted
by the injection rate, burstiness, and by how many passive stations can be
simultaneously activated by injecting packets into their empty queues.
We consider deterministic distributed broadcast algorithms, which are
further categorized by their properties. We investigate for which injection
rates can algorithms attain bounded packet latency, when adversaries are
restricted to be able to activate at most one station per round. The rates
of algorithms we present make the increasing sequence $\frac{1}{3}$, $\frac{3}{8}$ and $\frac{1}{2}$, re-
flecting the additional features of algorithms. We show that no injection
rate greater than $\frac{3}{4}$ can be handled with bounded packet latency.

Keywords: multiple access channel, adversarial queuing, distributed
broadcast, deterministic algorithm, stability, packet latency.

1 Introduction

Multiple access channels model shared-medium networks in which simultaneous
broadcast to all users is provided. They are an abstraction of the networking
technology of the popular implementation of local area networks by the Ethernet
suite of algorithms [18]. In a multiple access channel, transmissions by multiple
users that overlap in time result in interference so that none can be successfully
received. This makes it necessary either to avoid conflict for access to the channel
altogether or to have a mechanism to resolve conflict when it occurs. We consider
broadcasting in multiple-access channels in a dynamic scenario when there are
many stations attached to the channel, but only a few of them are active at any
time and the stations' status of active versus passive keeps changing.

Considering deterministic algorithms and their worst-case performance requires a methodological setting specifying worst-case bounds on how much traffic a network would need to handle. This can be accomplished formally through
suitable adversarial models of demands on network traffic. Another component

---

* The full version of this paper is available at http://arxiv.org/abs/1306.6109.
This work was supported by the NSF Grant 1016847.

T. Moscibroda and A.A. Rescigno (Eds.): SIROCCO 2013, LNCS 8179, pp. 237–248, 2013.

in a specification of a broadcast system is how much knowledge about the system can communicating agents use in their codes of algorithms. Historically, the first approach was to use the queue-free model, in which each injected packet is treated as if handled by an independent station without any name and no private memory for a queue. In such an ad hoc model, the number of stations is not set in any way, as stations come and go similarly as packets do; see [12] for the initial work on this model, and [6] for more recent one. An alternative approach to ad hoc channels is to have a system with a fixed number n of stations, each equipped with a private memory to store packets in a queue. An attractive feature of such fixed-size systems is that even simple randomized protocols like Aloha are stable under suitable traffic load [22], while in the queue-free model the binary exponential backoff is unstable for any arrival rate [1].

The popular assumptions used in the literature addressing distributed deterministic broadcasting stipulate that there are some n stations attached to a channel and that each station is identified by its name in the interval $[0, n - 1]$, with each station knowing the number n and its own name; see [2,4,3,9,10]. Our goal is to explore deterministic broadcasting on multiple-access channels when there are many stations attached to a channel but only a few stations use it at a time. In such a situation, using names permanently assigned to stations by deterministic distributed algorithms may create an unnecessarily large overhead measured as packet latency and queue size.

In this paper, we consider distributed deterministic broadcasting but we depart from the assumption about a fixed known size of the system. Instead, we view the system as consisting of a very large set of stations that are not individually identified in any way. The stations that want to use the channel to communicate join broadcasting activity. This needs to be coordinated with the other currently active stations by an algorithm, which could be associated with the medium-access control layer. The process of activating stations is modeled by a suitable adversarial model that we propose. This adversarial model is designed to represent a flexible system in which we relax the assumption that there is a finite fixed set of stations attached to the channel, and that their number is known to each participating station, and that each station has assigned a unique name which it knows. We call such channels *ad hoc* to emphasize the volatility of the system and the relative lack of knowledge of individual stations about themselves and the environment.

Our results. We propose an adversarial model of traffic demands for ad hoc multiple access channels, which represents *dynamic* environments in which stations freely join and leave broadcasting activity. To make an anonymous system able to break symmetry in a deterministic manner, we restrict adversaries by allowing them to activate at most one station per round. This is shown to be sufficient for *deterministic* distributed broadcast algorithms to exist. We categorize algorithms into acknowledgment based, activation based and full sensing. Independently from that, we differentiate algorithms by the property if they use control bits in messages or not, calling them adaptive and non-adaptive, respectively. We give a number of algorithms, for channels with and without collision detection, for which we assess injection rates they can handle with bounded packet latency. Our non-adaptive activation-based algorithm can handle injections rates smaller than $\frac{1}{3}$ on channels with collision detection, the non-adaptive full-sensing algorithm can handle injection rate $\frac{3}{8}$ on channels with collision detection, and the adaptive activation-based algorithm can handle injection rate $\frac{1}{2}$ on channels without collision detection. We show that no algorithm can provide bounded packet latency when injection rates are greater than $\frac{3}{4}$.

Related work. The adversarial queuing methodology was introduced by Borodin et al. [8] and Andrews et al. [5], who used it to study the stability of store-and-forward routing in wired networks. Adversarial queueing on multiple access channels was first studied by Bender et al. [6], who considered randomized algorithms for the queue-free model. A deterministic distributed broadcasting on multiple access channels with queues in adversarial settings was investigated by Chlebus et al. [9,10] and by Anantharamu et al. [2,4,3]. That work on deterministic distributed algorithms was about systems with a known number of stations attached to the channel and with stations using individual names. Acknowledgment-based algorithms include the first randomized algorithms studied on dynamic channels, as Aloha and binary exponential backoff fall into this category. The throughput of multiple access channels, understood as the maximum injection rate with Poisson traffic that can be handled by a randomized algorithm and make the system stable (ergodic), has been intensively studied in the literature. It was shown to be as low as 0.568 by Tsybakov and Likhanov [21]. Goldberg et al. [13] gave related bounds for backoff, acknowledgment-based and full-sensing algorithms. Håstad et al. [16] compared polynomial and exponential backoff algorithms in the queuing model with respect to bounds on their throughput. For early work on full-sensing algorithms in channels with collision detection in the queue-free model see the survey by Gallager [12]. Randomized algorithms of bounded packet latency were given by Raghavan and Upfal [19] in the queuing model and by Goldberg et al. [14] in the queue-free model. Upper bounds on packet latency in adversarial networks was studied by Anantharamu et al. [2,4] in the case of multiple access channels with injection rate less than 1 and by Rosén and Tsirkin [20] for general networks and adversaries of rate 1. Deterministic algorithms for collision resolution in static algorithmic problems on multiple access channels were first considered by Greenberg and Winograd [15] and Komlós and Greenberg [17]. Algorithmic problems of distributed-computing flavor in systems in which multiple access channels provide the underlying communication infrastructure were considered by Bieńkowski et al. [7] and Czyżowicz et al. [11].

2   Technical Preliminaries

A multiple-access channel consists of a shared communication medium and stations attached to it. We consider dynamic broadcasting, in which packets are injected into stations continually and the goal is to have them successfully transmitted on the channel. A *message* transmitted by a station includes at most one packet and some control bits, if any. Every station receives a transmitted message successfully, including the transmitting station, when the transmission of this message does not overlap with transmissions by other stations of their messages; in such a case we say that the message is *heard* on the channel. We consider synchronous channels which operate in rounds. Rounds and messages are calibrated such that transmitting one message takes the duration of one round. A message transmitted in a round is delivered to every station in the same round. When at least two messages are transmitted in the same round then this creates a *collision*, which prevents any station from hearing any of the transmitted messages. When no station transmits in a round, then the round is called *silent*. A channel is said to be *with collision detection* when the feedback from the channel in a collision round is different from the feedback received during a silent round, otherwise the channel is *without collision detection*. For a channel without collision detection, a collision round and a silent one are perceived the same. A round is *void* when no station hears a message; such a round is either silent or a collision one.

Ad hoc channels. A station is said to be *active*, at a point in time, when it has pending packets that have not been heard on the channel yet. A station is *passive*, at a point in time, if either it has never had any packets to broadcast or all the packets it has ever received to broadcast have already been heard on the channel. These "points in time" are understood as real-number time coordinates, which are finer than the discrete partitioning of time into rounds. This is needed to avoid ambiguity in a situation when a station begins a round with just one pending packet, this packet is heard on the channel in this round, and new packets are injected into this station in this very round. We assume that there is an unbounded supply of passive stations. A passive station is said to get *activated* when a packet or multiple packets are injected into it. We impose quantitative restrictions on how passive stations may be activated in a round, which results in finitely many stations being active in any round. There is no upper bound on the number of active stations in a round in an infinite execution, since there is an unbounded supply of passive stations. Stations are *anonymous* when there are no individual names assigned to them. We consider channels that are *ad hoc* which means that (1) every station is anonymous, (2) an execution starts with every station initialized as passive, and (3) there is an unbounded supply of passive stations.

Adversarial model of packet injection. Packets are injected by leaky-bucket adversaries. For a number $0 < \rho \le 1$ and integer $b > 0$, the *adversary of type* (ρ, b) may inject at most $\rho|\tau| + b$ packets in any time interval τ of $|\tau|$ rounds. In such a context, the number ρ is called the *rate of injection*. The maximum number of packets that an adversary may inject in one round is called the *burstiness* of this adversary. The adversary of type (ρ, b) has burstiness equal to $\lfloor \rho + b \rfloor$. The adversaries we consider are constrained by how many stations they can activate in a round. An adversary is *k-activated*, for an integer $k > 0$, if at most k stations may be activated in a round. We consider 1-activated adversaries, unless explicitly stated otherwise.

Broadcast algorithms. We consider deterministic distributed broadcast algorithms. In the context of communication algorithms, the "knowledge" of properties of a system means using such properties as a part of code of an algorithm. The algorithms we consider do not know the names of stations and the number of stations in the system. This is in contrast with previous work on deterministic distribute algorithms, see [2,4,3,9,10], where the names of stations and the number of stations could be used in a code. No information about adversaries is reflected in the code executed by stations. Every station has a private memory to store data relevant to executing a communication algorithm. This memory is considered to be unbounded, in the sense that it may store an arbitrary amount of data. The part of a private memory of a station used to store packets pending transmission is organized as a queue operating in a first-in-first-out manner. Successfully broadcast packets are removed from their queues and discarded. Packets are never dropped unless just after a successful broadcast. The *state* of a station is determined by the values of its private variables, with the exception of the queue to store packets, which is not a part of a state. One state is distinguished as *initial.* An execution begins with every station in the initial state and with empty queue. The algorithms we consider are distributed in the sense that they are "event driven." An *event*, in which a station participates, consists of everything that happens to the station in a round, including what the station receives as feedback from the channel and how many packets are injected into it. An event is structured as the following sequence of actions occurring in a round in the order given: (i) transmitting a packet, (ii) receiving feedback from the channel, (iii) having new packets injected, (iv) making a state transition. Some among the actions (i) and (iii) may be void in a station in a round. A state transition depends on the current state, the feedback from the channel, and on whether new packets were injected in the round. In particular, the following actions occur during a state transition. If a packet has just been successfully transmitted then it is dequeued and discarded. If new packets have just been injected then they are all enqueued. If a message is to be transmitted in the next round, possibly subject to packet availability, then a message to be transmitted is prepared. Such a message may include the packet from the top of the queue, when the queue is nonempty, but a message may consist only of some control bits. A station that begins a round as active becomes passive when it successfully transmits its only pending packet. More precisely, such a station becomes passive at the point in time when this station receives the transmitted message as the feedback from the channel. When new packets are injected into this station in this very round, then it means that this passive station gets activated again. A station's status, of active versus passive, is dynamic in the course of an execution. In particular, an active station may eventually be relegated to passive and stay such forever, or it may stay active forever, or it may change its status between active and passive any number of times.

Classes of algorithms. We define subclasses of algorithms by specifying what can be sent in messages and how state transitions occur. We begin with the categorizations into full-sensing, activation-based and acknowledgment-based algorithms.

General algorithms are called *full sensing*. This means that stations may have state transitions occur in each round, according to the state-transition rules represented by the code. This term "full sensing" is to indicate that every station is sensing the channel in every round. This encompasses passive stations, which means that when a full-sensing algorithm is executed, then passive stations undergo state transitions from the beginning of the execution. Algorithms such that every station stays in the initial state while passive and it resets itself to the initial state when it becomes passive again, that is, in a round in which its last pending packet is heard on the channel, are called *activation based*. These algorithms have stations ignore the feedback from the channel when they do not have any packets to broadcast. Finally, algorithms such that a station stays in the initial state while passive and it resets itself to the initial state in a round in which a packet that it transmitted was heard on the channel are called *acknowledgment based*. This definition is correct due to the stipulation that the contents of queues do not belong to what constitutes a state; in particular, a station may be in the initial state when its queue is nonempty. A station executing a full-sensing algorithm may (in principle) remember the whole history of the feedback from the channel, unless the size of its private memory restricts it in this respect, which is not the case in our considerations. An active station executing an activation-based algorithm may remember the history of the feedback from the channel since the activation. An active station executing an acknowledgment-based algorithm may remember the history of the feedback from the channel since the latest successful transmission or the latest activation, whichever occurred later. We understood these categorizations so that an acknowledgment-based algorithm is activation based, and an activation-based algorithm is full sensing. This is because a station executing an activation-based algorithm could be considered as receiving feedback from the channel but idling in the initial state when not having pending packets. When control bits are used in messages then we say that a algorithm is *adaptive*, otherwise the algorithm is *nonadaptive*. The categorization of adaptive versus non-adaptive is independent of the other three categorizations, into full sensing and activation based and acknowledgment based, so we have six categories of algorithms overall. This categorization of algorithms holds independently for channels with and without collision detection. The strongest algorithms are full sensing adaptive for channels with collision detection, while the weakest ones are acknowledgment-based non-adaptive for channels without collision detection.

The quality of broadcasting. An execution of an algorithm is said to be *fair* when each packet injected into a station is eventually heard on the channel. An algorithm is *fair* against an adversary when each of its executions is fair when packets are injected subject to the constrains of the type of the adversary. An execution of an algorithm has *at most Q packets queued* when in each round the number of packets stored in the queues of the active stations is at most Q. We say that an algorithm has *at most Q packets queued*, against an adversary of a given type, when at most Q packets queued in each execution of the algorithm against such an adversary. An algorithm is *stable*, against an adversary of a given type, when there exist an integer Q such that at most Q packets are queued in any execution against this adversary. When an algorithm is unstable then the queues may grow unbounded in some executions, but no packet is ever dropped unless heard on the channel. The semantics of multiple access channels allows at most one packet to be heard on the channel in a round. This means that when injection rate of an adversary is greater than 1 then for any algorithm some of its executions produce unbounded queues. In this paper, we consider only injection rates that are at most 1. An execution of an algorithm has *packet latency* t when each packet spends at most t rounds in the queue before it is heard on the channel. We say that an algorithm has packet latency t against an adversary of a given type when each execution of the algorithm against such an adversary has packet latency t.

3 Limitations on Deterministic Broadcasting

In this section we consider what limitations on deterministic distributed broadcasting are inherent in the properties of ad-hoc multiple access channels and the considered classes of algorithms.

Proposition 1. *No deterministic distributed algorithm is fair against a 2-activated adversary of burstiness at least 2.*

In the light of Proposition 1, we will restrict our attention to 1-activated adversaries in what follows. For 1-activated adversaries, we may refer to stations participating in an execution by the round numbers in which they get activated. So when we refer to the *station* v, for an integer $v \geq 0$, then we mean the station that got activated in the round v. If no station got activated in a round v, then a station bearing the number v does not exist.

Proposition 2. *No acknowledgment-based algorithm is fair against a 1-activated adversary of type (ρ, b) such that $2\rho + b \geq 3$.*

Theorem 1. *No deterministic distributed algorithm can provide bounded packet latency against a 1-activated adversary of injection rate greater than $\frac{3}{4}$ and with burstiness at least 2.*

Theorem 1 demonstrates a difference between the adversarial model of ad-hoc channels with the model of channels in which stations know the fixed number of stations attached to the channel and their names. In that latter model, a bounded packet latency can be attained for any injection rate less than 1, see [2,4], and a mere stability can be obtained even for the injection rate 1, as it was demonstrated in [9].

4 A Non-adaptive Activation Based Algorithm

We propose a non-adaptive activation-based algorithm COUNTING-BACKOFF. It is designed for channels with collision detection. The underlying paradigm of algorithm COUNTING-BACKOFF is that active stations maintain a global virtual stack, that is, a last-in-first-out queue. Each station needs to remember its position on the stack, which is maintained as a counter with the operations of incrementing and decrementing by one. A passive or newly activated station has the counter equal to zero. The station at the top of the stack has the counter equal to one. The algorithm applies the rule that if a collision of two concurrent transmissions occurs then the station activated earlier gives up temporarily, which is understood as giving up the position at the top of the stack, while the station activated later persists in transmissions, which is interpreted as claiming the top position on the stack. Every station has a private integer-valued variable `backoff_counter`, which is set to zero when the station is passive. The private instantiations of the variable `backoff_counter` are manipulated by the active stations according to the following general rules. An active station transmits a packet in a round when it `backoff_counter` is at most one. When a collision occurs, then each active station increments its `backoff_counter` by one. When a silent round occurs, then each active station decrements its `backoff_counter` by one. When a message is heard then the counters `backoff_counter` are not modified, with the possible exception of a station activated in the previous round which changes this variable from zero to one. A station that gets activated initially keeps its `backoff_counter` equal to zero, so the station transmits in the round just after the activation. Such a station increments its `backoff_counter` in the next round, unless its only packet got heard, in which case the station becomes passive without ever modifying its `backoff_counter`. A station that transmits and its packet is heard withholds the channel and keeps transmitting in the following rounds, unless it does not have any other pending packets or a collision occurs. The variables `backoff_counter` are manipulated such that they implement positions on a stack, and thereby serve as dynamic transient names for the stations that are otherwise nameless. This prevents conflicts for access among the stations that are already on the stack.

Theorem 2. *When algorithm* COUNTING-BACKOFF *is executed against an adversary of type* (ρ, b), *where* $\rho < \frac{1}{3}$ *and* $b > 1$, *then the packet latency is at most* $\frac{3b-3}{1-3\rho}$ *and there are at most* $\frac{3b-5}{2}$ *packets queued in any round.*

The bound on packet latency of algorithm COUNTING-BACKOFF given in Theorem 2 is tight. It follows that packet latency grows unbounded when the injection rate ρ approaches $\frac{1}{3}$. On the other hand, the bound on queue size given in Theorem 2 depends only on the burstiness of the adversary. The upper bound $\frac{3b-5}{2}$ on queues holds also when injection rate equals $\frac{1}{3}$, so algorithm COUNTING-BACKOFF is stable but not fair when injection rate equals $\frac{1}{3}$.

5   A Non-adaptive Full Sensing Algorithm

Stations executing a full-sensing algorithm can listen to the channel at all times and so they may have a sense of time by maintaining common references to the past rounds. This makes it possible to process consecutive past rounds to give stations activated in them an opportunity to transmit. This, just by itself, may result in unbounded packet latency, if we spend at least one round to examine any past round for a possible activation in it, because the repeated case of active stations with multiple packets would accrue unbounded delays. To prevent this from occurring, one may consider groups of rounds and have stations activated in these rounds transmit simultaneously. If at most one station got activated in a group then we save at least one round of examination, which compensates for delay due to some stations holding more than one packet and for occasional collisions. To implement this approach, a channel needs to be with collision detection, which is assumed in this section. We refer to active stations by the respective rounds of their activation. A round gets *verified* when either all the packets of the station activated in this round have been heard or when it becomes certain that no station got activated in this round.

We present a non-adaptive full-sensing algorithm which we call QUADRUPLE-ROUND. The rounds of an execution of the algorithm are partitioned into disjoint groups of four consecutive rounds, each called a *segment*. The first and second rounds of a segment make its *left pair*, while the third and fourth rounds make the *right pair* of the segment. The rounds of execution spent on processing the rounds in a segment make the *phase* corresponding to this segment. The purpose of a phase is to verify the stations in the corresponding segment.

A phase is organized as a loop, which repeats actions that we collectively refer to as an *iteration* of the loop. It takes at most four rounds to perform an iteration. An iteration is executed as follows. All the stations activated in the rounds of the phase's segment, if there are any, transmit together in the first round of an iteration. A station, that is scheduled to transmit, transmits a packet from its private queue, unless the queue is empty. This results in either a silence or a message heard or a collision, as a feedback from the channel. This creates the three corresponding cases which we consider next.

When the first round of an iteration is silent, then this ends the iteration and also the loop. This is because such a silence confirms that there are no outstanding packets in the active stations in the segment. When a message is heard in the first round of an iteration, then this ends the iteration but not the loop. The reason of continuing the loop is that the station, which transmitted the packet heard on the channel, may have more packets. If a collision occurs in the first round of an iteration, then the stations of the left pair transmit together in the second round. This leads to the three sub-cases presented next.

The first sub-case is of silence in the second round, which means that no station in the left pair is active. As the first round produced a collision, this means that each station in the right pair holds a pending packet. In this sub-case, the third and fourth rounds of the iteration are spend by the third and fourth stations of the segment transmitting one packet each in order, which concludes the iteration but not the loop. The second sub-case is of a message heard in the second round, which concludes the iteration but not the loop. The third case occurs when there is a collision in the second round of the iteration, which means that each station in the left pair of the segment holds an outstanding packet. In this case, the third and fourth rounds are spend by the first and second stations of the segment transmitting one packet each in order, which concludes the iteration but not the loop.

Theorem 3. *When algorithm* QUADRUPLE-ROUND *is executed against an adversary of type* $(\frac{3}{8}, b)$, *then packet latency is at most* $2b + 4$ *and there are at most* $b + \mathcal{O}(1)$ *packets queued in any round.*

6   An Adaptive Activation Based Algorithm

Adaptive algorithms may use control bits in messages. We present an adaptive activation-based algorithm which we call QUEUE-BACKOFF. The underlying paradigm is that active stations maintain a global virtual first-in-first-out queue. This approach is implemented so that if a collision occurs, caused by two concurrent transmissions, then the station activated earlier persists in transmitting while the station activated later gives up temporarily. This is a dual alternative to the rule used in algorithm COUNTING-BACKOFF. Assume first that the channel is with collision detection. Every station has three private integer-valued variables: queue_size, queue_position, and collision_count, which are all set to zero in a passive station. The values of these variables represent a station's knowledge about the global distributed virtual queue of stations. A message transmitted on the channel includes a packet and the value of the sender's variable queue_size; if this is the last packet from the sender's queue then a marker bit "over" is also set on in the message. In a round, an active station transmits a message when its queue_position equals either zero or one. The private variables are manipulated according to the following rules. When a collision occurs, then each active station with a positive value of queue_size increments its queue_size by one while an active station with queue_size $= 0$ increments its collision_count by one and sets queue_position $\leftarrow -1$. When a message with some value $K > 0$ of queue_size is heard and an active station has queue_position $= -1$, then the station sets queue_size $\leftarrow K$ and queue_position $\leftarrow K - (\text{collision\_count} - 1)$. When a message with the "over" bit is heard, then each active station decrements its variables queue_position and queue_size by one. When a station is still active, it has just heard its own message and its queue_size equals zero, then the station sets its variable queue_size $\leftarrow 1$ and queue_position $\leftarrow 1$; this occurs when the global virtual queue is empty.

Some of the underlying ideas of this algorithm are similar to those used in the design of algorithm COUNTING-BACKOFF, they are as follows. A station that becomes activated transmits in the next round after activation, as then its queue_position is still zero. A station that transmits and the transmitted message is heard withholds the channel by transmitting in the following rounds, subject to packet availability. This works because the first transmission is with queue_position equal to either zero or one and the following ones with queue_position equal to one. A collision in a round means that some new station got activated in the previous round. This is because the station that has transmitted multiple times, with no other station successfully intervening, has its `queue_position` equal to one, while the other option is to have this variable equal to zero, which is only possible when this value is inherited from the state when still being a passive station.

Theorem 4. *When algorithm* QUEUE-BACKOFF *is executed against an adversary of type* $(\frac{1}{2}, b)$, *then there are at most* $2b - 3$ *packets queued in any round and packet latency is at most* $4b - 4$.

Algorithm QUEUE-BACKOFF was presented as implemented for channels with collision detection. When the global queue is nonempty then each round contributes either a collision or a message heard on the channel. This means that when the channel is without collision detection, then collisions can be detected as void rounds by any involved active station, while passive stations do not participate anyway. It follows that this algorithm can be executed on channels without collision detection with minor modifications in code only and with the same performance bounds.

7   Conclusion

We introduced ad hoc multiple access channels along with an adversarial model of packet injection in which deterministic distributed algorithms can handle nontrivial injection rates. These rates make the increasing sequence of $\frac{1}{3}$, $\frac{3}{8}$ and $\frac{1}{2}$. To improve beyond the rate $\frac{1}{3}$ attained by an activation-based non-adaptive algorithm, we designed a full sensing algorithm that handles injection rate $\frac{3}{8}$ and an adaptive one that handles the injection rate $\frac{1}{2}$. The optimality of these algorithms, in terms of the magnitude of the injection rate that the algorithms in the respective class of algorithms can handle with bounded packet latency against 1-activated adversaries, is open. We showed that no algorithm can handle the injection rate higher that $\frac{3}{4}$. It is an open question if any injection rate in the interval $(\frac{1}{2}, \frac{3}{4})$ can be handled with bounded packet latency by deterministic distributed algorithms against 1-activated adversaries.

References

1. Aldous, D.J.: Ultimate instability of exponential back-off protocol for acknowledgment-based transmission control of random access communication channels. IEEE Transactions on Information Theory 33(2), 219–223 (1987)
2. Anantharamu, L., Chlebus, B.S., Kowalski, D.R., Rokicki, M.A.: Deterministic broadcast on multiple access channels. In: Proceedings of the 29th IEEE International Conference on Computer Communications (INFOCOM), pp. 1–5 (2010)
3. Anantharamu, L., Chlebus, B.S., Rokicki, M.A.: Adversarial multiple access channel with individual injection rates. In: Abdelzaher, T., Raynal, M., Santoro, N. (eds.) Principles of Distributed Systems. LNCS, vol. 5923, pp. 174–188. Springer, Heidelberg (2009)

4. Anantharamu, L., Chlebus, B.S., Kowalski, D.R., Rokicki, M.A.: Medium access control for adversarial channels with jamming. In: Kosowski, A., Yamashita, M. (eds.) SIROCCO 2011. LNCS, vol. 6796, pp. 89–100. Springer, Heidelberg (2011)
5. Andrews, M., Awerbuch, B., Fernández, A., Leighton, F.T., Liu, Z., Kleinberg, J.M.: Universal-stability results and performance bounds for greedy contention-resolution protocols. Journal of the ACM 48(1), 39–69 (2001)
6. Bender, M.A., Farach-Colton, M., He, S., Kuszmaul, B.C., Leiserson, C.E.: Adversarial contention resolution for simple channels. In: Proceedings of the 17th Annual ACM Symposium on Parallel Algorithms (SPAA), pp. 325–332 (2005)
7. Bieńkowski, M., Klonowski, M., Korzeniowski, M., Kowalski, D.R.: Dynamic sharing of a multiple access channel. In: Proceedings of the 27th International Symposium on Theoretical Aspects of Computer Science (STACS), Schloss Dagstuhl–Leibniz-Zentrum fuer Informatik. Leibniz International Proceedings in Informatics, vol. 5, pp. 83–94 (2010)
8. Borodin, A., Kleinberg, J.M., Raghavan, P., Sudan, M., Williamson, D.P.: Adversarial queuing theory. Journal of the ACM 48(1), 13–38 (2001)
9. Chlebus, B.S., Kowalski, D.R., Rokicki, M.A.: Maximum throughput of multiple access channels in adversarial environments. Distributed Computing 22(2), 93–116 (2009)
10. Chlebus, B.S., Kowalski, D.R., Rokicki, M.A.: Adversarial queuing on the multiple access channel. ACM Transactions on Algorithms 8(1), 5:1–5:31 (2012)
11. Czyżowicz, J., Gąsieniec, L., Kowalski, D.R., Pelc, A.: Consensus and mutual exclusion in a multiple access channel. EEE Transaction on Parallel and Distributed Systems 22(7), 1092–1104 (2011)
12. Gallager, R.G.: A perspective on multiaccess channels. IEEE Transactions on Information Theory 31(2), 124–142 (1985)
13. Goldberg, L.A., Jerrum, M., Kannan, S., Paterson, M.: A bound on the capacity of backoff and acknowledgment-based protocols. SIAM Journal on Computing 33(2), 313–331 (2004)
14. Goldberg, L.A., MacKenzie, P.D., Paterson, M., Srinivasan, A.: Contention resolution with constant expected delay. Journal of the ACM 47(6), 1048–1096 (2000)
15. Greenberg, A.G., Winograd, S.: A lower bound on the time needed in the worst case to resolve conflicts deterministically in multiple access channels. Journal of the ACM 32(3), 589–596 (1985)
16. Håstad, J., Leighton, F.T., Rogoff, B.: Analysis of backoff protocols for multiple access channels. SIAM Journal on Computing 25(4), 740–774 (1996)
17. Komlós, J., Greenberg, A.G.: An asymptotically fast nonadaptive algorithm for conflict resolution in multiple-access channels. IEEE Transactions on Information Theory 31(2), 302–306 (1985)
18. Metcalfe, R.M., Boggs, D.R.: Ethernet: Distributed packet switching for local computer networks. Communications of the ACM 19(7), 395–404 (1976)
19. Raghavan, P., Upfal, E.: Stochastic contention resolution with short delays. SIAM Journal on Computing 28(2), 709–719 (1998)
20. Rosén, A., Tsirkin, M.S.: On delivery times in packet networks under adversarial traffic. Theory of Computing Systems 39(6), 805–827 (2006)
21. Tsybakov, B.S., Likhanov, N.B.: Upper bound on the capacity of a random multiple-access system. Problemy Peredachi Informatsii 23(3), 64–78 (1987)
22. Tsybakov, B.S., Mikhailov, V.A.: Ergodicity of a slotted ALOHA system. Problemy Peredachi Informatsii 15(4), 301–312 (1979)

Profit Maximization
in Flex-Grid All-Optical Networks

Mordechai Shalom[1], Prudence W.H. Wong[2], and Shmuel Zaks[3]

[1] TelHai Academic College, Upper Galilee, 12210, Israel
`cmshalom@telhai.ac.il`
[2] Department of Computer Science, University of Liverpool, UK
`pwong@liverpool.ac.uk`
[3] Department of Computer Science, Technion, Haifa, Israel
Visiting Professor, Departamento de Ingenieria Telematica, Universidad Carlos III de
Madrid, Spain, and Visiting Researcher, Institute IMDEA Networks, Madrid, Spain
`zaks@cs.technion.ac.il`

Abstract. All-optical networks have been largely investigated due to
their high data transmission rates. The key to the high speeds in all-
optical networks is to maintain the signal in optical form, to avoid the
overhead of conversion to and from electrical form at the intermediate
nodes. In the traditional WDM technology the spectrum of light that can
be transmitted through the optical fiber has been divided into frequency
intervals of fixed width with a gap of unused frequencies between them.
In this context the term wavelength refers to each of these predefined
frequency intervals.

An alternative architecture emerging in very recent studies is to move
towards a flexible model in which the usable frequency intervals are of
variable width. Every lightpath is assigned a frequency interval which
remains fixed through all the links it traverses. Two different lightpaths
using the same link have to be assigned disjoint sub-spectra. This technology is termed *flex-grid* or *flex-spectrum*.

The introduction of this technology requires the generalization of
many optimization problems that have been studied for the fixed-grid
technology. Moreover it implies new problems that are irrelevant or trivial in the current technology. In this work we focus on bandwidth utilization in path toplogy and consider two *wavelength assignment*, or in
graph theoretic terms *coloring*, problems where the goal is to maximize
the total profit. We obtain bandwidth maximization as a special case.

Keywords: all-optical networks, flex-grid, approximation algorithms,
network design.

1 Introduction

The WDM Technology: All-optical networks have been largely investigated in
recent years due to the promise of high data transmission rates. Its major applications are in video conferencing, scientific visualization, real-time medical imag-
ing, high-speed super-computing, cloud computing, distributed computing, and

T. Moscibroda and A.A. Rescigno (Eds.): SIROCCO 2013, LNCS 8179, pp. 249–260, 2013.
© Springer International Publishing Switzerland 2013

media-on-demand. The key to high speeds in all-optical networks is to maintain the signal in optical form, thereby avoiding the prohibitive overhead of conversion to and from the electrical form at the intermediate nodes.

In modern optical networks, high-speed signals are sent through optical fibers using WDM (Wavelength Division Multiplexing) technology: several signals connecting different source - destination pairs may share a link, provided they are transmitted on carriers having different wavelengths of light. These signals are routed at intermediate nodes by optical cross-connects (OXCs) that can route an incoming signal arriving from an incident edge to another, based on the signal's wavelength. A signal transmitted optically from some source node to some destination node over a wavelength is termed a *lightpath*.

Fixed-Grid and Flex-Grid DWDM Networks: Traditionally the spectrum of light that can be transmitted through the fiber has been divided into frequency intervals of fixed width with a gap of unused frequencies between them. In this context the term wavelength refers to each of these predefined frequency intervals. This technology is termed WDM, DWDM or UDWDM depending on the gap of unused frequencies between the wavelengths.

An alternative architecture emerging in very recent studies is to move away from this rigid DWDM model towards a flexible model in which the usable frequency intervals are of variable width (even within the same link). Every lightpath has to be assigned a frequency interval (*sub-spectrum*), which remains fixed through all the links it traverses. As in the traditional model, two different lightpaths using the same link have to be assigned disjoint sub-spectra. This technology is termed *flex-grid* or *flex-spectrum*, as opposed to *fixed-grid* or *fixed-spectrum* current technology. Specifically this new technology is feasible due to gridless wavelength selective switches (WSS), based on a very large number of pixels. This *sliceable* transceiver technology is not as mature, but is critical to the economic viability of flex-grid.

The introduction of the flex-grid technology requires the generalization of most of the many optimization problems that have been studied under the fixed-grid technology. For instance, as a result of the variability of the width of the sub-spectra, lightpaths have different transmission impairments, thus different regeneration needs. Another major difference is that in the fixed-grid it is assumed that lightpath requests are for one wavelength's bandwidth because otherwise it can be treated as multiple independent requests. In the flex-grid technology this assumption does not hold because two lightpaths assigned two arbitrary wavelengths are not equivalent to one lightpath assigned two consecutive colors. This assignments differ both in terms of regeneration needs, and in terms of bandwidth utilization.

In this work we focus on the bandwidth utilization in path topology as a basic network to analyze in this introductory work. Results on path topology may extend to rings and trees that are other natural topologies in optical networks. Such results often have applications in the scheduling context in which the path network becomes the time axis. For problems that are provably hard in the general case we consider special cases such as bounded load and proper intervals.

We assume that the lightpath requests have bandwidth requirements that are multiples of some basic unit. This unit is smaller than the traditional wavelength bandwidth. The entire bandwidth of the fiber is W units. We consider two *wavelength assignment*, or in graph theoretic terms *coloring*, problems. In both problems every lightpath request consists of a path P, with minimum and maximum bandwidth requirements $\mathbf{a}_P$ and $\mathbf{b}_P$ respectively, and a unit profit $\mathbf{w}_P$ (i.e., the profit for each color assigned). In the first problem such a lightpath P has to be assigned a set $w(P)$ of colors such that $\mathbf{a}_p \leq |w(P)| \leq \mathbf{b}_P$ where color is a number between 0 and $W-1$. In the second problem, in addition, the set $w(P)$ of colors assigned to a lightpath P constitutes an interval of colors from some color λ to some color $\lambda' \geq \lambda$ so that the loss, due to the otherwise unused gap between the colors, is avoided. We term these colorings as *non-contiguous colorings* (or just *colorings*), and *contiguous colorings* respectively. Note that these colorings correspond to ordinary colorings and to *interval colorings* of the intersection graph of the paths.

The profit obtained from a lightpath is proportional to the number of colors it is assigned and its unit profit, i.e. $\mathbf{w}_P \cdot |w(P)|$. Our goal is to maximize the total profit. We have an important special case when $\mathbf{w}_P$ is equal to the length of the path P. In this case the profit is the total bandwidth utilization of the network.

Related Work: [1] is a general reference for optical networks. For a discussion of their data transmission rates see [2]. [3,4] suggest flex-grid DWDM as an alternative emerging architecture. The network implications of this new architecture are explained in detail in [5], which refers to the key enabling technologies for the flex-spectrum.

Closely related to our work is the coloring and interval coloring of interval graphs. [6] is an excellent reference book on these subjects. To find an interval coloring with minimum colors in an interval graph is known as the *shipbuilding* problem, and also as the *dynamic storage allocation* problem. The problem is stated in [7] as NP-Complete under the latter name (problem [SR2]). In [8] it is conjectured to be in APX-Hard. Interval coloring of interval graphs with different optimization functions have also been studied in the literature (see for instance [9]).

Our Contribution: In this paper we consider three profit maximization problems, PMC is for non-contiguous coloring, PMCC is for contiguous coloring and PMCCC is for circularly contiguous coloring. Circularly contiguous coloring means that the interval of colors assigned can be wrapped around from $W-1$ back to 0. For PMC, we show a polynomial time optimal algorithm for arbitrary $\mathbf{a}$ and $\mathbf{b}$ when the network is a path. For PMCC, we derive an algorithm that converts a circularly contiguous coloring to a contiguous coloring with a small loss in the profit. We observe that PMCC is NP-Hard for path networks and study special cases. We study the case when the number of paths that using any given edge is bounded by some constant and give a polynomial time optimal algorithm. We further consider the case when the input set of paths is proper, i.e., no path properly contains another, and show an approximation algorithm with approximation ratio $4/3$ for some special values of $\mathbf{a}$ and $\mathbf{b}$.

2   Preliminaries

Graphs and Paths: In path (multi) coloring problems we are given a network modeled by a graph G and a set of lightpaths modeled by a set $\mathcal{P}$ of non-trivial paths of G. $V(G)$ and $E(G)$ denote the vertex set and edge set of G, respectively. We denote by $\delta_G(v)$ the set of edges incident to a vertex v in G, i.e. $\delta_G(v) = \{e \in E(G) | v \in e\}$, and $d_G(v) = |\delta_G(v)|$ is the degree of v in G. For a directed graph G, $A(G)$ denotes the arc set of G. We denote by $\delta_G^-(v)$ and $\delta_G^+(v)$ the sets of incoming arcs and outgoing arcs of a vertex v, respectively. Similarly $d_G^-(v) = |\delta_G^-(v)|$ (resp. $d_G^+(v) = |\delta_G^+(v)|$) denotes the in-degree (resp. out-degree) of v in G.

We consider paths as sets of edges, e.g. for two paths P, P' we denote by $P \cap P'$ the set of their common edges, and by $|P|$ the length of P. For an edge e of G, we denote by $\mathcal{P}_e$ the subset of $\mathcal{P}$ consisting of the paths containing e, i.e. $\mathcal{P}_e \stackrel{def}{=} \{P \in \mathcal{P} : e \in P\}$. The number of these paths is termed the *load* on the edge e, and denoted by $L_e(\mathcal{P}) \stackrel{def}{=} |\mathcal{P}_e|$. An important parameter we consider is the maximum load over all the edges of G. We denote it by $L_{max}(\mathcal{P}) \stackrel{def}{=} \max\{L_e(\mathcal{P}) : e \in E(G)\}$. Note that in the intersection graph of the paths $\mathcal{P}$, the subset of vertices corresponding to $\mathcal{P}_e$ is a clique. Therefore $L_{max}(\mathcal{P})$ is a lower bound to the size of the maximum clique of the intersection graph.

In this work we focus on the case where G is a path, i.e. the intersection graph of $\mathcal{P}$ is an *interval graph*. It is well known that every clique of an interval graph corresponds to some $\mathcal{P}_e$, therefore $L_{max}(\mathcal{P})$ is equal to the size of the maximum clique. A set of paths that no two of them intersect is an independent set of the intersection graph. When we say that a set of paths is a clique (or an independent set) we implicitly refer to their intersection graph.

Colors and Colorings: In addition to the graph G and the set $\mathcal{P}$ of paths, we are given an integer W that denotes the number of colors available. For two integers i, j such that $i \leq j$, $[i, j] \stackrel{def}{=} \{k \in \mathbb{N} : i \leq k \leq j\}$. The set of available colors is $\Lambda = [0, W - 1]$. A set $[i, j] \subseteq \Lambda$ is said to be an *interval* of colors. When $0 \leq j < i \leq W - 1$ we define $[i, j] \stackrel{def}{=} [i, W - 1] \cup [0, j]$. In both cases $[i, j]$ is termed a *circular interval* of colors, i.e. colors that are consecutive on a ring (in which 0 is the successor of $W - 1$).

A *(multi)coloring* is a function $w : \mathcal{P} \mapsto 2^{\Lambda}$ that assigns to each path $P \in \mathcal{P}$ a subset of the set Λ of colors. A coloring w is *valid* if for any two paths $P, P' \in \mathcal{P}$ such that $P \cap P' \neq \emptyset$ we have $w(P) \cap w(P') = \emptyset$. For a color $\lambda \in \Lambda$, $\mathcal{P}_\lambda^w$ denotes the set of paths assigned the color λ by w, i.e. $\mathcal{P}_\lambda^w = \{P \in \mathcal{P} : \lambda \in w(P)\}$. If w is a valid coloring, then for any two paths $P, P' \in \mathcal{P}_\lambda^w$ we have $P \cap P' = \emptyset$. In other words, $\mathcal{P}_\lambda^w$ is an independent set of $\mathcal{P}$. When there is no ambiguity, we omit the superscript w and denote $\mathcal{P}_\lambda^w$ as $\mathcal{P}_\lambda$.

A coloring is *contiguous* (resp. *circularly contiguous*), if for every $P \in \mathcal{P}$, $w(P)$ is an interval (resp. circular interval) of colors.

Vector Notation and Profits: Throughout the paper we use *vectors* of integers indexed by the elements of $\mathcal{P}$. We denote vectors with bold typeface, e.g. $\mathbf{v} =$

$\{\mathbf{v}_P : P \in \mathcal{P}\}$. The vector $\mathbf{0}$ is the zero vector, $\mathbf{1}$ is the vector consisting of a 1 in every index.

The size vector of a coloring w is a vector $\mathbf{s}(w)$ such that $\mathbf{s}(w)_P = |w(P)|$ for every $P \in \mathcal{P}$, i.e. the entries of $\mathbf{s}(w)$ are the number of colors assigned to each path. We say that a coloring w is a $(\mathbf{a} - \mathbf{b})$-coloring if $\mathbf{a} \le \mathbf{s}(w) \le \mathbf{b}$, and w is a $\mathbf{v}$-coloring if it is a $(\mathbf{v} - \mathbf{v})$-coloring. An ordinary coloring in which every path is assigned one color corresponds to a $\mathbf{1}$-coloring, and clearly any coloring is a $(\mathbf{0} - W \cdot \mathbf{1})$-coloring.

Given a real vector $\mathbf{w}$ of weights, the profit $p^w(P, \mathbf{w})$ obtained by a coloring w, from a path P is $p^w(P, \mathbf{w}) \overset{def}{=} \mathbf{w}_P \cdot |w(P)|$. The total profit due to a coloring w is $p^w(\mathcal{P}, \mathbf{w}) \overset{def}{=} \sum_{P \in \mathcal{P}} p^w(P, \mathbf{w})$.

In this work we use the term *maximum independent set* to mean an independent set with maximum profit, and denote the profit obtained from such a set as $\alpha(\mathcal{P}, \mathbf{w})$. Usually the weight function under consideration will be clear from the context and we will use $p^w(\mathcal{P})$ (resp. $\alpha(\mathcal{P})$) as a shorthand for $p^w(\mathcal{P}, \mathbf{w})$ (resp. $\alpha(\mathcal{P}, \mathbf{w})$).

We note that $p^w(\mathcal{P}) = \sum_{P \in \mathcal{P}} p^w(P) = \sum_{P \in \mathcal{P}} \mathbf{w}_P \cdot |w(P)| = \mathbf{w} \cdot \mathbf{s}(w)$. We can write the profit of a valid coloring w, from a path P as the sum of the profits obtained from every color of P, i.e. $p^w(P) = \mathbf{w}_P \cdot |w(P)| = \sum_{\lambda \in w(P)} \mathbf{w}_P$ and therefore

$$p^w(\mathcal{P}) = \sum_{P \in \mathcal{P}} \sum_{\lambda \in w(P)} \mathbf{w}_P = \sum_{\lambda \in \Lambda} \sum_{P \in \mathcal{P}_\lambda^w} \mathbf{w}_P \le \sum_{\lambda \in \Lambda} \alpha(\mathcal{P}) = W \cdot \alpha(\mathcal{P})$$

where the inequality follows from the fact that $\mathcal{P}_\lambda^w$ is an independent set.

The Problem(s): In this work we consider the following problem and its variants.

PROFIT MAXIMIZING COLORING(PMC)
Input: A tuple $(G, \mathcal{P}, W, \mathbf{a}, \mathbf{b}, \mathbf{w})$ where G is a graph, $\mathcal{P}$ is a set of paths on G, W is an integer, $\mathbf{a}$ and $\mathbf{b}$ are two integer vectors and $\mathbf{w}$ is a real vector indexed by $\mathcal{P}$.
Output: A valid $(\mathbf{a} - \mathbf{b})$-coloring w.
Objective: Maximize $p^w(\mathcal{P}, \mathbf{w})$.

The problems Profit Maximizing Contiguous Coloring (PMCC) and Profit Maximizing Circularly Contiguous Coloring (PMCCC) problems are variants of PMC in which the coloring w has to be contiguous, and circularly contiguous, respectively.

We denote the optimum of an instance $(G, \mathcal{P}, W, \mathbf{a}, \mathbf{b}, \mathbf{w})$ of a problem PRB $\in$ {PMC, PMCC, PMCCC} by $\text{OPT}_{\text{PRB}}(G, \mathcal{P}, W, \mathbf{a}, \mathbf{b}, \mathbf{w})$. A contiguous coloring is a circularly contiguous coloring, which is in turn a coloring. Therefore we have:

$$\text{OPT}_{\text{PMCC}}(G, \mathcal{P}, W, \mathbf{a}, \mathbf{b}, \mathbf{w}) \le \text{OPT}_{\text{PMCCC}}(G, \mathcal{P}, W, \mathbf{a}, \mathbf{b}, \mathbf{w})$$
$$\le \text{OPT}_{\text{PMC}}(G, \mathcal{P}, W, \mathbf{a}, \mathbf{b}, \mathbf{w}). \qquad (1)$$

Any coloring, and in particular an optimal one that we denote by w^*, satisfies
$p^{w^*}(\mathcal{P}) \leq W \cdot \alpha(\mathcal{P})$. Therefore we have

$$\text{OPT}_{\text{PMC}}(G, \mathcal{P}, W, \mathbf{a}, \mathbf{b}, \mathbf{w}) \leq W \cdot \alpha(\mathcal{P}).$$

We now observe that the above inequalities are tight when the lower and upper
bounds $\mathbf{a}$ and $\mathbf{b}$ are trivial. In other words, in this case all the three problems
equivalent to the problem of finding $\alpha(\mathcal{P})$.

Proposition 1. *If* $\mathbf{a} = \mathbf{0}, \mathbf{b} = W \cdot \mathbf{1}$ *then*

$$\text{OPT}_{\text{PMCC}}(G, \mathcal{P}, W, \mathbf{a}, \mathbf{b}, \mathbf{w}) = \text{OPT}_{\text{PMCCC}}(G, \mathcal{P}, W, \mathbf{a}, \mathbf{b}, \mathbf{w})$$
$$= \text{OPT}_{\text{PMC}}(G, \mathcal{P}, W, \mathbf{a}, \mathbf{b}, \mathbf{w}) = W \cdot \alpha(\mathcal{P}).$$

Proof. It suffices to show that $\text{OPT}_{\text{PMCC}}(G, \mathcal{P}, W, \mathbf{a}, \mathbf{b}, \mathbf{w}) \geq W \cdot \alpha(\mathcal{P})$. Indeed,
let $\mathcal{I}$ be a maximum independent set of $\mathcal{P}$. The coloring that assigns Λ to ever
path of $\mathcal{I}$ and $\emptyset$ to all the rest is a valid contiguous $(\mathbf{0} - W \cdot \mathbf{1})$-coloring with
profit $W \cdot \alpha(\mathcal{P})$.   □

Path Networks: When G is a path we assume without loss of generality that
the vertex set of G is $[1, n]$ where the vertices are numbered according to their
order in G. We sometimes refer to the vertices and edges of G as drawn on the
real line where 1 is the leftmost vertex and n is the rightmost one. Given this
numbering, $s(P)$ and $t(P)$ denote the endpoints of a path P with $s(P) < t(P)$.
We term these vertices as the *start* and *termination* vertices of P, respectively.
We denote a sub-path of G with endpoints $i < j$ as $[i, j]$, i.e. $P = [s(P), t(P)]$.
Given a sub-path δ of G, $\mathcal{P}_\delta$ denotes the set of all paths of $\mathcal{P}$ that are contained
in δ.

3   Profit Maximizing Colorings

A maximum independent set can be calculated in polynomial time when the
network is a path [6]. By Proposition 1 this implies an algorithm for all three
problems for the case where G is a path and $\mathbf{a} = \mathbf{0}$ and $\mathbf{b} = W \cdot \mathbf{1}$. In this
section we extend the study to path networks for arbitrary $\mathbf{a}$ and $\mathbf{b}$, and provide
a polynomial-time optimal algorithm.

We first introduce notations and definitions that we use in this section. Let w
be a coloring of a set $\mathcal{Q}$ of paths, and $\mathcal{Q}' \subseteq \mathcal{Q}$. $w' = w\big|_{\mathcal{Q}'}$ denotes the coloring w
restricted to $\mathcal{Q}'$, i.e. $w'(P) = w(P)$ whenever $P \in \mathcal{Q}'$, and $w'(P) = \emptyset$ otherwise.

We reduce PMC to the Minimum Cost Maximum Flow (MINCOSTMAXFLOW)
problem that is well known to be solvable in polynomial time [10]. Instances of
MINCOSTMAXFLOW are tuples $(H, s, t, \kappa, \kappa', c)$ where H is a directed graph,
$s \in V(H)$ (resp. $t \in V(H)$) is the source (resp. sink) vertex, $\kappa : A(H) \mapsto \mathbb{R}$
(resp. $\kappa' : A(H) \mapsto \mathbb{R}$) determines the lower (resp. upper) bounds of the flow on
every arc, and finally $c : A(H) \mapsto \mathbb{R}$ determines the cost of a unit flow on every
arc. The goal is to find a flow $f : A(H) \mapsto \mathbb{R}$ from s to t that has a minimum

cost among all maximum flows, i.e. among all flows of maximum amount, as follows. Recall that the amount of a flow f is the amount of flow entering t, i.e. $\sum_{e \in \delta_H^-(t)} f(e)$ and its cost $c(f)$ is $\sum_{e \in A(H)} f(e) \cdot c(e)$.

Given an instance $I = (G, \mathcal{P}, W, \mathbf{a}, \mathbf{b}, \mathbf{w})$ of PMC, we build a flow network $N(I) = (H, s, t, \kappa, \kappa', c)$. For convenience we introduce two additional semi-infinite (i.e. having one endpoint) paths $P^{(-)} = [-\infty, 1]$ and $P^{(+)} = [n, \infty]$ with zero profit, and we define $\mathcal{P}' = \mathcal{P} \cup \{P^{(-)}, P^{(+)}\}$. $V(H) = S \cup T$ where $T = \{t_P : P \in \mathcal{P}'\}$, $S = \{s_P : P \in \mathcal{P}'\}$. $A(H) = A_1 \cup A_2$ where $A_1 = \{(s_P, t_P) : P \in \mathcal{P}'\}$ and $A_2 = \{(t_P, s_{P'}) : s(P') \geq t(P)\}$. We proceed with the bounds and costs of the arcs. For every path $P \in \mathcal{P}$ the bounds and costs on the corresponding arc $a = (s_P, t_P) \in A_1$ are $\kappa(a) = \mathbf{a}_P$, $\kappa'(a) = \mathbf{b}_P$ and $c(a) = -\mathbf{w}_P$. For each one of the two arcs a corresponding to the two semi-infinite paths we set $\kappa(a) = 0$, $\kappa'(a) = W$ and $c(a) = 0$. For an arc $a = (t_P, s_{P'})$ of A_2 we set $\kappa(a) = 0$, $\kappa'(a) = \infty$ and $c(a) = 0$. Finally we set $s = s_{P^{(-)}}$ and $t = t_{P^{(+)}}$.

Lemma 1. *For every feasible coloring w of an instance I of PMC, there is a maximum flow $f^{(w)}$ of $N(I)$, such that $c(f^{(w)}) = -p^w(\mathcal{P})$. Moreover, given a maximum flow f of $N(I)$ a coloring w such that $f^{(w)} = f$ can be found in polynomial-time.*

Proof. We first observe that the maximum flow of $N(I)$ is W. Indeed a flow of amount W can be pushed from $s_{P^{(-)}}$ via $t_{P^{(-)}}$ and $s_{P^{(+)}}$ to $t_{P^{(+)}}$. On the other hand this is a maximum flow because the arc $(s_{P^{(-)}}, t_{P^{(-)}})$ constitutes a cut of weight W.

Given a feasible coloring w of I we define the flow $f^{(w)}$ as the sum of W flows $f_1^{(w)}, f_2^{(w)}, \ldots, f_W^{(w)}$. For each color $\lambda \in \Lambda$, $f_\lambda^{(w)}$ corresponds to the independent set $\mathcal{P}_\lambda^w$. $f_\lambda^{(w)}$ pushes one unit of flow from $s_{P^{(-)}}$ to $t_{P^{(+)}}$ over the path that consists of the arcs of A_1 corresponding to the paths of $\mathcal{P}_\lambda^w$ and the arcs of A_2 connecting two consecutive paths of $\mathcal{P}_\lambda^w$. The cost of an A_2 arc is zero, and the cost of an A_1 arcs corresponding to a path P is $-\mathbf{w}_P$. Therefore the cost of f_λ is $c(f_\lambda^{(w)}) = -\sum_{P \in \mathcal{P}_\lambda^w} \mathbf{w}_P$. Summing up over all colors λ we get

$$c(f^{(w)}) = \sum_{\lambda \in \Lambda} c(f_\lambda^{(w)}) = -\sum_{\lambda \in \Lambda} \sum_{P \in \mathcal{P}_\lambda^w} \mathbf{w}_P = -p^w(\mathcal{P}).$$

$f^{(w)}$ satisfies the bounds κ and κ'. Indeed, for an arc a of A_1 corresponding to a path $P \in \mathcal{P}'$ we have $f^{(w)}(a) = |w(P)|$ and $\kappa(a) = \mathbf{a}_P \leq |w(P)| \leq \mathbf{b}_P = \kappa'(a)$. For the arcs of A_2 we have $\kappa(a) = 0 \leq f^{(w)}(a) \leq \infty = \kappa'(a)$.

Any maximum flow f of $N(I)$ can be split, in polynomial time, into W unit flows $f_1, f_2, \ldots, f_W$. Each unit flow uses a path from $s_{P^{(-)}}$ to $t_{P^{(+)}}$. Such a path starts with an A_1 arc, and alternates between A_1 and A_2 arcs. The set of odd arcs corresponds to an independent set paths of $\mathcal{P}$. w is defined such that $\mathcal{P}_\lambda^w$ is the independent set corresponding to f_λ. $\square$

Corollary 1. *The profit $p^w(\mathcal{P})$ is maximized when $c(f^{(w)})$ is minimum.*

This implies the following a polynomial time algorithm for PMC: Given an instance I, calculate a minimum cost maximum flow f of $N(I)$ and return a coloring w such that $f^{(w)} = f$.

4 Profit Maximizing Contiguous Colorings

In this section we consider contiguous colorings. We first observe that the problem is NP-Hard even if the graph is a path. In Section 4.1 we compare circularly contiguous colorings to contiguous colorings and we provide an algorithm that transforms a circularly contiguous coloring to a contiguous coloring with a small loss in the profit. In Section 4.2 we consider the case where the load on the edges is bounded by some constant and provide a polynomial-time algorithm for this case. In Section 4.3 we provide an approximation algorithm for another special case where the paths constitute a proper set.

Let G be a graph and f a weight function $f : V(G) \to \mathbb{N}$ on its vertices. An interval coloring w of G, f assigns an interval $w(v)$ of $f(v)$ integers to every vertex v of G, such that $f(v) \cap f(v') = \emptyset$ whenever v and v' are adjacent in G. The weight $f(K)$ of a clique $K \subseteq V(G)$ is the sum $\sum_{v \in K} f(v)$ of the individual weights of its vertices. The clique number $\omega(G, f)$ of the weighted graph (G, f) is the maximum weight of its cliques. The interval chromatic number of $\chi(G, f)$ is the minimum number of colors used by an interval coloring of (G, f) [6]. Clearly $\chi(G, f) \geq \omega(G, f)$. The problem of finding the interval chromatic number of a weighted interval graph is also known as the *shipbuilding* problem, and also as the *dynamic storage allocation* problem. This problem is known to be NP-Complete [7]. Therefore

Lemma 2. PMCC *is* NP-*Hard even when G is a path.*

Proof. Let (G, f) be a weighted interval graph, and $\mathcal{P}$ the set of paths on a path H which represent G. Let $\mathbf{w}$ be any weight function on $\mathcal{P}$. The instance $(H, \mathcal{P}, W, f, f, \mathbf{w})$ is feasible if and only if the interval chromatic number of (G, f) is at most W.   □

4.1 Comparison with Circularly Contiguous Colorings

In this section we present the algorithm CIRCULARTOCONTIGUOUS that converts a circularly contiguous $(\mathbf{a} - \mathbf{b})$-*coloring* w^{cc} to a contiguous $(\lceil \mathbf{a}/2 \rceil - \mathbf{b})$-*coloring* w^c such that $p^{w^c}(\mathcal{P}) \geq \frac{3}{4} p^{w^{cc}}(\mathcal{P})$.

A circularly contiguous interval $[i, j]$ is either contiguous or the disjoint union of two contiguous intervals $[j, W-1], [0, i]$. The size of one of these sub-intervals is at least half of the size of the entire interval. CIRCULARTOCONTIGUOUS chooses a color $\bar{\lambda}$ uniformly at random and renames all the colors such that $\bar{\lambda}$ becomes 0, $(\bar{\lambda} + 1) \mod W$ becomes 1, and so on. Then to every path P for which the obtained coloring is not contiguous it assigns the biggest among the two corresponding contiguous colorings.

w^c is clearly a contiguous ($\lceil \mathbf{a}/\mathbf{2} \rceil - \mathbf{b}$)-*coloring*. For a given path P we now calculate the expected value of $|w^c(P)|$. Let $\ell = |w^{cc}(P)|$, and $[i,j] = w^{cc}(P)$. We consider three cases: (a) $\bar{\lambda}$ is not in $[i+1,j]$. In this case, after the renaming phase, $w^c(P)$ is contiguous. Therefore $|w^c(P)| = \ell$. (b) $\bar{\lambda} = i+k$ and $k <= \ell/2$. In this case $|w^c(P)| = \ell - k$. (c) $\bar{\lambda} = i+k$ and $l/2 < k < l$. In this case $|w^c(P)| = k$. The probability that $\bar{\lambda}$ gets any given value is $1/W$. We consider only the case that ℓ is even which leads to a smaller expected value. We have

$$E[|w^c(P)|] = \frac{1}{W} \left(\sum_{k=1}^{\ell/2}(\ell - k) + \sum_{k=\ell/2+1}^{l-1} k + (W - \ell + 1)\ell \right)$$

$$= \frac{1}{W}\left(\frac{3}{4}\ell^2 - \ell + (W - \ell + 1)\ell \right) = \ell - \frac{\ell}{W}\frac{\ell}{4} \geq \frac{3}{4}\ell = \frac{3}{4}|w^{cc}(P)|.$$

We use the above inequality and linearity of expectation to calculate the expected value of the solution.

$$E[p^{w^c}(\mathcal{P})] = E[\mathbf{w} \cdot \mathbf{s}(w^c)] = \mathbf{w} \cdot E[\mathbf{s}(w^c)] \geq \frac{3}{4}E[\mathbf{s}(w^{cc})] = \frac{3}{4}p^{w^{cc}}(\mathcal{P}).$$

Therefore

Lemma 3. *There is a randomized polynomial-time algorithm that converts a valid circularly contiguous* ($\mathbf{a}-\mathbf{b}$)-*coloring* w^{cc} *to a valid contiguous* ($\lceil \mathbf{a}/\mathbf{2} \rceil - \mathbf{b}$)-*coloring* w^c *satisfying* $p^{w^c}(\mathcal{P}) \geq \frac{3}{4}p^{w^{cc}}(\mathcal{P})$.

The above randomized algorithm can be de-randomized by trying every possible value of W and picking up the best result. Clearly at least one solution is at least as good as the expected value. This de-randomization does not lead to a polynomial-time algorithm whenever the value of W is exponential in the input size. An efficient de-randomization can be obtained by guessing each one bit of $\bar{\lambda}$ at a time. We conclude

Lemma 4. *There is a deterministic polynomial-time algorithm that converts a valid circularly contiguous* ($\mathbf{a}-\mathbf{b}$)-*coloring* w^{cc} *to a valid contiguous* ($\lceil \mathbf{a}/\mathbf{2} \rceil - \mathbf{b}$)-*coloring* w^c *satisfying* $p^{w^c}(\mathcal{P}) \geq \frac{3}{4}p^{w^{cc}}(\mathcal{P})$.

4.2   Bounded Load

Let $I = (G, \mathcal{P}, W, \mathbf{a}, \mathbf{b}, \mathbf{w})$ be an instance of PRB $\in \{\text{PMC}, \text{PMCC}, \text{PMCCC}\}$, and let $v \in [1,n]$. We denote by $I^{(v+)}$ the instance obtained from I by restricting the paths set to ones that start at vertex v or before. Formally $I^{(v+)} = (G, \mathcal{P}^{(v+)}, W, \mathbf{a}^{(v)}, \mathbf{b}^{(v+)}, \mathbf{w}^{(v+)})$ where $\mathcal{P}^{(v+)} = \{P \in \mathcal{P} : s(P) \leq v\}$, $\mathbf{a}^{(v)} = \mathbf{a}\big|_{\mathcal{P}^{(v+)}}$, $\mathbf{b}^{(v+)} = \mathbf{b}\big|_{\mathcal{P}^{(v+)}}$ and $\mathbf{w}^{(v+)} = \mathbf{w}\big|_{\mathcal{P}^{(v+)}}$.

We say that two colorings w, w' of two subsets $\mathcal{Q}, \mathcal{Q}'$ of $\mathcal{P}$ *agree* if $w(P) = w'(P)$ whenever $P \in \mathcal{Q} \cap \mathcal{Q}'$, and we denote this by $w \sim w'$. Let $\bar{w}$ be a coloring of the paths $\mathcal{P}_{e_v}$ where e_v denotes the edge $\{v-1, v\}$. We denote by $\text{OPT}_{\text{PRB}}(I, v, \bar{w})$ the optimum of problem PRB for the instance $I^{(v)}$ when the feasible colorings are restricted to colorings that *agree* with $\bar{w}$. As our goal in this section is to provide an optimal algorithm for PMCC, and this implies an optimal algorithm for all problems, in the sequel we refer only to this problem, although the arguments hold for all three problems. Clearly

$$\mathrm{OPT_{PMCC}}(I) = \mathrm{OPT_{PMCC}}(I^{(n)}) = \max\left\{\mathrm{OPT_{PMC}}(I, n, \bar{w}) : \bar{w} \text{ is a cont. coloring of} \mathcal{P}_{e_n}\right\}.$$

Consider a contiguous coloring w of $\mathcal{P}^{(v+)}$, and a contiguous coloring w^- of $\mathcal{P}^{(v-1)}$ that agrees w. We have

$$p^w(\mathcal{P}^{(v+)}) = p^{w^-}(\mathcal{P}^{(v-1)}) + \sum_{P \text{ s.t. } s(P)=v-1} |w(P)| \cdot \mathbf{w}_P.$$

We note that the second term depends only on $w\big|_{\mathcal{P}_{e_v}}$. Among all contiguous colorings w that agree with a given contiguous coloring $\bar{w}$ of $\mathcal{P}_{e_v}$, the second term is a constant. Therefore the maximum is obtained at the maximum of the first term. We conclude

$$\mathrm{OPT_{PMCC}}(I, v, \bar{w}) = \max_{\bar{w}^- \sim \bar{w}} \left\{\mathrm{OPT_{PMCC}}(I, v-1, \bar{w}^-)\right\} + \sum_{P \text{ s.t. } s(P)=v-1} |w(P)| \cdot \mathbf{w}_P.$$

These equations imply the dynamic programming algorithm CONTCOLOR-DYNPROG. For simplicity CONTCOLORDYNPROG calculates the optimum of the instance without explicitly finding an optimal coloring. It can be easily extended to return an optimal coloring.

Algorithm 1. CONTCOLORDYNPROG $I = (G, \mathcal{P}, W, \mathbf{a}, \mathbf{b}, \mathbf{w})$

1: $\mathrm{OPT_{PMC}}(I, 1, w_{empty}) \leftarrow 0$.   ▷ w_{empty} is the empty coloring.
2: **for** $v = 2$ to $v = n$ **do**
3:     **for all** Contiguous colorings $\bar{w}$ of $\mathcal{P}_{e_v}$ **do**
4:         $C \leftarrow \sum_{P \text{ s.t. } s(P)=v-1} |w(P)| \cdot \mathbf{w}_P$.
5:         $M \leftarrow 0$.
6:         **for all** Contiguous colorings $\bar{w}^-$ of $\mathcal{P}_{e_{v-1}}$ s.t. $\bar{w}^- \sim \bar{w}$ **do**
7:             **if** $\mathrm{OPT_{PMCC}}(I, v-1, \bar{w}^-) > M$ **then**
8:                 $M \leftarrow \mathrm{OPT_{PMCC}}(I, v-1, \bar{w}^-)$.
9:             **end if**
10:         **end for**
11:         $\mathrm{OPT_{PMCC}}(I, v, \bar{w}) \leftarrow M + C$.
12:     **end for**
13: **end for**
14: **return** $\max\left\{\mathrm{OPT_{PMC}}(I, n, \bar{w}) : \bar{w} \text{ is a contiguous coloring of } \mathcal{P}_{e_n}\right\}$.

The loops at lines 3 and 6 constitute the dominant part in the running time of the algorithm. A contiguous coloring of $\mathcal{P}_{e_v}$ can be found by fixing a permutation of the $\ell = L_{e_v}$ paths, and assigning to each path a positive number so that their sum does not exceed W. The number of permutations is $\ell!$ and the number of possible assignments of the numbers is $\binom{W}{\ell}$. Therefore each one of the loops iterates at most $\ell! \binom{W}{\ell} \leq W^\ell$ times, and the total number of iterations is at most $W^{2 \cdot \ell} \leq W^{2 \cdot L_{max}(\mathcal{P})}$. Therefore

Lemma 5. *There is a polynomial-time algorithm that solves* PMC $(G, \mathcal{P}, W, \mathbf{a}, \mathbf{b}, \mathbf{w})$ *when G is a path network and $L_{max}(\mathcal{P})$ is bounded by a constant.*

4.3   Proper Sets of Paths

A set of paths is *proper* if no path in the set properly contains another. The intersection graph of a proper set of paths on a path graph is a *proper interval graph*. Let P, P' be two paths in a proper set $\mathcal{P}$ of paths. $s(P) \leq s(P')$ if and only of $t(P) \leq t(P')$.

We present a simple algorithm PROPERTOCIRCULAR that converts any coloring w of a proper set of paths can to a circularly contiguous coloring w^{cc} with the same profit. PROPERTOCIRCULAR iterates over the paths according to the total order implied by their start vertices. Every path is assigned a circular interval $[\lambda, \lambda + |w(P)| - 1]$ where $\lambda = 0$ for the first path, and for each subsequent path λ is the last color of the previous path, plus one. Clearly w^{cc} is a circularly contiguous $(\mathbf{a} - \mathbf{b})$-*coloring* and $p^{w^{cc}}(\mathcal{P}) = p^w(\mathcal{P})$. It remains to show that w^{cc} is valid.

Assume, by way of contradiction, that w^{cc} is not valid. Then there are two intersecting paths $P, P' \in \mathcal{P}$ and a color λ such that $\lambda \in w^{cc}(P) \cap w^{cc}(P')$. Assume without loss of generality that $s(P) \leq s(P')$, and let e be the last edge of P, i.e. $e = \{t(P) - 1, t(P)\}$. As $\mathcal{P}$ is a proper set of paths and $P \cap P' \neq \emptyset$, we have $e \in P'$. Moreover any path P'' such that $s(P) \leq s(P'') \leq s(P')$ contains the edge e. Therefore the set Q of all paths whose start vertices are between $s(P)$ and $s(P')$ (inclusive) is a subset of $\mathcal{P}_e$. As PROPERTOCIRCULAR considers the paths in the order of their start vertices, and λ was used in both P and P', this means that the number of colors assigned by w^{cc} to the paths of Q exceeds W. However, this is exactly the number of colors assigned to these paths by w. Then w assigns more than W colors to the paths of $\mathcal{P}_e$, therefore invalid, contradicting our assumption.

Combining with (1) we conclude

Lemma 6. *When G is a path and $\mathcal{P}$ is a proper set of paths*

$$\text{OPT}_{\text{PMCCC}}(G, \mathcal{P}, W, \mathbf{a}, \mathbf{b}, \mathbf{w}) = \text{OPT}_{\text{PMC}}(G, \mathcal{P}, W, \mathbf{a}, \mathbf{b}, \mathbf{w}).$$

Moreover there is a polynomial-time algorithm solving PMCCC$(G, \mathcal{P}, W, \mathbf{a}, \mathbf{b}, \mathbf{w})$ *optimally.*

Combining this with Lemma 4 we obtain the following two corollaries.

Corollary 2. *There is a deterministic polynomial-time $4/3$-approximation algorithm for* PMCC $(G, \mathcal{P}, W, \mathbf{a}, \mathbf{b}, \mathbf{w})$ *when G is a path, $\mathcal{P}$ is a proper set of paths, $\mathbf{b}$ is a valid coloring and $\mathbf{a} \leq \lceil \mathbf{b}/2 \rceil$.*

5   Future Work

A few open problems regarding contiguous colorings in path networks that are closely related to our results, namely: a) to find an approximation algorithm for PMCC, b) to obtain prove APX-hardness of PMCC, c) To determine if PMCC is polynomial time solvable for proper intervals.

Another research direction is to extend the results to other topologies, especially those that are relevant in optical networks, such as rings, trees, grids, bounded treewidth. Finally, as stated in the introduction, the flex-grid technology opens a wide range of problems, such as regenerator placement, traffic grooming etc., that have been studied in the fixed-grid context, to be reconsidered in the flex-grid context.

References

1. Ramaswami, R., Sivarajan, K.N., Sasaki, G.H.: Optical Networks: A Practical Perspective. Kaufmann Publisher Inc., San Francisco (2009)
2. Klasing, R.: Methods and problems of wavelength-routing in all-optical networks. In: Proceeding of the MFCS 1998 Workshop on Communication, Brno, Czech Republic, August 24-25, pp. 1–9 (1998)
3. Jinno, M., Takara, H., Kozicki, B., Tsukishima, Y., Sone, Y., Matsuoka, S.: Spectrum-efficient and scalable elastic optical path network: architecture, benefits, and enabling technologies. Comm. Mag. 47, 66–73 (2009)
4. Gerstel, O.: Realistic approaches to scaling the IP network using optics. In: Optical Fiber Communication Conference and Exposition and the National Fiber Optic Engineers Conference (OFC/NFOEC), pp. 1–3 (March 2011)
5. Gerstel, O.: Flexible use of spectrum and photonic grooming. In: Photonics in Switching, OSA (Optical Society of America) Technical Digest, page paper PMD3 (2010)
6. Golumbic, M.C.: Algorithmic Graph Theory and Perfect Graphs. Annals of Discrete Mathematics, vol. 57. North-Holland Publishing Co., Amsterdam (2004)
7. Garey, M., Johnson, D.S.: Computers and Intractability, A Guide to the Theory of NP-Completeness. Freeman (1979)
8. Halldórsson, M.M., Kortsarz, G.: Multicoloring: Problems and techniques. In: Fiala, J., Koubek, V., Kratochvíl, J. (eds.) MFCS 2004. LNCS, vol. 3153, pp. 25–41. Springer, Heidelberg (2004)
9. Buchsbaum, A.L., Karloff, H., Kenyon, C., Reingold, N., Thorup, M.: Opt versus load in dynamic storage allocation. SIAM Journal of Computing 33(3), 632–646 (2004)
10. Goldberg, A.V., Tarjan, R.E.: Finding minimum-cost circulations by canceling negative cycles. J. ACM 36(4), 873–886 (1989)

Measuring the Impact of Adversarial Errors on Packet Scheduling Strategies*

Antonio Fernández Anta[1], Chryssis Georgiou[2], Dariusz R. Kowalski[3,**],
Joerg Widmer[1], and Elli Zavou[1,4,***]

[1] Institute IMDEA Networks
[2] University of Cyprus
[3] University of Liverpool
[4] Universidad Carlos III de Madrid

Abstract. In this paper we explore the problem of achieving efficient packet transmission over unreliable links with worst case occurrence of errors. In such a setup, even an omniscient offline scheduling strategy cannot achieve stability of the packet queue, nor is it able to use up all the available bandwidth. Hence, an important first step is to identify an appropriate metric for measuring the efficiency of scheduling strategies in such a setting. To this end, we propose a *relative throughput* metric which corresponds to the *long term competitive ratio* of the algorithm with respect to the optimal. We then explore the impact of the error detection mechanism and feedback delay on our measure. We compare instantaneous error feedback with deferred error feedback, that requires a faulty packet to be fully received in order to detect the error. We propose algorithms for worst-case adversarial and stochastic packet arrival models, and formally analyze their performance. The relative throughput achieved by these algorithms is shown to be close to optimal by deriving lower bounds on the relative throughput of the algorithms and almost matching upper bounds for any algorithm in the considered settings. Our collection of results demonstrate the potential of using instantaneous feedback to improve the performance of communication systems in adverse environments.

1 Introduction

Motivation. Packet scheduling [8] is one of the most fundamental problems in computer networks. As packets arrive, the sender (or scheduler) needs to continuously make scheduling decisions. Typically, the objective is to maximize the *throughput* of the link or to achieve stability. Furthermore, the sender needs to take decisions without knowledge of future packet arrivals. Therefore, many times this problem is treated as an *online* scheduling problem [4,10] and *competitive analysis* [1,13] is used to evaluate the performance of proposed solutions: the worst-case performance of an online algorithm is compared with the performance of an offline optimal algorithm that has a priori knowledge of the problem's input.

---

* This research was supported in part by the Comunidad de Madrid grant S2009TIC-1692, Spanish MICINN/MINECO grant TEC2011-29688-C02-01, and NSF of China grant 61020106002.
** This work was performed during the visit of D. Kowalski to Institute IMDEA Networks.
*** Partially supported by FPU Grant from MECD.

T. Moscibroda and A.A. Rescigno (Eds.): SIROCCO 2013, LNCS 8179, pp. 261–273, 2013.
© Springer International Publishing Switzerland 2013

In this work we focus on online packet scheduling over *unreliable* links, where packets transmitted over the link might be corrupted by bit errors. Such errors may, for example, be caused by an increased noise level or transient interference on the link, that in the worst case could be caused by a malicious entity or an attacker. In the case of an error the affected packets must be retransmitted. To investigate the impact of such errors on the scheduling problem under study and provide *provable guarantees*, we consider the worst case occurrence of errors, that is, we consider errors caused by an omniscient and adaptive *adversary* [12]. The adversary has full knowledge of the protocol and its history, and it uses this knowledge to decide whether it will cause errors on the packets transmitted in the link at a certain time or not. Within this general framework, the packet arrival is continuous and can either be controlled by the adversary or be stochastic.

Contributions. Packet scheduling performance is often evaluated using throughput, measured in absolute terms (e.g., in bits per second) or normalized with respect to the bandwidth (maximum transmission capacity) of the link. This throughput metric makes sense for a link without errors or with random errors, where the full capacity of the link can be achieved under certain conditions. However, if adversarial bit errors can occur during the transmission of packets, the full capacity is usually not achievable by any protocol, unless restrictions are imposed on the adversary [2,12]. Moreover, since a bit error renders a whole packet unusable (unless costly techniques like PPR [5] are used), a throughput equal to the capacity minus the bits with errors is not achievable either. As a consequence, in a link with adversarial bit errors, a fair comparison should compare the throughput of a specific algorithm to the maximum achievable amount of traffic that *any* protocol could send across the link. This introduces the challenge of identifying an appropriate metric to measure the throughput of a protocol over a link with adversarial bit errors.

Relative throughput: Our first contribution is the proposal of a *relative throughput* metric for packet scheduling algorithms under unreliable links (Section 2). This metric is a variation of the competitive ratio typically considered in online scheduling. Instead of considering the ratio of the performance of a given algorithm over that of the optimal offline algorithm, we consider the limit of this ratio as time goes to infinity. This corresponds to the *long term competitive ratio* of the algorithm with respect to the optimal.

Problem outline: We consider a sender that transmits packets to a receiver over an unreliable link, where the errors are controlled by an adversary. Regarding packet arrivals (at the sender), we consider two models: (a) the arrival times and their sizes follow a stochastic distribution, and (b) the arrival times and their sizes are also controlled by an adversary. The general offline version of our scheduling problem, in which the scheduling algorithm knows a priori when errors will occur, is NP-hard. This further motivates the need for devising simple and efficient online algorithms for the problem we consider.

Feedback mechanisms: Then, moving to the online problem requires detecting the packets received with errors, in order to retransmit them. The usual mechanism [7], which we call *deferred feedback*, detects and notifies the sender that a packet has suffered an error after the whole packet has been received by the receiver. It can be shown that, even when the packet arrivals are stochastic and packets have the same length, no online scheduling algorithm with deferred feedback can be competitive with respect to

Table 1. Summary of results presented. The results for deferred feedback are for one packet length, while the results for instantaneous feedback are for 2 packet lengths ℓ_{min} and ℓ_{max}. Note that $\gamma = \ell_{max}/\ell_{min}$, $\overline{\gamma} = \lfloor \gamma \rfloor$, λp is the arrival rate of ℓ_{min} packets, and p and $q = 1 - p$ are the proportions of ℓ_{min} and ℓ_{max} packets, respectively.

Arrivals	Feedback	Upper Bound	Lower Bound
	Deferred	0	0
Adversarial	Instantaneous	$T_{Alg} \leq \overline{\gamma}/(\gamma + \overline{\gamma})$ $T_{LL} = 0, T_{SL} \leq 1/(\gamma + 1)$	$T_{SL-Pr} \geq \overline{\gamma}/(\gamma + \overline{\gamma})$
	Deferred	0	0
Stochastic	Instantaneous	$T_{Alg} \leq \overline{\gamma}/\gamma$ $T_{Alg} \leq \max\left\{\lambda p\ell_{min}, \frac{\overline{\gamma}}{(\gamma+\overline{\gamma})}\right\}$ if $p < q$ $T_{LL} = 0, T_{SL} \leq 1/(\gamma + 1)$	$T_{CSL-Pr} \geq \overline{\gamma}/(\gamma + \overline{\gamma})$, if $\lambda p\ell_{min} \leq \overline{\gamma}/(2\gamma)$ $T_{CSL-Pr} \geq \min\{\lambda p\ell_{min}, \overline{\gamma}/\gamma\}$, otherwise

the offline one. Hence, we center our study a second mechanism, which we call *instantaneous feedback*. It detects and notifies the sender of an error the moment this error occurs. This mechanism can be thought of as an abstraction of the emerging Continuous Error Detection (CED) framework [11] that uses arithmetic coding to provide continuous error detection. The difference between deferred and instantaneous feedback is drastic, since for the instantaneous feedback mechanism, and for packets of the same length, it is easy to obtain optimal relative throughput of 1, even in the case of adversarial arrivals. However, the problem becomes substantially more challenging in the case of non-uniform packet lengths. Hence, we analyze the problem for the case of packets with two different lengths, ℓ_{min} and ℓ_{max}, where $\ell_{min} < \ell_{max}$.

Bounds for adversarial arrivals: We show (Section 3), that an online algorithm with instantaneous feedback can achieve at most almost half the relative throughput with respect to the offline one. It can also be shown that two basic scheduling policies, giving priority either to short *(SL – Shortest Length)* or long *(LL – Longest Length)* packets, are not efficient under adversarial errors. Therefore, we devise a new algorithm, called SL-Preamble, and show that it achieves the optimal online relative throughput. Our algorithm, transmits a "sufficiently" large number of short packets while making sure that long packets are transmitted from time to time.

Bounds for stochastic arrivals: In the case of stochastic packet arrivals (Section 4), as one might expect, we obtain better relative throughput in some cases. The results are summarized in Table 1. We propose and analyze an algorithm, called CSL-Preamble, that achieves relative throughput that is optimal. This algorithm schedules packets according to SL-Preamble, giving preference to short packets depending on the parameters of the stochastic distribution of packet arrivals. (If the distribution is not known, then one needs to use the algorithm developed for the case of adversarial arrivals that needs no knowledge a priori.) We show that the performance of algorithm CSL-Preamble is optimal for a wide range of parameters of stochastic distributions of packets arrivals, by proving a matching upper bound for the relative throughput of any algorithm in this setting.(Analyzing algorithms yields lower bounds on the relative throughput, while analyzing adversarial strategies yields upper bounds on the relative throughput.)

A note on randomization: All the proposed algorithms are deterministic. Interestingly, it can be shown that using randomization does not improve the results; the upper bounds already discussed hold also for the randomized case.

To the best of our knowledge, this is the first work that investigates in depth the impact of adversarial worst-case link errors on the throughput of the packet scheduling problem. Collectively, our results (see Table 1) show that instantaneous feedback can achieve a significant relative throughput under worst-case adversarial errors (almost half the relative throughput that the offline optimal algorithm can achieve). Furthermore, we observe that in some cases, stochastic arrivals allow for better performance.

Omitted results, proofs and discussion can be found in the full version [3].

Related Work. A vast amount of work exists for online (packet) scheduling. Here we focus only on the work that is most related to ours. For more information the reader can consult [9] and [10]. The work in [6] considers the packet scheduling problem in wireless networks. Like our work, it looks at both stochastic and adversarial arrivals. Unlike our work though, it considers only *reliable* links. Its main objective is to achieve maximal throughput guaranteeing *stabiliy*, meaning bounded time from injection to delivery. The work in [2] considers online packet scheduling over a wireless channel, where both the channel conditions and the data arrivals are governed by an adversary. Its main objective is to design scheduling algorithms for the base-station to achieve stability in terms of the size of queues of each mobile user. Our work does not focus on stability, as we assume errors controlled by an unbounded adversary that can always prevent it. The work in [12] considers the problem of devising local access control protocols for wireless networks with a single channel, that are provably robust against *adaptive adversarial jamming*. At certain time steps, the adversary can jam the communication in the channel so that the wireless nodes do not receive messages (unlike our work, where the receiver might receive a message, but it might contain bit errors). Although the model and the objectives of this line of work is different from ours, it shares the same concept of studying the impact of adversarial behavior on network communication.

2   Model

Network Setting. We consider a sending station transmitting packets over a link. Packets arrive at the sending station continuously and may have different lengths. Each packet that arrives is associated with a length and its arrival time (based on the station's local clock). We denote by ℓ_{min} and ℓ_{max} the smallest and largest lengths, respectively, that a packet may have. We use the notation $\gamma = \ell_{max}/\ell_{min}$, $\overline{\gamma} = \lfloor \gamma \rfloor$ and $\hat{\gamma} = \lceil \gamma \rceil - 1$. The link is unreliable, that is, transmitted packets might be corrupted by bit errors. We assume that all packets are transmitted at the same bit rate, hence the transmission time is proportional to the packet's length.

Arrival Models. We consider two models for packet arrivals.

Adversarial: The packets' arrival time and length are governed by an adversary. We define an adversarial arrival pattern as a collection of packet arrivals caused by the adversary.

Stochastic: We consider a probabilistic distribution D_a, under which packets arrive at the sending station and a probabilistic distribution D_s, for the length of the packets. In particular, we assume packets arriving according to a Poisson process with parameter $\lambda > 0$. When considering two packet lengths, ℓ_{min} and ℓ_{max}, each packet that arrives is assigned one of the two lengths independently, with probabilities $p > 0$ and $q > 0$ respectively, where $p + q = 1$.

Packet Bit Errors. We consider an adversary that controls the bit errors of the packets transmitted over the link. An adversarial error pattern is defined as a collection of error events on the link caused by the adversary. More precisely, an error event at time t specifies that an instantaneous error occurs on the link at time t, so the packet that happens to be on the link at that time is corrupted with bit errors. A corrupted packet transmission is unsuccessful, therefore the packet needs to be retransmitted in full. As mentioned before, we consider an *instantaneous feedback* mechanism for the notification of the sender about the error. The instant the packet suffers a bit error the sending station is notified (hence it can stop transmitting the remainder of the packet, if any).

The Power of the Adversary. Adversarial models are typically used to argue about the algorithm's behavior in worst-case scenarios. In this work we assume an adaptive adversary that knows the algorithm and the history of the execution up to the current point in time. In the case of stochastic arrivals, this includes all stochastic packet arrivals up to this point, and the length of the packets that have arrived. However it only knows the distribution but neither the exact timing nor the length of the packets arriving beyond the current time.

Note that in the case of deterministic algorithms, in the model of adversarial arrivals the adversary has full knowledge of the computation, as it controls both packet arrivals and errors, and can simulate the behavior of the algorithm in the future (there are no random bits involved in the computation). This is not the case in the model with stochastic arrivals, where the adversary does not control the timing of future packet arrivals, but knows only about the packet arrival and length distributions.

Efficiency Metric: *Relative Throughput.* Due to dynamic packet arrivals and adversarial errors, the real link capacity may vary throughout the execution. Therefore, we view the problem of packet scheduling in this setting as an online problem and we pursue long-term competitive analysis. Specifically, let A be an arrival pattern and E an error pattern. For a given deterministic algorithm Alg, let $L_{Alg}(A, E, t)$ be the total length of all the successfully transferred (i.e., non-corrupted) packets by time t under patterns A and E. Let OPT be the offline optimal algorithm that knows the exact arrival and error patterns before the start of the execution. We assume that OPT devises an optimal schedule that maximizes at each time t the successfully transferred packets $L_{OPT}(A, E, t)$. Observe that, in the case of stochastic arrivals, the worst-case adversarial error pattern may depend on stochastic injections. Therefore, we view E as a function of an arrival pattern A and time t. In particular, for an arrival pattern A we consider a function $E(A, t)$ that defines errors at time t based on the behavior of a given algorithm Alg under the arrival pattern A up to time t and the values of function $E(A, t')$ for $t' < t$.

Let $\mathcal{A}$ denote a considered arrival model, i.e., a set of arrival patterns in case of adversarial, or a distribution of packet injection patterns in case of stochastic, and let $\mathcal{E}$ denote the corresponding adversarial error model, i.e., a set of error patterns derived by the adversary, or a set of functions defining the error event times in response to the arrivals that already took place in case of stochastic arrivals. In case of adversarial arrivals, we require that any pair of patterns $A \in \mathcal{A}$ and $E \in \mathcal{E}$ occurring in an execution must allow non-trivial communication, i.e., the value of $L_{\mathrm{OPT}}(A, E, t)$ in the execution is unbounded with t going to infinity. In case of stochastic arrivals, we require that any adversarial error function $E \in \mathcal{E}$ applied in an execution must allow non-trivial communication for any stochastic arrival pattern $A \in \mathcal{A}$.

For arrival pattern A, adversarial error function E and time t, we define the *relative throughput* $T_{\mathrm{Alg}}(A, E, t)$ *of a deterministic algorithm Alg by time* t as:

$$T_{\mathrm{Alg}}(A, E, t) = \frac{L_{\mathrm{Alg}}(A, E, t)}{L_{\mathrm{OPT}}(A, E, t)}.$$

For completeness, $T_{\mathrm{Alg}}(A, E, t)$ equals 1 if $L_{\mathrm{Alg}}(A, E, t) = L_{\mathrm{OPT}}(A, E, t) = 0$.

We define the *relative throughput* of Alg in the adversarial arrival model as:

$$T_{\mathrm{Alg}} = \inf_{A \in \mathcal{A}, E \in \mathcal{E}} \lim_{t \to \infty} T_{\mathrm{Alg}}(A, E, t),$$

while in the stochastic arrival model it needs to take into account the random distribution of arrival patterns in $\mathcal{A}$, and is defined as follows:

$$T_{\mathrm{Alg}} = \inf_{E \in \mathcal{E}} \lim_{t \to \infty} \mathbb{E}_{A \in \mathcal{A}}[T_{\mathrm{Alg}}(A, E, t)].$$

To prove lower bounds on relative throughput, we compare the performance of a given algorithm with that of OPT. When deriving upper bounds, it is not necessary to compare the performance of a given algorithm with that of OPT, but instead, with the performance of some carefully chosen offline algorithm OFF. As we demonstrate later, this approach leads to accurate upper bound results.

Finally, we consider *work conserving* online scheduling algorithms, in the following sense: as long as there are pending packets, the sender does not cease to schedule packets. Note that it does not make any difference whether one assumes that offline algorithms are work-conserving or not, since their throughput is the same in both cases (a work conserving offline algorithm always transmits, but stops the ongoing transmission as soon as an error occurs and then continues with the next packet). Hence for simplicity we do not assume offline algorithms to be work conserving.

3 Adversarial Arrivals

This section focuses on adversarial packet arrivals. First, observe that it is relatively easy and efficient to handle packets of only one length.

Proposition 1. *Any work conserving online scheduling algorithm with instantaneous feedback has optimal relative throughput of 1 when all packets have the same length.*

3.1   Upper Bound for at Least Two Packet Lengths

Let Alg be any deterministic algorithm for the considered packet scheduling problem. In order to prove upper bounds, Alg will be competing with an offline algorithm OFF. The scenario is as follows. We consider an infinite supply of packets of length ℓ_{max} and initially assume that there are no packets of length ℓ_{min}. We define as a *link error event*, the point in time when the adversary corrupts (causes an error to) any packet that happens to be in the link at that specific time. We divide the execution in *phases*, defined as the periods between two consecutive link error events. We distinguish 2 types of phases as described below and give a description for the behavior of the adversarial models $\mathcal{A}$ and $\mathcal{E}$. The adversary controls the arrivals of packets at the sending station and error events of the link, as well as the actions of algorithm OFF. The two types of phases are as follows:

1. A phase in which Alg starts by transmitting an ℓ_{max} packet (the first phase of the execution belongs to this class). Immediately after Alg starts transmitting the ℓ_{max} packet, a set of $\hat{\gamma}\,\ell_{min}$ packets arrive, that are scheduled and transmitted by OFF. After OFF completes the transmission of these packets, a link error occurs, so Alg cannot complete the transmission of the ℓ_{max} packet (more precisely, the packet undergoes a bit error, so it needs to be retransmitted). Here we use the fact that $\hat{\gamma} < \gamma$.

2. A phase in which Alg starts by transmitting an ℓ_{min} packet. In this case, OFF transmits an ℓ_{max} packet. Immediately after this transmission is completed, a link error occurs. Observe that in this phase Alg has transmitted successfully several ℓ_{min} packets (up to $\overline{\gamma}$ of them).

Let A and E be the specific adversarial arrival and error patterns in an execution of Alg. Let us consider any time t (at the end of a phase for simplicity) in the execution. Let v_1 be the number of phases of type 1 executed by time t. Similarly, let $v_2(j)$ be the number of phases of type 2 executed by time t in which Alg transmits j ℓ_{min} packets, for $j \in [1, \overline{\gamma}]$. Then, the relative throughput can be computed as follows.

$$T_{\text{Alg}}(A, E, t) = \frac{\ell_{min} \sum_{j=1}^{\overline{\gamma}} j v_2(j)}{\ell_{max} \sum_{j=1}^{\overline{\gamma}} v_2(j) + \ell_{min}\hat{\gamma}v_1}. \tag{1}$$

From the arrival pattern A, the number of ℓ_{min} packets injected by time t is exactly $\hat{\gamma}v_1$. Hence, $\sum_{j=1}^{\overline{\gamma}} j v_2(j) \leq \hat{\gamma}v_1$. It can be easily observed from Eq. 1 that the relative throughput increases with the average number of ℓ_{min} packets transmitted in the phases of type 2. Hence, the throughput would be maximal if all the ℓ_{min} packets are used in phases of type 2 with $\overline{\gamma}$ packets. With the above we obtain the following theorem.

Theorem 1. *The relative throughput of Alg under adversarial patterns A and E and up to time t is at most $\frac{\overline{\gamma}}{\gamma+\overline{\gamma}} \leq \frac{1}{2}$ (the equality holds iff γ is an integer).*

3.2   Lower Bound and SL-Preamble Algorithm

Two natural scheduling policies one could consider are the *Shortest Length* (SL) and *Longest Length* (LL) algorithms; the first gives priority to ℓ_{min} packets, whereas the second gives priority to the ℓ_{max} packets. However, these two policies are not efficient in the considered setting; LL cannot achieve a relative throughput more than 0 while SL achieves at most $T = \frac{1}{\gamma+1}$. Therefore, we present algorithm SL-Preamble that tries to combine, in a graceful and efficient manner, these two policies.

Algorithm description: At the beginning of the execution and whenever the sender is (immediately) notified by the instantaneous feedback mechanism that a link error occurred, it checks the queue of pending packets to see whether there are at least $\overline{\gamma}$ packets of length ℓ_{min} available for transmission. If there are, then it schedules $\overline{\gamma}$ of them — this is called a *preamble* — and then the algorithm continues to schedule packets using the LL policy. Otherwise, if there are not enough ℓ_{min} packets available, it simply schedules packets following the LL policy.

Algorithm analysis (sketch): We show that algorithm SL-Preamble achieves a relative throughput that matches the upper bound shown in the previous subsection, and hence, it is optimal. According to the algorithm there are four types of phases that may occur.

1. Phase starting with ℓ_{min} packet and has length $L < \overline{\gamma}\ell_{min}$
2. Phase starting with ℓ_{min} packet and length $L \geq \overline{\gamma}\ell_{min}$
3. Phase starting with ℓ_{max} packet and has length $L < \ell_{max}$
4. Phase starting with ℓ_{max} packet and length $L \geq \ell_{max}$

For phases of type 1, SL-Preamble is not able to transmit successfully the $\overline{\gamma}$ packets ℓ_{min} of the preamble, but clearly OPT is only able to complete at most as much *work* (understood as the total length of sent packets). For phases of type 2 and 4, the amount of work completed by OPT can be at most the work completed by SL-Preamble plus ℓ_{max} (and hence the former is at most twice the latter). In the case of phases of type 3, SL-Preamble is not able to successfully transmit any packet, whereas OPT might transmit up to $\hat{\gamma}\ell_{min}$ packets. Amortizing the work completed by OPT in these phases with those completed in the preambles of types 1 and 2 by algorithm SL-Preamble is the most challenging part of the proof. This process is divided into two cases, depending on whether the number of type 3 phases is bounded or not, leading to the following:

Theorem 2. *The relative throughput of Algorithm SL-Preamble is at least* $\frac{\overline{\gamma}}{\gamma+\overline{\gamma}}$.

4   Stochastic Arrivals

We now turn our attention to stochastic packet arrivals.

4.1   Upper Bounds for at least Two Packet Lengths

In order to find the upper bound of the relative throughput, we consider again an arbitrary work conserving algorithm Alg. Recall that we assume that $\lambda p > 0$ and $\lambda q > 0$, which implies that there are in fact injections of packets of both lengths ℓ_{min} and ℓ_{max} (recall the definitions of λ, p and q from Section 2). We define the following adversarial error model $\mathcal{E}$.

1. When Alg starts a phase by transmitting an ℓ_{max} packet then,
   (a) If OFF has ℓ_{min} packets pending, then the adversary extends the phase so that OFF can transmit successfully as many ℓ_{min} packets as possible, up to $\hat{\gamma}$. Then, it ends the phase so that Alg does not complete the transmission of the ℓ_{max} packet (since $\hat{\gamma}\ell_{min} < \ell_{max}$).
   (b) If OFF does not have any ℓ_{min} packets pending, then the adversary inserts a link error immediately (say after infinitesimally small time ϵ).
2. When Alg starts a phase by transmitting an ℓ_{min} packet then,
   (a) If OFF has a packet of length ℓ_{max} pending, then the adversary extends the phase so OFF can transmit an ℓ_{max} packet. By the time this packet is successfully transmitted, the adversary inserts an error and finishes the phase. Observe that in this case Alg was able to successfully transmit up to $\overline{\gamma}$ packets ℓ_{min}.
   (b) If OFF has no ℓ_{max} packets pending, then the adversary inserts an error immediately and ends the phase.

Observe that in phases of type 1b and 2b, neither OFF nor Alg are able to transmit any packet. These phases are just used by the adversary to wait for the conditions required by phases of type 1a and 2a to hold. In those latter types some packets are successfully transmitted (at least by OFF). Hence we call them *productive* phases. Analyzing a possible execution, in addition to the concept of phase that we have already used, we define *rounds*. There is a round associated with each productive phase. The round ends when its corresponding productive phase ends, and starts at the end of the prior round (or at the start of the execution if no prior round exists). Depending on the type of productive phase they contain, rounds can be classified as type 1a or 2a.

Let us fix some (large) time t. We denote by $r_{1a}^{(j)}$ the number of rounds of type 1a in which $j \leq \hat{\gamma} \, \ell_{min}$ packets are sent by OFF completed by time t. The value $r_{2a}^{(j)}$ with $j \leq \overline{\gamma} \, \ell_{min}$ packets sent by Alg, is defined similarly for rounds of type 2a. (Here rounding effects do not have any significant impact, since they will be compensated by the assumption that t is large.) We assume that t is a time when a round finishes. Let us denote by r the total number or rounds completed by time t, i.e., $\sum_{j=1}^{\overline{\gamma}} r_{2a}^{(j)} + \sum_{j=1}^{\hat{\gamma}} r_{1a}^{(j)} = r$. The relative throughput by time t can be computed as

$$T_{\text{Alg}}(A, E, t) = \frac{\ell_{min} \sum_{j=1}^{\overline{\gamma}} j \cdot r_{2a}^{(j)}}{\ell_{max} \sum_{j=1}^{\overline{\gamma}} r_{2a}^{(j)} + \ell_{min} \sum_{j=1}^{\hat{\gamma}} j \cdot r_{1a}^{(j)}}. \tag{2}$$

From this expression, we can show the following result.

Theorem 3. *No algorithm Alg has relative throughput larger than $\frac{\overline{\gamma}}{\gamma}$.*

Proof. It can be observed in Eq. 2 that, for a fixed r, the lower the value of $r_{1a}^{(j)}$ the higher the relative throughput. Regarding the values $r_{2a}^{(j)}$, the throughput increases when there are more rounds in the larger values of j. E.g., under the same conditions, a configuration with $r_{2a}^{(j)} = k_1$ and $r_{2a}^{(j+1)} = k_2$, has lower throughput than one with $r_{2a}^{(j)} = k_1 - 1$ and $r_{2a}^{(j+1)} = k_2 + 1$. Then, the throughput is maximized when $r_{2a}^{(\overline{\gamma})} = r$ and the rest of values $r_{1a}^{(j)}$ and $r_{2a}^{(j)}$ are 0, which yields the bound. ∎

To provide tighter bounds for some special cases, we prove the following lemma.

Lemma 1. *Consider any two constants η, η' such that $0 < \eta < \lambda < \eta'$. Then:*

(a) *there is a constant $c > 0$, dependent only on λ, p, η, such that for any time $t \geq \ell_{min}$, the number of packets of length ℓ_{min} (resp., ℓ_{max}) injected by time t is at least $t\eta p$ (resp., $t\eta q$) with probability at least $1 - e^{-ct}$;*

(b) *there is a constant $c' > 0$, dependent only on λ, p, η', such that for any time $t \geq \ell_{min}$, the number of packets of length ℓ_{min} (resp., ℓ_{max}) injected by time t is at most $t\eta' p$ (resp., $t\eta' q$) with probability at least $1 - e^{-c't}$.*

Now we can show the following result.

Theorem 4. *Let $p < q$. Then, the relative throughput of any algorithm Alg is at most $\min\left\{\max\left\{\lambda p\ell_{min}, \frac{\overline{\gamma}}{\gamma+\overline{\gamma}}\right\}, \frac{\overline{\gamma}}{\gamma}\right\}$.*

Proof. The claim has two cases. In the first case, $\lambda p\ell_{min} \geq \frac{\overline{\gamma}}{\gamma}$. In this case, the upper bound of $\frac{\overline{\gamma}}{\gamma}$ is provided by Theorem 3. In the second case $\lambda p\ell_{min} < \frac{\overline{\gamma}}{\gamma}$. For this case, define two constants η, η' such that $0 < \eta < \lambda < \eta'$ and $\eta' p < \eta q$. Observe that these constants always exist. Then, we prove that the relative throughput of any algorithm Alg in this case is at most $\max\left\{\eta' p\ell_{min}, \frac{\overline{\gamma}}{\gamma+\overline{\gamma}}\right\}$.

Let us introduce some notation. We use a_t^{min} and a_t^{max} to denote the number of ℓ_{min} and ℓ_{max} packets, respectively, injected up to time t. Let r_t^{off} and s_t^{off} be the number of ℓ_{max} and ℓ_{min} packets respectively, successfully transmitted by OFF by time t. Similarly, let s_t^{alg} be the number of ℓ_{min} packets transmitted by algorithm Alg by time t. Observe that $s_t^{alg} \geq r_t^{off} \geq \lfloor \frac{s_t^{alg}}{\overline{\gamma}} \rfloor$.

Let us consider a given execution and the time instants at which the queue of OFF is empty of ℓ_{min} packets in the execution. We consider two cases.

Case 1: For each time t, there is a time $t' > t$ at which OFF has the queue empty of ℓ_{min} packets. Let us fix a value $\delta > 0$ and define time instants $t_0, t_1, \ldots$ as follows. t_0 is the first time instant no smaller than ℓ_{min} at which OFF has no ℓ_{min} packet and such that $a_{t_0}^{min} > \ell_{max}$. Then, for $i > 0$, t_i is the first time instant not smaller than $t_{i-1}+\delta$ at which OFF has no ℓ_{min} packets. The relative throughput at time t_i can be bounded as

$$T_{Alg}(A, E, t_i) \leq \frac{s_{t_i}^{alg}\ell_{min}}{r_{t_i}^{off}\ell_{max} + a_{t_i}^{min}\ell_{min}} \leq \frac{s_{t_i}^{alg}\ell_{min}}{\lfloor \frac{s_{t_i}^{alg}}{\overline{\gamma}} \rfloor \ell_{max} + a_{t_i}^{min}\ell_{min}}.$$

This bound grows with $s_{t_i}^{alg}$ when $a_{t_i}^{min} > \ell_{max}$, which leads to a bound on the relative throughput as follows:

$$T_{Alg}(A, E, t_i) \leq \frac{a_{t_i}^{min}\ell_{min}}{a_{t_i}^{min}(\frac{\ell_{max}}{\overline{\gamma}} + \ell_{min}) - \ell_{max}} = \frac{a_{t_i}^{min}\overline{\gamma}}{a_{t_i}^{min}(\gamma + \overline{\gamma}) - \gamma\overline{\gamma}}.$$

Which as i goes to infinity yields a bound of $\frac{\overline{\gamma}}{\gamma+\overline{\gamma}}$.

Case 2: There is a time t_* after which OFF never has the queue empty of ℓ_{min} packets. Recall that for any $t \geq \ell_{min}$, from Lemma 1, we have that the number of ℓ_{min} packets injected by time t satisfy $a_t^{min} > \eta' pt$ with probability at most $\exp(-c't)$ and the injected max packets satisfy $a_t^{max} < \eta qt$ with probability at most $\exp(-ct)$. By the assumption of the theorem and the definition of η and η', $\eta' p < \eta q$. Let us define

$t^* = 1/(\eta q - \eta' p)$. Then, for all $t \geq t^*$ it holds that $a_t^{\max} \geq a_t^{\min} + 1$, with probability at least $1 - \exp(-c't) - \exp(-ct)$. If this holds, it implies that OFF will always have ℓ_{max} packets in the queue.

Let us fix a value $\delta > 0$ and define $t_0 = \max(t_*, t^*)$, and the sequence of instants $t_i = t_0 + i\delta$, for $i = 0, 1, 2, \ldots$. By the definition of t_0, at all times $t > t_0$ OFF is successfully transmitting packets. Using Lemma 1, we can also claim that in the interval $(t_0, t_i]$ the probability that more than $\eta' p i \delta$ packets ℓ_{min} are injected is no more than $\exp(-c'' i \delta)$.

With the above, the relative throughput at any time t_i for $i \geq 0$ can be bounded as

$$T_{\text{Alg}}(A, E, t_i) \leq \frac{(a_{t_0}^{\min} + \eta' p \cdot i\delta)\ell_{min}}{r_{t_0}^{off}\ell_{max} + s_{t_0}^{off}\ell_{min} + i\delta}$$

with probability at least $1 - \exp(-ct_i) - \exp(-c't_i) - \exp(-c''t_i)$. Observe that as i goes to infinity the above bound converges to $\eta' p \ell_{min}$, while the probability converges exponentially fast to 1. ∎

4.2   Lower Bound and Algorithm CSL-Preamble

In this section we consider algorithm CSL-Preamble (stands for Conditional SL-Preamble), which builds on algorithm SL-Preamble presented in Section 3.2, in order to solve packet scheduling in the setting of stochastic packet arrivals. The algorithm, depending on the arrival distribution, either follows the SL policy (giving priority to ℓ_{min} packets) or algorithm SL-Preamble. More precisely, algorithm CSL-Preamble acts as follows:

If $\lambda p \ell_{min} > \frac{\overline{\gamma}}{2\gamma}$ then algorithm SL is run, otherwise algorithm SL-Preamble is executed.

Theorem 5. *The relative throughput of algorithm CSL-Preamble is not smaller than* $\frac{\overline{\gamma}}{\gamma+\overline{\gamma}}$ *for* $\lambda p \ell_{min} \leq \frac{\overline{\gamma}}{2\gamma}$, *and not smaller than* $\min\left\{\lambda p \ell_{min}, \frac{\overline{\gamma}}{\gamma}\right\}$ *otherwise.*

Proof. (Sketch) We break the analysis of the algorithm into cases according to the probability of ℓ_{min} packet arrivals and consider the time line of executions ignoring any OPT-unproductive periods.

Case $\lambda p \ell_{min} \leq \frac{\overline{\gamma}}{2\gamma}$. In this case algorithm CSL-Preamble runs algorithm SL-Preamble, achieving, per Theorem 2, relative throughput of at least $\frac{\overline{\gamma}}{\gamma+\overline{\gamma}}$ under *any* error pattern.

Case $\frac{\overline{\gamma}}{2\gamma} \leq \lambda p \ell_{min} \leq 1$. It can be proved that the relative throughput is not smaller than $\min\left\{\eta p \ell_{min}, \frac{\overline{\gamma}}{\gamma}\right\}$, for any η satisfying $\lambda/2 < \eta < \lambda$. To prove it, we consider time points t_i being multiples of ℓ_{max} and show that with high probability, at those points there have already arrived at least $t_i \eta p$ packets. Using this property, we show that the relative throughput at time t_j is at least $\min\left\{\eta p \ell_{min} - \frac{\overline{\gamma}\ell_{min}}{t_j}, (1 - 1/\sqrt{j}) \cdot \frac{\overline{\gamma}}{\gamma}\right\}$ with probability at least $1 - c' \exp\left(-ct_{\sqrt{j}}\right)$, for some constant $c, c' > 0$ dependent only on λ, η, p. It follows that if j grows to infinity, we obtain the desired relative throughput.

Case $\lambda p \ell_{min} > 1$. In this case we simply observe that we get at least the same relative throughput as in case $\lambda p \ell_{min} = 1$, because we are dealing with executions saturated with packets of length ℓ_{min} with probability converging to 1 exponentially fast. (Recall that we use the same algorithm SL in the specification of CSL-Preamble, both for $\lambda p\ell_{min} = 1$ and for $\lambda p\ell_{min} > 1$.) Consequently, the relative throughput in this case is at least $\min\{\eta p\ell_{min}, \overline{\gamma}/\gamma\}$, for any $\lambda/2 < \eta < \lambda$, and thus it is at least $\min\{\lambda p\ell_{min}, \overline{\gamma}/\gamma\} \geq \min\{1, \overline{\gamma}/\gamma\} = \overline{\gamma}/\gamma$.

Combining the three cases, we get the claimed result.   ∎

Observe that if we compare the upper bounds on relative throughput shown in the previous subsection with the lower bounds of the above theorem, then we may conclude that in the case where γ is an integer, algorithm CSL-Preamble is optimal (wrt relative throughput). In the case where γ is not an integer, there is a small gap between the upper and lower bound results.

5   Conclusions

This work has considered packet scheduling with dynamic packet arrivals and adversarial bit errors. We studied scenarios with two different packet lengths, developed efficient algorithms, and proved upper and lower bounds for relative throughput in average-case (i.e., stochastic) and worst-case (i.e., adversarial) online packet arrivals. These results demonstrate that exploring instantaneous feedback mechanisms (and developing more effective implementations of it) has the potential to significantly increase the performance of communication systems.

Several future research directions emanate from this work. Some of them concern the exploration of variants of the model considered, for example, assuming that packets that suffer errors are not retransmitted (which applies when Forward Error Correction [11] is used), considering packets of more than two lengths, or assuming bounded buffers. Other lines of work deal with adding QoS requirements to the problem, such as requiring fairness in the transmission of the packets from different flows or imposing deadlines to the packets. In the considered adversarial setting, it is easy to see that even an omniscient offline solution cannot achieve stability: for example, the adversary could prevent any packet from being transmitted correctly. Therefore, an interesting extension of our work is to study conditions (e.g., restrictions on the adversary) under which an online algorithm could maintain stability, and still be efficient with respect to relative throughput. Finally, we believe that the definition of relative throughput as proposed here can be adapted, possibly in a different context, to other metrics and problems.

References

1. Ajtai, M., Aspnes, J., Dwork, C., Waarts, O.: A theory of competitive analysis for distributed algorithms. In: 1994 Proceedings of the 35th Annual Symposium on Foundations of Computer Science, pp. 401–411. IEEE (1994)
2. Andrews, M., Zhang, L.: Scheduling over a time-varying user-dependent channel with applications to high-speed wireless data. J. ACM 52(5), 809–834 (2005)
3. Anta, A.F., Georgiou, C., Kowalski, D.R., Widmer, J., Zavou, E.: Measuring the impact of adversarial errors on packet scheduling strategies. ArXiv (2013)

4. Awerbuch, B., Kutten, S., Peleg, D.: Competitive distributed job scheduling. In: Proceedings of the Twenty-fourth Annual ACM Symposium on Theory of Computing, pp. 571–580. ACM (1992)
5. Jamieson, K., Balakrishnan, H.: Ppr: partial packet recovery for wireless networks. In: Proceedings of the 2007 Conference on Applications, Technologies, Architectures, and Protocols for Computer Communications, SIGCOMM 2007, pp. 409–420. ACM, New York (2007)
6. Kesselheim, T.: Dynamic packet scheduling in wireless networks. In: PODC, pp. 281–290 (2012)
7. Lin, S., Costello, D.J.: Error control coding, vol. 123. Prentice-Hall, Englewood Cliffs (2004)
8. Meiners, C., Torng, E.: Mixed criteria packet scheduling. In: Algorithmic Aspects in Information and Management, pp. 120–133 (2007)
9. Pinedo, M.L.: Scheduling: theory, algorithms, and systems. Springer (2012)
10. Pruhs, K., Torng, E., et al.: Online scheduling (2007)
11. Raghavan, A., Ramchandran, K., Kozintsev, I.: Continuous error detection (ced) for reliable communication. IEEE Transactions on Communications 49(9), 1540–1549 (2001)
12. Richa, A., Scheideler, C., Schmid, S., Zhang, J.: Competitive throughput in multi-hop wireless networks despite adaptive jamming. In: Distributed Computing, pp. 1–13 (2012)
13. Sleator, D.D., Tarjan, R.E.: Amortized efficiency of list update and paging rules. Communications of the ACM 28(2), 202–208 (1985)

Optimal Buffer Management
for 2-Frame Throughput Maximization

Jun Kawahara[1] and Koji M. Kobayashi[2]

[1] Nara Institute of Science and Technology
jkawahara@is.naist.jp
[2] National Institute of Informatics
kobaya@nii.ac.jp

Abstract. We consider a variant of the online buffer management problem in network switches, called the *k-frame throughput maximization* problem (*k*-FTM). Large data, called *frames*, carried on the Internet are split into small k packets by a sender, and the receiver can reconstruct each frame only if he/she accepts all the k constituent packets of the frame. Packets pass through network switches on the Internet, and each switch is equipped with a FIFO buffer to temporarily store arriving packets. Since the size of the buffer is bounded, some packets must be discarded if it is full. It is impossible to reconstruct frames including discarded packets any more. Our goal is to maximize the number of reconstructed frames. Kesselman et al. proposed this problem, and showed that any online algorithm has an unbounded competitive ratio even when $k = 2$. Hence, they considered the "order-respecting" variant of *k*-FTM. They showed that the competitive ratio of their algorithm is at most $(\frac{2kB}{\lfloor B/k \rfloor} + k)$ for any $B \geq k$, where B is the size of the buffer. Also, they gave a lower bound of $\frac{B}{\lfloor 2B/k \rfloor}$ on the competitive ratio when $2B \geq k$ and k is a power of 2. Furthermore, they proved that the competitive ratio of a greedy algorithm is at most $(11 + \frac{8}{B-1})$ for any $B(\geq 2)$ and $k = 2$. We analyze a greedy algorithm for $k = 2$, and show that its competitive ratio is at most 3 for any B, improving the previous upper bound of $\frac{4B}{\lfloor B/2 \rfloor} + 2(\geq 10)$. Moreover, we show that the competitive ratio of any deterministic algorithm is at least 3 for any B in $k = 2$, which matches our upper bound.

1 Introduction

Large and sequential data are currently used by real-time multimedia applications on the Internet. The data are called *frames*, which are too large to be transferred over the Internet. (For example, each frame in the case of video data corresponds to each picture of the video.) Thus, they are fragmented into small packets. When all the packets arrive at the receiver, each frame is reconstructed from the packets. Then, each packet has to pass through many switches (routers) on the Internet. Buffer management in the switches can become a bottleneck for transferring packets to the receiver. In particular, each switch is equipped with a buffer to store packets arriving at a burst. However, if the number of arriving

T. Moscibroda and A.A. Rescigno (Eds.): SIROCCO 2013, LNCS 8179, pp. 274–285, 2013.
© Springer International Publishing Switzerland 2013

packets surpasses the size of the buffer, the switch has to decide which packets can be accepted for insertion into its buffer.

Recently, this kind of problem was modeled as online problems, and a great amount of work has been done. Many models have been proposed, of which the most basic one is as follows [1]: A switch is equipped with a buffer (FIFO queue) of bounded size B. An input consists of a sequence of events. Each event is an arrival event or a send event. At an arrival event, one packet arrives at an input port. Each packet is of unit size and has a value that represents its priority. A buffer can store packets provided that the total size of stored packets does not exceed B, namely, a switch can store up to B packets at the same time. Stored packets are delivered in the FIFO order. At an arrival event, if the buffer is full, the new packet is rejected. If there is room for the new packet, an online policy determines, without knowledge of the future, whether to accept the packet. At each send event, the packet at the head of the queue is transmitted. The goal of the problem is to maximize the sum of the values of the transmitted packets. The performance of an online algorithm is evaluated by competitive analysis [5,16]. If, for any input σ, a deterministic online algorithm ALG gains the benefit, which is at least $1/c$ of the optimal offline policy for σ, then we say that ALG is c-competitive.

Kesselman et al. [11] focused on the buffer management with frame reconstruction, and formulated the k-frame throughput maximization problem (k-FTM), where $k\,(\geq 2)$ is an integer. Each arriving packet belongs to some frame, and every frame consists of exactly k packets. We say that a frame f is completed if all the packets constituting f are transmitted. Otherwise, we say that f is incompleted. Our goal is to maximize the number of completed frames.

Previous Results.   Kesselman et al. [11] showed that the competitive ratio of any algorithm for k-FTM is unbounded even when $k = 2$. The order of arrival of each packet in the instance used in the proof does not have the relation between packets in different frames. However, such an instance does not reflect the actual situation of networks since each packet generally arrives in the order of departure in a network such as a TCP/IP network. Hence, the authors introduced the order-respecting setting as follows: For any frame f, and the $i(\in [1, k])$th arriving packet p which is included in f, we call p the i-packet in f. The arrival order of the j-packets of frames f_i and $f_{i'}$ must obey the arrival order of the j'-packets of f_i and $f_{i'}$ $(j' < j)$ (a formal definition will be given later). We call the k-FTM problem in the order-respecting setting the order-respecting k-frame throughput maximization problem (k-OFTM). For the k-OFTM problem, Kesselman et al. showed a lower bound of $\frac{B}{\lceil 2B/k \rceil}$ on the competitive ratio for any deterministic algorithm, where $B \geq k/2$ and k is a power of 2. Also, they presented a $(\frac{2kB}{\lfloor B/k \rfloor} + k)$-competitive deterministic algorithm, called STATICPARTITIONING (SPA), when $B \geq k$. The authors proved that for $k \geq 3$, a greedy algorithm for k-OFTM is not competitive. They also showed that the competitive ratio of the preemptive greedy algorithm for 2-OFTM is at most $11 + 8/(B - 1)$ for any $B(\geq 2)$.

Our Results.   In this paper, we analyze a greedy algorithm (GR) for 2-OFTM, and improve the upper bound from $\frac{2kB}{\lfloor B/k \rfloor} + k = \frac{4B}{\lfloor B/2 \rfloor} + 2 \geq 10$ to 3 for any B. Furthermore, we prove a lower bound of 3 for any deterministic algorithm for any B in 2-OFTM, which matches our upper bound. In computational complexity theory, it is common to evaluate the performance of algorithms by its asymptotic behavior, e.g., when k approaches infinity. However, from a practical point of view, it is natural to assume that the number of packets constructing each frame, namely k, is bounded. Thus, it is significant to analyze the competitive ratio when k is constant.

Let us briefly explain our idea of improvement. Our main idea is to "assign" packets in frames completed by an online algorithm ALG to the 1-packet in each frame completed by an optimal offline algorithm OPT at the end of the input. Suppose that any 1-packet (2-packet, respectively) in ALG's completed frame is assigned at most x times (y times, respectively). Then, it can be shown that the competitive ratio of ALG is at most $x + y$. We can consider that the authors in [11] showed that $x = y = \frac{B/2 + 2B}{B/2} = 5$ for their online algorithm SPA and $x = 1, y = 2 + 8(1 + \frac{1}{B-1})$ for some greedy algorithm. (Note that they showed the results NOT using "assignments" of packets transmitted by an online algorithm.) We prove that the assignment such that $x = 1$ and $y = 2$ can be constructed in order to show that the competitive ratio of GR is at most 3. We try to assign each GR's packet at the moment when it arrives at the buffer. However, when we assign GR's 1-packet p_1 to OPT's 1-packet, we do not know whether or not to complete the frame f including p_1 in the future. Specifically, the 2-packet in f can be discarded by GR after p_1 is transmitted. If p_1 is assigned to some 1-packet which arrives at OPT's buffer, the competitive ratio of GR cannot correctly be evaluated. To that end, if we assign the 1-packet in an incompleted frame, (that is, the 2-packet corresponding to the 1-packet is discarded by GR,) then we assign a packet in a completed frame to the discarded 2-packet. Hence, we can bound the competitive ratio of GR from above.

Related Results.   Scalosub et al. [15] proposed the *max frame goodput* problem which is a generalized problem of k-FTM. A set of frames is called a *stream* in this problem, and a constraint is imposed on the arrival order of each packet in frames which belong to the same stream. They established an $O((kMB + M)^{k+1})$-competitive deterministic algorithm, where k is the number of packets in each frame, B is the size of a buffer, and M is the number of streams. Furthermore, they showed that the competitive ratio of any deterministic algorithm is $\Omega(kM/B)$.

Many studies concentrate on buffer management. The most basic model is that by Aiello et al. [1] consisting of a single FIFO queue as mentioned above. In this model, each packet can take one of two values: 1 or $\alpha(> 1)$. Andelman et al. [4] generalized the values of packets to any value between 1 and α. The goal of these problems is to maximize the sum of the values of packets transmitted by an algorithm. Results of the competitiveness on these models are given in [9,17,10,3,2,7].

In addition, some research focuses on reconstructing frames from packets arriving at buffers. In the *online set packing* problem by Emek et al. [6], elements of each set are given in an online fashion, and the goal of this problem is to maximize the number of sets whose elements are completed. Sets and elements in each set are regarded as frames and packets in each frame, respectively. The numbers of elements in any two given sets are not necessarily equivalent. Let k_{max} denote the maximum number of elements which belong to one set, and let σ_{max} denote the maximum number of elements that may arrive simultaneously. Emek et al. designed a $k_{max}\sqrt{\sigma_{max}}$-competitive randomized algorithm. Moreover, some variants of this problem have been studied [8,13,14].

2 Preliminaries

2.1 Order-Respecting k-Frame Throughput Maximization Problem (k-OFTM)

In this section, we give a formal description of k-OFTM. Arriving packets, each of size 1, are stored in a buffer. The buffer, which is a FIFO queue, can store B packets at the same time. An input is a sequence of *phases* starting from the 0th phase. Each phase consists of an *arrival subphase* and a *delivery subphase*. At an arrival subphase, some packets arrive at the buffer, and the task of an algorithm for each arriving packet p is to decide whether to *accept* or *reject* it. An algorithm can discard a packet p' existing in the current buffer, and make space in order to accept an arriving packet (*preempt p'*). If a packet p'' is rejected or preempted, we say that p'' is *dropped*. If there is no space to accept arriving packets in the buffer, they have to be rejected. If a packet is accepted, it is stored at the tail of the queue. Packets accepted at the same arrival subphase can be inserted into the queue in an arbitrary order. If a packet p arrives at the ith arrival subphase, we write $\text{arr}(p) = i$. Next, a delivery subphase follows the arrival subphase. At this subphase, the first packet of the queue is transmitted if the buffer is not empty. Each frame f consists of k packets $p_1, \ldots, p_k$, where $k(\geq 2)$ is an integer. For any frame $f = \{p_1, \ldots, p_k\}$ such that p_i $(i \in [1, k])$ is a packet constructing f and $\text{arr}(p_1) \leq \cdots \leq \text{arr}(p_k)$, we call p_i the *i-packet* in f for any $i \in [1, k]$. For any frames $f_i = \{p_{i,1}, \ldots, p_{i,k}\}$ and $f_{i'} = \{p_{i',1}, \ldots, p_{i',k}\}$ such that for any $\ell \in [1, k]$, $p_{i,\ell}$ and $p_{i',\ell}$ are the ℓ-packets, and for any $j, j' = 1, \ldots, k$, $\text{arr}(p_{i,j}) \leq \text{arr}(p_{i',j}) \Leftrightarrow \text{arr}(p_{i,j'}) \leq \text{arr}(p_{i',j'})$, which is called *order-respecting*. For any algorithm ALG, and any frame $f = \{p_1, \ldots, p_k\}$ such that they arrive at the buffer of ALG, we say that f of ALG is *completed* if all the packets which construct f, namely $p_1, \ldots, p_k$, are transmitted by ALG. The *benefit* of an algorithm is the number of completed frames. Therefore, our goal is to maximize the number of completed frames. The benefit gained by an algorithm ALG for an input σ is denoted by $V_{ALG}(\sigma)$.

Without loss of generality, we can assume that OPT never preempts packets. In addition, we assume that OPT never accepts packets in an incompleted frame. For ease of presentation, for any i-packet and $j(\neq i)$-packet such that they belong to the same frame, we say that the i-packet *corresponds to* the j-packet. Also, for any algorithm ALG and any packet p which arrives at ALG's buffer, we say that p is *a packet of ALG*.

2.2   Greedy Algorithm

In this section, we give the definition of a preemptive greedy algorithm (GR) for $k = 2$ which is analyzed in this paper. In brief, GR's favorite packets are as follows: (A 2-packet for which the 1-packet corresponding to it has been transmitted) > (A 2-packet for which the 1-packet corresponding to it has already arrived, but has not been transmitted) = (A 1-packet for which the 2-packet corresponding to it has already arrived, but has not been transmitted) > (A 1-packet for which the 2-packet corresponding to it has not yet arrived).

Next, we discuss the formal definition. When $k = 2$, packets stored in GR's buffer can be classified into four categories as follows: (B1-1) A 1-packet for which the decision to accept the 2-packet corresponding to it has not been made yet, (B1-2) a 1-packet for which the 2-packet corresponding to it is stored in the buffer of GR, (B2-2) a 2-packet for which the 1-packet corresponding to it is stored in the buffer of GR, and (B2-3) a 2-packet for which the 1-packet corresponding to it has already been transmitted. Furthermore, arriving packets at GR's buffer can be classified into six categories as follows: (A1-1) A 1-packet for which the 2-packet corresponding to it has not yet arrived, (A1-4) a 1-packet for which the 2-packet corresponding to it arrives simultaneously, (A2-2) a 2-packet for which the 1-packet corresponding to it is stored in the buffer of GR, (A2-3) a 2-packet for which the 1-packet corresponding to it has already been transmitted, (A2-4) a 2-packet for which the 1-packet corresponding to it also arrives at the same time, and (A2-5) a 2-packet for which the 1-packet corresponding to it has already been dropped.

The priority of packets for GR is as follows: (B2-3)>(A2-3)>(B2-2)=(B1-2)>(A2-2)>(A2-4)=(A1-4)>(B1-1)>(A1-1)>(A2-5). It is easy to see that any reasonable algorithm (including GR) never accepts 2-packets in (A2-5). GR decides whether to accept each arriving packet in order of the priority, namely, (A2-3)>(A2-2)>(A2-4)=(A1-4)>(A1-1). If there exist some packets in (A2-2) (or (A2-3)) at a time, GR deals with them in order of the arrival time of the corresponding 1-packet. If GR's buffer has space to store an arriving packet p, GR accepts p. Otherwise, (that is, GR's buffer has no space to accept p,) if there exists a packet whose priority is lower than that of p, GR preempts one of the packets with the lowest priority in its buffer, and accepts p. If several packets have the same priorities, GR preempts the packet with the same priority which is the closest to the tail of GR's queue. Otherwise, (that is, in the case where any packet in GR's buffer has a higher or the same priority,) p is rejected. However, for any 1-packet p' in (A1-4), GR decides whether to accept both p' and the 2-packet p'' corresponding to p' in (A2-4) together. In addition, when they are accepted, p'' is inserted into GR's queue after it first inserts p'. For any arriving 2-packet $\tilde{p}$ in (A2-2), (that is, $\tilde{p}$ arrives, and there exists the 1-packet $\hat{p}$ corresponding to $\tilde{p}$ in GR's buffer,) $\hat{p}$ is preempted if GR rejects $\tilde{p}$.

3   Upper Bound

3.1   Overview of the Analysis

We give some definitions for our analysis. Let ALG denote either GR or OPT throughout Sec. 3. For any algorithm ALG, and any $j(\in \{1,2\})$, $V_{ALG,j}$ denotes the number of j-packets transmitted by ALG by the end of the input. For a given input, $\overline{V}_{GR,2}$ denotes the total number of frames, each of which includes the 1-packet transmitted by GR and the 2-packet dropped by GR. Then, $V_{OPT}(\sigma) = V_{OPT,1}$ since OPT does not accept any packet in an incompleted frame. Also, $V_{GR}(\sigma) = V_{GR,2} = V_{GR,1} - \overline{V}_{GR,2}$.

At the end of the input σ, packets of GR are "assigned" to both all the 1-packets in completed frames of OPT and some 2-packets in incompleted frames of GR according to the assignment routine which is defined later. (GR's packets used by the routine are not always in completed frames.) We can show a connection between packets being assigned and assigned ones, and have the following properties from (i) to (iv). (i) For any 1-packet p in OPT's completed frame, either one 1-packet transmitted by GR or one 2-packet in GR's completed frame is assigned to p, (ii) for any 2-packet p' dropped by GR for which the 1-packet corresponding to p' is assigned to some packet, either one 1-packet transmitted by GR or one 2-packet in GR's completed frame is assigned to p', (iii) the number of packets to which one 1-packet transmitted by GR is assigned is at most one, and (iv) the number of packets to which the 2-packet in one completed frame of GR is assigned is at most two.

Using these properties, we have that $V_{OPT,1} + \overline{V}_{GR,2} \leq V_{GR,1} + 2V_{GR,2}$. (See Lemma 1.) By the above inequalities, $V_{OPT}(\sigma) = V_{OPT,1} \leq -\overline{V}_{GR,2} + V_{GR,1} + 2V_{GR,2} = V_{GR,2} + 2V_{GR,2} = 3V_{GR}(\sigma)$ holds. Therefore, we have the following theorem:

Theorem 1. *The competitive ratio of GR is at most 3 when $k = 2$.*

3.2   Notation for Analysis

We use the following notation, because dealing with packets one by one in our analysis when they arrive. Suppose that n packets $p_1, p_2, \ldots, p_n$ arrive at the ith arrival subphase, and GR deals with the n packets one by one in this order. Let t_{p_i} denote the moment when GR decides whether or not to accept p_i ($1 \leq i \leq n$). We call this moment the *decision time* of p_i. Let t_{d_i} denote the moment when the ith delivery subphase occurs. We call this moment the *delivery time*. Also, a delivery time or a decision time is called an *event time*. A moment which is not any event time is called a *non-event time*. For all the event times at the ith phase, we define $t_{p_1} < t_{p_2} < \cdots < t_{p_n} < t_{d_i}$. However, when a 1-packet p and the 2-packet p' corresponding to p arrive at the same phase, GR decides whether or not to accept them at the same time by the definition of GR. Hence, we define $t_p = t_{p'}$. Furthermore, for any integers i and $i'(> i)$, and any event time e (e', respectively) at the ith phase (i'th phase, respectively), let us define

$e < e'$. For any event time e, any non-event time d, and the event time e' just before d (just after d, respectively), we define $e < d$ ($d < e$, respectively) if $e \leq e'$ ($e' \leq e$, respectively). Moreover, for any event time e, $e+$ denotes the non-event time after e and before the next event time. Also, $e-$ denotes the non-event time before e and after the previous event time. To analyze the performance of algorithms, we introduce the above notation to specify the location of each packet in a buffer shortly before and after the moment when a packet is accepted or transmitted. For each arriving packet, we assume that OPT decides whether or not to accept it at the same time as GR does.

3.3   Assignment Routine

Next, we introduce some notation to give the definition of our assignment routine. For any non-event time d, and any 2-packet p of GR, we say that p is *1-completed* at d if GR transmits the 1-packet corresponding to p before d. For any non-event time d, and any 2-packet p of GR such that p is 1-completed at d and the 1-packet corresponding to p is assigned to some packet at d, we say that p is a *key* 2-packet at d.

For any non-event time d, we define *candidate packets* at d as those packets that satisfy one of the following conditions: (i) Any 1-packet p is stored in GR's buffer at d such that p is not assigned to any packet at d, and (ii) any 2-packet p' is stored in GR's buffer at d such that p' is assigned to at most one packet at d. Especially, we define a *candidate 2-packet* at d as a packet satisfying the condition (ii).

At each time when any packet to be assigned arrives, we assign one candidate packet at this time to the arriving packet. (However, there is one exception in Case 3.2.1.) As a result, the assignments mentioned in the previous section are achieved sequentially. Also, if a packet p is assigned to some packet p', and GR preempts p at some time, then we unassign p, and assign another packet p'' of GR to p'. Hence, for any packet $\hat{p}$ to be assigned, some packet of GR is always assigned to $\hat{p}$ when $\hat{p}$ is stored in a buffer. Furthermore, when a packet p is assigned to some packet p', the assignment never changes after p is transmitted. Therefore, we obtain the assignments which satisfy the properties mentioned in Sec. 3.1 after the final packet is transmitted.

The assignments described above are realized by AssignmentRoutine which is defined later. For all arriving packets, this routine is executed at each decision time. First, we sketch the actions of AssignmentRoutine at the decision time e. The actions are categorized into three types: (i) If OPT accepts a 1-packet p at e, then the routine assigns one candidate packet at $e-$ to p. (Cases 1.1 and 1.2) (ii) If a packet p in GR's buffer is preempted at e, and p is assigned to some packet q at $e-$, then the routine unassigns p, and assigns some packet in GR's buffer at $e+$ to q. Moreover, if there exist some packets p' and q' such that p' is assigned to q' at $e-$, p' is stored in GR's buffer at $e-$, and the position of p is closer to the head of GR's queue than that of p' at $e-$, then the routine unassigns p', and instead assigns some packet in GR's buffer at $e+$ to q'. (Cases 2.2.1, 2.3.1, 3.4.1, and 3.5.1) This is because each assignment needs to satisfy the *position condition* which is described below. Also, note that packets used to assign are not only candidate packets in GR's buffer at $e-$ but also packets which the routine unassigns at e and the packet inserted into GR's buffer at e. The routine might assign some packets in the actions in (ii), and calls the subroutine REASSIGNMENTSUBROUTINE, which has two parameters: the names of the packet which GR decides to accept at e and the packet p preempted at e. (iii) If GR rejects a key 2-packet p at e, then the routine assigns one packet at $e-$ to p. (Case 3.2) The packet used in Case 3.2.1 is not a candidate packet. In this case, notice that GR's buffer is full, and all the packets in its buffer are only 2-packets.

Now let us explain the position condition mentioned above. For any non-event time d, any algorithm ALG, and any packet p of ALG, we define $\ell_{ALG}(d,p) = x$ if p is located at the xth position from the top in ALG's buffer at d. Also, $\ell_{ALG}(d,p) = \infty$ if either p arrives after d or p is dropped by ALG before d. Furthermore, $\ell_{ALG}(d,p) = 0$ if ALG transmits p before d. At each decision time e, when the routine assigns some packet p' of GR to a packet p of $A'(\in \{GR, OPT\})$, it certainly selects p' such that $\ell_{GR}(e+, p') \leq \ell_{A'}(e+, p)$, which is called the *position condition*. (It will be shown in the full version of this paper.) By satisfying the condition, we can prove that each packet in GR's buffer is not assigned to any packet which has already been transmitted. (For details, see the full version.) As a result, we can show that there exist enough packets to assign at each time when the routine needs to assign packets.

Note that the case where both a 1-packet and the corresponding 2-packet may arrive at the same time is included in Case 1. Also, note that the case in which GR accepts an arriving 1-packet, and preempts another packet in its buffer (in Case 1) never occurs. For ease of presentation, we give the notation. For any algorithm ALG, any non-event time d, any $x \in [1, B]$, and any $y \in [x, B]$, let $P_{ALG}(d, x, y)$ denote the set of packets p such that p is located at a position in $[x, y]$ in ALG's buffer at d. In other words, $P_{ALG}(d, x, y) = \{p \mid \ell_{ALG}(d,p) \in [x, y]\}$. For any non-event time d, any $x \in [1, B]$ and any $y \in [x, B]$, we define $A(d, x, y) = \{q \mid p \in P_{GR}(d, x, y)$ is assigned to q at $d\}$. That is, $A(d, x, y)$ denotes the set of packets p' such that some packet located at a position in $[x, y]$ in GR's buffer at d is assigned to p' at d.

ASSIGNMENTROUTINE

Consider the decision time of a packet p.

Case 1: p is a 1-packet.

    Case 1.1: Both OPT and GR accept p.

        Case 1.1.1: $\ell_{GR}(t_p+, p) \leq \ell_{OPT}(t_p+, p)$.
        Assign p of GR to p of OPT.

        Case 1.1.2: $\ell_{GR}(t_p+, p) > \ell_{OPT}(t_p+, p)$.
        Let p' be a candidate packet at t_p- such that $\ell_{GR}(t_p+, p') \leq \ell_{OPT}(t_p+, p)$.
Assign p' to p of OPT.

    Case 1.2: OPT accepts p, but GR rejects p.

Let p' be a candidate packet at t_p- such that $\ell_{GR}(t_p+, p') \leq \ell_{OPT}(t_p+, p)$. Assign p' to p of OPT.

Case 1.3: OPT rejects p.

Do nothing.

Case 2: p is a 2-packet to which the corresponding 1-packet p' is stored in GR's buffer.

Case 2.1: GR accepts p.

Do nothing.

Case 2.2: GR preempts only one packet p'', and accepts p.

Case 2.2.1: p'' is assigned to some packet at t_p-.

Call REASSIGNMENTSUBROUTINE(p, p'').

Case 2.2.2: p'' is not assigned to any packet at t_p-.

Do nothing.

Case 2.3: GR preempts p', and rejects p.

Case 2.3.1: p' is assigned to some packet at t_p-.

Call REASSIGNMENTSUBROUTINE(p, p').

Case 2.3.2: p' is not assigned to any packet at t_p-.

Do nothing.

Case 3: p is a 1-completed 2-packet at t_p-.

Case 3.1: GR accepts p.

Do nothing.

Case 3.2: GR rejects p and p is key at t_p-.

Case 3.2.1: There exists a 2-packet $\tilde{p}$ in GR's buffer at t_p- such that the 1-packet p' corresponding to $\tilde{p}$ is not assigned at t_p-.

Assign p' to p of GR.

Case 3.2.2: Otherwise, i.e., there does not exist such $\tilde{p}$ at Case 3.2.1.

Let p'' be a candidate 2-packet at t_p-. Assign p'' to p of GR.

Case 3.3: GR rejects p and p is not key at t_p-.

Do nothing.

Case 3.4: GR accepts p, and preempts a 1-packet p'.

Case 3.4.1: p' is assigned to some packet at t_p-.

Call REASSIGNMENTSUBROUTINE(p, p').

Case 3.4.2: p' is not assigned to any packet at t_p-.

Do nothing.

Case 3.5: GR accepts p, and preempts both a 1-packet p' and the 2-packet p'' corresponding to p'.

Case 3.5.1: Either p' or p'' is assigned to some packet at t_p-.

Call REASSIGNMENTSUBROUTINE(p, p').

Case 3.5.2: Neither p' nor p'' is assigned to any packet at t_p-.

Do nothing.

Case 4: Otherwise.

Do nothing.

REASSIGNMENTSUBROUTINE(p, p')
  for each packet $\tilde{p} \in P_{GR}(t_p-, \ell_{GR}(t_p-, p'), B)$ **do**
    if $\tilde{p}$ is assigned to some packet at t_p-, **then** unassign $\tilde{p}$.
  end for
  $Q := \{p\} \cup P_{GR}(t_p-, \ell_{GR}(t_p-, p') + 1, B)$.
/* Q is the set of packets which can be used for assignments by this subroutine.
*/
  $R := A(t_p-, \ell_{GR}(t_p-, p'), B)$ /* R is the set of packets to be assigned. */
  while $R \neq \emptyset$ **do**
    $q \in \arg\min_{q''} \{\ell_{OPT}(t_p-, q'') \mid q'' \in R\}$.
    $q' := $ (the packet located at the lowest position in $\{q'' \in Q \mid \ell_{GR}(t_p+, q'') \leq \ell_{OPT}(t_p+, q)\}$).
    Assign q' to q.
    $R := R \setminus \{q\}$.
    if either q' is a 1-packet or q' is a 2-packet, and the number of packets to which q' is assigned is 2, **then** $Q := Q \setminus \{q'\}$.
  end for

If this routine can be executed at each time, we can show the following lemma. We then obtain Theorem 1 using the lemma.

Lemma 1. *Suppose that the assignment routine can be executed at any decision time. Then, $V_{OPT,1} + \overline{V}_{GR,2} \leq V_{GR,1} + 2V_{GR,2}$.*

Proof. We have the following facts by the definitions of candidate packets and the routine. For any 1-packet p accepted by OPT, either one 1-packet transmitted by GR or one 2-packet in a completed frame of GR is assigned to p. (By the assumption of OPT, OPT does not preempt any packet.) Also, for any rejected 2-packet p' which is key just before GR rejects p', either one 1-packet transmitted by GR or one 2-packet in a completed frame of GR is assigned to p'. (We call these facts the property (i).) On the other hand, the number of packets to which a 1-packet transmitted by GR is assigned is at most one. Also, the number of packets to which a 2-packet included in a completed frame of GR is assigned is at most two. (We call these facts the property (ii).)

Then, let $V_{GR,11}$ denote the number of 1-packets, each of which is transmitted by GR and is assigned to a 1-packet transmitted by OPT, and let $V_{GR,12}$ denote the number of 2-packets, each of which is in a completed frame of GR and is assigned to a 1-packet transmitted by OPT. Let $V_{GR,21}$ denote the number of 1-packets, each of which is transmitted by GR and is assigned to a rejected 2-packet p' which is key just before GR rejects p', and let $V_{GR,22}$ denote the number of 2-packets, each of which is in a completed frame of GR and is assigned to a rejected 2-packet p' which is key just before GR rejects p'. Furthermore, let $\hat{V}_{GR,2}$ ($\check{V}_{GR,2}$, respectively) denote the number of 2-packets, each of which is key (not key, but only 1-completed, respectively) just before GR rejects the 2-packet.

By the property (i), we have $V_{OPT,1} = V_{GR,11} + V_{GR,12}$, $\tilde{V}_{GR,2} = V_{GR,21} + V_{GR,22}$, and $\overline{V}_{GR,2} = \tilde{V}_{GR,2} + \hat{V}_{GR,2}$. Also, using the property (ii), $V_{GR,11} + V_{GR,21} + \hat{V}_{GR,2} \le V_{GR,1}$ and $V_{GR,12} + V_{GR,22} \le 2V_{GR,2}$. By the above equalities and inequalities, $V_{OPT,1} + \overline{V}_{GR,2} = V_{GR,11} + V_{GR,12} + \tilde{V}_{GR,2} + \hat{V}_{GR,2} \le V_{GR,11} + V_{GR,12} + V_{GR,21} + V_{GR,22} + V_{GR,1} - V_{GR,11} - V_{GR,21} = V_{GR,12} + V_{GR,22} + V_{GR,1} \le 2V_{GR,2} + V_{GR,1}$.

Due to page limitation, the proof for executability of each case will be shown in the full version of this paper.

4 Lower Bound

In this section, we present a lower bound of 3 on the competitive ratio of any deterministic algorithm for $k = 2$.

Theorem 2. *When $k = 2$, the competitive ratio of any deterministic algorithm is at least 3 for any B.*

Proof. Fix an online algorithm ALG. We consider the following input σ: At the 0th phase, $2B$ 1-packets arrive. Then, ALG accepts $x(\le B)$ 1-packets, and OPT accepts B 1-packets which are not accepted by ALG. Let C (D, respectively) be the set of the x packets (the B packets, respectively) accepted by ALG (OPT, respectively).

After B delivery subphases occur, $2B$ 1-packets arrive in the same manner as the first $2B$ 1-packets. We suppose that ALG accepts $y(\le B)$ packets. Then, OPT accepts B packets which are not accepted by ALG. Let E (F, respectively) be the set of the y packets (the B packets, respectively) accepted by ALG (OPT, respectively).

This process is repeated one more time. Namely, $2B$ 1-packets arrive at the $2B$th phase after B delivery subphases occur. Then, ALG accepts $z(\le B)$ packets and OPT accepts B packets which are not accepted by ALG. Let G (H, respectively) be the set of the z packets (the B packets, respectively) accepted by ALG (OPT, respectively). Afterward, no additional 1-packets arrive in this instance. After the $2B$th phase, each 2-packet arrives.

First, the B 2-packets corresponding to B 1-packets in D arrive after B delivery subphases happen. OPT accepts them, and they are transmitted at the B delivery subphase. On the other hand, there is no point for ALG to accept them.

Then, $x + y + z$ 2-packets corresponding to $x + y + z$ 1-packets in C, E and G arrive at a burst. ALG can accept at most B packets.

On the other hand, at the same time, the B 2-packets corresponding to B 1-packets in F arrive. OPT accepts and transmits them at B delivery subphases. After OPT transmits them, the B 2-packets corresponding to B 1-packets in H arrive. Similarly, OPT accepts them and they are transmitted at B delivery subphases.

By the above argument, we have $V_{ALG}(\sigma) \le B$ and $V_{OPT}(\sigma) = 3B$. Therefore, $\frac{V_{OPT}(\sigma)}{V_{ALG}(\sigma)} \ge \frac{3B}{B} = 3$.

References

1. Aiello, W., Mansour, Y., Rajagopolan, S., Rosén, A.: Competitive queue policies for differentiated services. Journal of Algorithms 55(2), 113–141 (2005)
2. Andelman, N.: Randomized queue management for DiffServ. In: Proc. of the 17th ACM Symposium on Parallel Algorithms and Architectures, pp. 1–10 (2005)
3. Andelman, N., Mansour, Y.: Competitive management of non-preemptive queues with multiple values. In: Proc. of the 17th International Symposium on Distributed Computing, pp. 166–180 (2003)
4. Andelman, N., Mansour, Y., Zhu, A.: Competitive queueing policies for QoS switches. In: Proc. of the 14th ACM-SIAM Symposium on Discrete Algorithms, pp. 761–770 (2003)
5. Borodin, A., El-Yaniv, R.: Online computation and competitive analysis. Cambridge University Press (1998)
6. Emek, Y., Halldórsson, M., Mansour, Y., Patt-Shamir, B., Radhakrishnan, J., Rawitz, D.: Online set packing and competitive scheduling of multi-part tasks. In: Proc. of the 29th ACM Symposium on Principles of Distributed Computing, pp. 440–449 (2010)
7. Englert, M., Westermann, M.: Lower and upper bounds on FIFO buffer management in QoS switches. In: Proc. of the 14th European Symposium on Algorithms, pp. 352–363 (2006)
8. Halldórsson, M., Patt-Shamir, B., Rawitz, D.: Online set packing and competitive scheduling of multi-part tasks. In: Proc. of the 28th Symposium on Theoretical Aspects of Computer Science, pp. 472–483 (2011)
9. Kesselman, A., Lotker, Z., Mansour, Y., Patt-Shamir, B., Schieber, B., Sviridenko, M.: Buffer overflow management in QoS switches. SIAM Journal on Computing 33(3), 563–583 (2004)
10. Kesselman, A., Mansour, Y., van Stee, R.: Improved competitive guarantees for QoS buffering. In: Di Battista, G., Zwick, U. (eds.) ESA 2003. LNCS, vol. 2832, pp. 361–372. Springer, Heidelberg (2003)
11. Kesselman, A., Patt-Shamir, B., Scalosub, G.: Competitive buffer management with packet dependencies. In: Proc. of the 23rd IEEE International Parallel and Distributed Processing Symposium, pp. 1–12 (2009)
12. Kobayashi, K., Miyazaki, S., Okabe, Y.: A tight bound on online buffer management for two-port shared-memory switches. In: Proc. of the 19th ACM Symposium on Parallel Algorithms and Architectures, pp. 358–364 (2007)
13. Mansour, Y., Patt-Shamir, B., Rawitz, D.: Overflow management with multipart packets. In: Proc. of the 31st IEEE Conference on Computer Communications, pp. 2606–2614 (2011)
14. Mansour, Y., Patt-Shamir, B., Rawitz, D.: Competitive router scheduling with structured data. In: Solis-Oba, R., Persiano, G. (eds.) WAOA 2011. LNCS, vol. 7164, pp. 219–232. Springer, Heidelberg (2012)
15. Scalosub, G., Marbach, P., Liebeherr, J.: Buffer management for aggregated streaming data with packet dependencies. In: Proc. of the 29th IEEE Conference on Computer Communications, pp. 1–5 (2010)
16. Sleator, D., Tarjan, R.: Amortized efficiency of list update and paging rules. Communications of the ACM 28(2), 202–208 (1985)
17. Sviridenko, M.: A lower bound for on-line algorithms in the FIFO model (2001) (unpublished manuscript)

Dynamically Maintaining Shortest Path Trees under Batches of Updates*

Annalisa D'Andrea, Mattia D'Emidio, Daniele Frigioni, Stefano Leucci,
and Guido Proietti

Department of Information Engineering, Computer Science and Mathematics,
University of L'Aquila, Italy
{annalisa.dandrea,mattia.demidio,daniele.frigioni,stefano.leucci,
guido.proietti}@univaq.it

Abstract. In this paper we focus on dynamic *batch* algorithms for single source shortest paths in graphs with positive real edge weights. A dynamic algorithm is called *batch* if it is able to handle graph changes that consist of multiple edge updates at a time, i.e. a *batch*. We propose a new algorithm to process a decremental batch (containing only delete and weight increase operations), a new algorithm to process an incremental batch (containing only insert and weight decrease operations), and a combination of these algorithms to process arbitrary sequences of incremental and decremental batches. These algorithms are *update-sensitive*, namely they are efficient w.r.t. to the number of nodes in the shortest paths tree that change the parent and/or the distance from the source as a consequence of the changes.

1 Introduction

The problem of updating shortest paths in real networks whose topology dynamically changes over time is a core functionality in many real-world scenarios as Internet routing, routing in road networks, timetabling in railways networks. In these scenarios, shortest-path trees are stored and have to be updated when the underlying graph undergoes dynamic updates. For example, in communication networks, a faulty network device or congestion phenomena may cause several links to become slower or unavailable. In order to preserve the quality of service, shortest paths need to be efficiently updated to reflect the underlying changes.

In general, the typical update operations that can occur on a network can be modelled as insertions and deletions of edges and edge weight changes (weight decrease or weight increase) in the underlying graph. When arbitrary sequences of these operations are allowed, we refer to the *fully dynamic problem*, otherwise we refer to the *partially dynamic problem*; if only insert and weight decrease (delete and weight increase) operations are allowed, then the partially dynamic problem is called *incremental* (*decremental*). A dynamic algorithm is a *batch*

---

* Research partially supported by the Research Grant 2010N5K7EB PRIN 2010 "ARS TechnoMedia" from the Italian Ministry of University and Research.

T. Moscibroda and A.A. Rescigno (Eds.): SIROCCO 2013, LNCS 8179, pp. 286–297, 2013.

algorithm if it is able to handle graph changes that consist of multiple edge updates at a time, i.e. a *batch*. If a batch consists of only delete and weight increase (insert and weight decrease) operations, then it is called *decremental batch* (*incremental batch*), otherwise it is called *full batch*.

Shortest paths can be computed by Dijkstra's algorithm [10]. Unfortunately, real-world networks are huge yielding unsustainable times to compute shortest paths. Over the last decade great efforts have been done to improve the practical performance of Dijkstra's algorithm. These efforts have led to the development of a number of so-called *speed-up techniques* (see, e.g., [1,2,9,14]), whose aim is to compute additional data in a preprocessing phase in order to accelerate the answer to shortest paths queries during an on-line phase. None of these techniques is theoretically better than Dijkstra's algorithm in the worst case, while some of them have been shown to be very effective in practice.

Related work. The problem of *updating* single-source shortest paths in dynamic scenarios has been widely studied in the literature. None of the proposed algorithms is better than the recomputation from scratch in the worst case ([6,7,8,12,13,15,16]). Some of these algorithms store a shortest-path tree [6,7,8,12,13,15] or the shortest-path subgraph [16]. Some of them are only able to cope with the update of one edge at a time [6,7,8,12,13], while others can perform *batch* updates [15,16]. In the original works these algorithms are theoretically analyzed with respect to different measures. These measures mostly depend on the size of the output information that changes, and in particular, given a graph G and an edge update μ, on the size of the set $\delta(G, \mu)$ containing the so-called *affected* vertices, i.e., vertices that change either their parent towards the source or their distance from the source in G as a consequence of μ. Recently, many works have been proposed to maintain approximate shortest path on digraphs (see [4,18] and references therein).

In [17] the authors provide a fully dynamic algorithm for general graphs. The worst case cost of a *single* edge update μ, is $O(||\mu||\cdot\log||\mu||)$. Here, $||\mu||$ represents the size of the change in the *input* and the *output*, and it is given by the sum of the number of affected vertices plus the number of *modified components* in the graph, i.e., the two vertices that are adjacent to the updated edge and the set of edges that are incident to affected vertices and/or to any of these two vertices. Note that, if n is the number of nodes of G, then $||\mu||$ might be n times larger than the number of affected vertices. In the case of batch updates, in [16] the same authors propose an algorithm running in $O(||\beta|| \cdot \log||\beta||)$ time, where β is a batch of update operations, $||\beta||$ is again the size of the change in the input and the output as a consequence of β.

In [11] the authors separately consider the *incremental* and the *decremental* problem. The decremental solution works only for planar graphs G, and each operation requires $O(|\delta(G, \mu)| \cdot \log n)$ time. The incremental solution works for any graph and its complexity depends on the existence of a *k-bounded accounting function* for G (*k-baf* from now on). In particular, for any incremental sequence of updates, if the final graph has a *k-baf*, then the complexity of the incremental algorithm is $O(k \cdot \log n)$ amortized time per affected vertex. An *accounting function* f for G is a function that, for each edge (x, y) of G, determines either vertex x or vertex y as the *owner* of the edge; f is k-bounded if k is the maximum over all vertices x of the cardinality of the set of edges owned by x. As shown in [12], several classes of graphs admit a k-baf. In general, if m is the number of edges of the graph, then $k = O(\sqrt{m})$ always holds. In [12] the results of [11] are extended to the fully dynamic case for general graphs as follows. If the input graph G admits a k-baf and only weight updates of edges are allowed, then each operation requires $O(k \cdot \log n)$ worst case time per affected vertex, i.e., a total $O(|\delta(G, \mu)| \cdot k \cdot \log n)$ time. If also insertions and deletions of edges are allowed, then the bound per affected vertex becomes amortized.

In [15] the authors propose a framework containing six single-edge update algorithms. The best of these algorithms requires $O(|\delta(G, \mu)| \cdot \log(|\delta(G, \mu)|) + \gamma \cdot D_{\max} \cdot |\delta(G, \mu)|)$ time, where $D_{\max}$ is the maximum vertex degree and γ can be as large as the number of nodes that change both their parent and their distance. These algorithms can be adapted to handle batches of updates, but doing so will not provide any tangible improvement on the computational complexity.

In [3] an experimental study of the algorithms in [12,15,16] have been proposed in the case of batch updates.

Contribution of the Paper. In this paper, we focus on the batch shortest path problem and extend the results of [11] to positively weighted general graphs, and to batch updates. In particular, if $\beta = (\mu_1, \ldots, \mu_h)$ is a batch of edge update operations, we define $\hat{\delta}(G, \beta)$ as the set of affected vertices caused by the execution of all the operations in β, simultaneously. Notice that $|\hat{\delta}(G, \beta)|$ can be much smaller than the sum over all i of $|\delta(G, \mu_i)|$, as vertices that are affected multiple times with respect to the single operations in β are considered only once in $\hat{\delta}(G, \beta)$, and a vertex affected several times may result as unaffected at the end of the batch. If the input graph admits a k-baf, then the results of this paper can be summarized as follows:

1. We propose a new algorithm which is able to process a *decremental* batch β in $O((|\beta| + |\hat{\delta}(G, \beta)| \cdot k) \cdot \log n)$ worst case time.
2. We propose a new algorithm which is able to process an *incremental* batch β of only weight decrease operations in $O(|\beta| + |\hat{\delta}(G, \beta)| \cdot k \cdot \log n)$ worst case time. If the incremental batch contains also edge insertions, then it can be processed in $O(|\beta| + |\hat{\delta}(G, \beta)| \cdot \max\{k, k^*\} \cdot \log n)$ worst case time, where k^* is the minimum integer such that a k^*-baf exists for the graph after β.
3. We combine these new algorithms to deal with a *mixed* sequence $B = (\beta_1, \ldots, \beta_h)$ of incremental and decremental batches in $O((|B| + |\hat{\Delta}(G, B)| \cdot k) \cdot \log n)$ overall time where $|B| = \sum_{\beta_i \in B} |\beta_i|$ and $|\hat{\Delta}(G, B)|$ is the sum over all batches of B of the number of vertices affected by each of these batches.

These results clearly improve over the results of [11]. Moreover, as far as the decremental and the incremental problem is concerned, we improve the corresponding results of [12]. Indeed, our solutions for the batch problem give basically the same bounds the solution of [12] gives in the case of a single update. As far as the mixed case is concerned, a comparison with the result provided in [16] for positively weighted graphs is quite unfeasible on a theoretical basis. This is due to the different parameters appearing in the respective time analysis. Notice indeed that even a single edge update may induce a linear number of modified components in the graph, which clearly makes non output-sensitive the algorithm given in [16]. However, if the updates in the batch induce a small number of modified components, then the result in [16] becomes very competitive.

2 Background

Let $G = (V, E, w)$ be a weighted undirected graph with n vertices and m edges, and let $s \in V$ be a fixed source. Each edge $(x, y) \in E$ has a real positive weight w_{xy} associated. Let $d : V \Rightarrow \mathbb{R}^+$ be a distance function giving, for each $x \in V$, the minimum distance of x from s, let $T(s) = (V_T, E_T)$ be a shortest paths tree of G rooted at s, and, for any $x \in V$, let $T(x)$ be the subtree of $T(s)$, rooted at x. Every $x \in V$ has one parent (except for the source s), denoted as $parent(x)$, and a set of children, denoted as $children(x)$, in $T(s)$. An edge (x, y) is a *tree edge* if $(x, y) \in E_T$; otherwise it is a *non-tree edge*.

If G is a graph and μ (resp., β) is an edge update (resp., a batch of edge updates) of G, we denote by G', $T'(s)$, $d'(x)$ and $parent'(x)$, the graph, the shortest paths tree, the distance of vertex x from s and the parent of vertex x in $T'(s)$, respectively, after the execution of μ (resp., β).

Complexity Model. In [11], the authors propose to measure the complexity of the single source shortest paths problem in a dynamic scenario as follows: given a graph $G = (V, E, w)$ with source s, the output information consists of $d(x)$, for any $x \in V$, and of $T(s)$. Let μ be an edge operation to be performed on G (insertion, deletion, or weight update), and G' be the new graph after that μ has been performed on G. The set of output updates $\delta(G, \mu)$ to be performed is given by the set of vertices that either change their distance from the source, or change their parent in the shortest paths tree, due to μ. The number of output updates caused by μ is the cardinality of $\delta(G, \mu)$. This notion has been also defined in [11] for sequences of updates as follows. Let $\sigma = (\mu_1, \mu_2, \ldots, \mu_h)$ be a sequence of input modifications (insertions, deletions or weight updates of edges) to be performed on G; each input modification $\mu_i \in \sigma$ is performed on graph G_{i-1}, with $G_0 \equiv G$, and gives the new graph G_i. The output information is required to be updated after each input modification $\mu_i \in \sigma$. Let $\delta(G_{i-1}, \mu_i)$ and $\Delta(G, \sigma)$ be the set of output updates caused by μ_i and by the whole sequence σ, respectively. The total number of output updates over the sequence σ is given by $|\Delta(G, \sigma)| = \sum_{\mu_i \in \sigma} |\delta(G_{i-1}, \mu_i)|$. Note that, no algorithm can process σ by performing explicit updates in less than $|\sigma| + |\Delta(G, \sigma)|$ time: this is the cost an ideal algorithm that carries out each update in constant time.

We extend the model of [11] to batches of updates as follows. Let $\beta = (\mu_1, \ldots, \mu_h)$ be a batch of operations on the edges of G, we call $\hat{\delta}(G, \beta)$ the set of output updates caused by all the operations in β, simultaneously. Notice that $|\hat{\delta}(G, \beta)|$ can be much smaller than $|\Delta(G, \beta)|$ as vertices that are affected multiple times with respect to $\Delta(G, \beta)$ are considered only once in $\hat{\delta}(G, \beta)$. In a similar way, if $B = (\beta_1, \beta_2, \dots, \beta_h)$ is a sequence of batches, then we define $|\hat{\Delta}(G, B)| = \sum_{\beta_i \in B} |\hat{\delta}(G_{i-1}, \beta_i)|$ where $G_0 \equiv G$ and G_i is the graph obtained after the first i batches of B. Moreover, with a little abuse of notation, let $|B|$ be the total number of update operations contained in B, i.e. $|B| = \sum_{\beta_i \in B} |\beta_i|$.

Data Structures. To implement our algorithms, we use the same technique of [12] to partition the edges incident to each vertex of G and to store them. In particular, any edge (x, y) has an *owner*, denoted as $owner(x, y)$, that is either x or y. For each vertex x, $ownership(x)$ denotes the set of edges owned by x, and $non\text{-}ownership(x)$ denotes the set of edges with one endpoint in x, but not owned by x. If G has a k-baf then, for each $x \in V$, $ownership(x)$ contains at most k edges. Notice that, a 2-approximation of an optimal accounting function for a generic graph G can be computed in linear time [13]. Furthermore, given an edge (z, q) of G, the backward level (forward level, respectively) of edge (z, q) and of vertex q, relative to vertex z, is the quantity $b\_level_z(q) = d(q) - w_{zq}$ (resp., $f\_level_z(q) = d(q) + w_{zq}$). The intuition behind these definition is that the level of an edge (z, q) provides information about the shortest available path from s to q passing through z. The edges in $non\text{-}ownership(x)$ are stored as follows. In the decremental batch algorithm, $non\text{-}ownership(x)$ is stored as a min-based priority queue F_x; the priority of edge (x, y) in F_x, denoted as $F_x(y)$, is the computed value of $f\_level_x(y)$. In the incremental batch algorithm, $non\text{-}ownership(x)$ is stored as a max-based priority queue B_x; the priority of edge (x, y) in B_x, denoted as $B_x(y)$, is the computed value of $b\_level_x(y)$. The priority queues are implemented by efficient heaps as those of [5], which support insert, findMin, findMax and decreaseKey operations in $O(1)$ worst-case time, and delete and increaseKey operations in $O(\log n)$ worst-case time. The described data structures require $O(|V| + |E|)$ space. Since distances and the shortest path tree are stored, shortest path queries are answered in optimal time.

In what follows, for any $x \in V$, we say that edges in $ownership(x)$ are scanned by ownership, edges in $non\text{-}ownership(x)$ are scanned by priority. Moreover, $D(x)$ (resp., $P(x)$) stores the distance from s to x (resp., parent of x in the current shortest paths tree) computed by the algorithms. We assume that, before the execution of any update procedure, $D(x) = d(x)$ for each $x \in V$, and we will prove that $D(x) = d'(x)$ upon termination of the procedures. We assume that before the execution of a batch of operations all data structures are correctly stored.

3   Decremental Batch Algorithm

In this section we provide an algorithm, named `decrementalBatch`, for the decremental batch single-source shortest path problem. Algorithm `decrementalBatch`, reported in Figure 1, computes the updated shortest path tree $T'(s)$ and distance function d' of G', where G' is the result of the application of a decremental batch $\beta = (\mu_1, \dots, \mu_h)$ on G. We use the following notion of coloring. Given a vertex $q \in V$, $color(q)$ is: white if q changes neither the distance from s nor the parent in $T(s)$; red if q increases the distance from s, i.e., $d'(q) > d(q)$; pink if q preserves its distance from s, but it must replace the old parent in $T(s)$, i.e., q is pink if $d'(q) = d(q)$, but $parent'(q) \neq parent(q)$.

The algorithm works in three phases. The first phase (Lines 1–10), analyzes each operation μ_i of the decremental batch β as follows. If μ_i is a weight increase operation on an edge (z, q), owned by z, then the priority of z needs to be updated in F_q. Otherwise, if μ_i is a delete operation, then the priority of z needs to be removed from F_q. Moreover, in both cases, if (z, q) is a tree edge and z is the parent of q in the shortest path tree, then q is inserted in a min-heap Q with priority equal to its distance from s in G.

Once all the affected edges have been analyzed, the second phase (Lines 11–17) is performed. In this phase, the algorithm assigns a color to each vertex as follows. It first extracts from Q the min-priority vertex x, that is the vertex which was closer to s in G. Now, two cases can occur. If vertex x has a neighbor y such that $color(y) \neq$ red and $d(y) + w_{xy} = d(x)$, that is a vertex that allows x to keep its distance unchanged, then the color of x is set to pink and its parent $P(x)$ is set to y (as shown in Procedure $checkPink$). Otherwise, the color of x is set to red. Furthermore, all the children v, in $T(s)$, of x are inserted in Q, if they have not been inserted yet, with priority equal to $d(v)$. The algorithm proceeds by extracting the next vertex from Q and by iterating the above strategy. When the heap becomes empty, the coloring phase terminates (Line 17).

Now, the algorithm performs a third phase (Lines 18–25) where it updates the distance from s of all the red vertices. Note that, the pink vertices do not change their distance and have already updated the parent in $T(s)$ during the second phase. The update phase proceeds as follows. For each red vertex x (as shown in Procedure $checkBestNonRed$), the algorithm tries to find what we call the best non-red neighbor, that is a vertex y such that $d(y) + w_{xy} = \min_{k \in N(x)} \{d(k) + w_{kx}\}$ and $color(y) \neq$ red. If such neighbor exists, it is inserted in a min-heap Q with priority equal to $d(y) + w_{xy}$, and the distance from s to x is set to the same value. Otherwise, x is inserted in Q with priority equal to infinity, the distance from s to x is set to infinity and the parent of x is set to null. In order to speed-up the search for the best non-red neighbor of a red vertex x, the priority of the edges owned by x, that links x to non-red vertices is updated during the first phase. In such way, the best non-red neighbor is always the best among the non-red neighbors j of x such that x owns (x, j) and the first non-red vertex in F_x.

Now, a relaxing step is performed. In particular, the min-priority vertex x is extracted from Q and, for each neighbor v of x, the distance from s to v is relaxed as shown in Procedure $processRed$. If $D(v) > D(x) + w_{xv}$, then $D(v)$ is set to $D(x) + w_{xv}$ and $P(v)$ is set to x. Moreover, the priority of v in Q is also updated to $D(v) + w_{xv}$. Note that, once a vertex is extracted from Q, its distance does not change anymore. The algorithm proceeds by extracting the next vertex from Q and by iterating the above strategy. When the heap becomes empty the algorithm and both the shortest path tree and the distance function are updated, that is for each $v \in V$, $D(v) = d'(v)$ and $P(v) = parent'(v)$.

Procedure: decrementalBatch(β)

   /* Phase 1:   */

1   **foreach** $\mu \in \beta$ **do**

2   **if** μ *is a weight increase on edge* (x, y) **then**

3   **if** $owner(x, y) = x$ **then** Update $F_y(x)$;

4   **else** Update $F_x(y)$;

5   **if** μ *is a deletion on edge* (x, y) **then**

6   **if** $owner(x, y) = x$ **then** Delete $F_y(x)$;

7   **else** Delete $F_x(y)$;

8   **if** (x, y) *is a tree edge* **then**

9   **if** $D(x) > D(y)$ **then** Enqueue($Q, \langle x, D(x) \rangle$);

10   **else** Enqueue($Q, \langle y, D(y) \rangle$);

   /* Phase 2:   */

11 **while** Q *is not empty* **do**

12   $\langle z, D(z) \rangle :=$ extractMin(Q);

13   $checkPink(z)$;

14   **if** $color(z) \neq pink$ **then**

15     $color(z) :=$ **red**;

16   **foreach** $v \in$ children(z) : $v \notin Q$ **do**

17   Enqueue($Q, \langle v, D(v) \rangle$);

   /* Phase 3:   */

18 **foreach** $z \in V$: $color(z) =$ **red** **do**

19   Create an empty list $list(z)$;

20 **foreach** $z \in V$: $color(z) =$ **red** **do**

21   $checkBestNonRed(z)$;

22   Enqueue($Q, \langle z, D(z) \rangle$);

23 **while** Q *is not empty* **do**

24   $\langle z, D(z) \rangle :=$ extractMin(Q);

25   $processRed(z)$;

26 Restore **white** color for all colored vertices

Fig. 1. Procedure decrementalBatch

Theorem 1. *If G has a k-baf, then* decrementalBatch *requires* $O(|\beta| \log n + |\hat{\delta}(G, \beta)| \cdot k \log n)$ *worst case time to process a decremental batch β.*

Proof. First of all, it is easy to see that Phase 1 requires time $O(|\beta| \log n)$, as each delete or increase operation of the batch β is processed in $O(\log n)$ time. Concerning Phase 2, consider that the overall cost of lines 1–4 of Procedure *checkPink* is proportional to the number of edges that are scanned by ownership while searching for the **non-red** neighbor, and requires at most $O(k)$ time per colored vertex. Note that if edge (z, q) is considered while searching for a **non-red** neighbor for z in Procedure *checkPink* then z will be colored either **pink** or **red**.

The rest of Procedure *checkPink* performs a scan by priority. All the edges scanned by priority but the last are such that both their endpoints are colored. As there are at most $O(k)$ of these edges per colored vertex, then the total cost of Phase 2 is bounded by $O(k \log n)$ times the number of **red** and **pink** vertices.

Procedure: $checkPink(z)$

1 **foreach** $(z, q) \in ownership(z) \; : \; color(q) \neq red$ **do**
2   **if** $D(q) + w_{zq} = D(z)$ **then**   /* Check by ownership */
3          $color(z) := $ pink;
4          $P(z) := q$;
5 **if** $color(z) \neq pink$ **then**   /* Check by ownership failed */
6   **while** F_z *is not empty* **do**
7          $q := $ extractMin(F_z);
8   **if** $D(q) + w_{zq} > D(z)$ **then break**;
9   **if** $color(q) \neq red$ **then**   /* Check by priority */
10              $color(z) := $ pink;
11              $P(z) := q$;
12   **break**;
13   Reinsert all the extracted vertices, with the same priority, into F_z;

Fig. 2. Procedure $checkPink$

Procedure: $checkBestNonRed(z)$

1 $D(z) := \infty$; $P(z) := $ **null**;
2 **foreach** $(z, q) \in ownership(z)$ **do**   /* Check by ownership */
3   **if** $color(q) \neq red$ **and** $D(q) + w_{zq} < D(z)$ **then**
4          $D(z) := D(q) + w_{zq}$;
5          $P(z) := q$;
6   **else**
7          $list(z).append(z, q)$;
8          $list(q).append(z, q)$;
9 **while** F_z *is not empty* **do**
10     $q := $ extractMin(F_z);
11   **if** $color(q) \neq red$ **and** $D(q) + w_{zq} < D(z)$ **then** /* Check by priority */
12         $D(z) := D(q) + w_{zq}$;
13         $P(z) := q$;
14   **break**;
15 Reinsert all the extracted vertices, with the same priority, into F_z;

Fig. 3. Procedure $checkBestNonRed$

Procedure: $processRed(z)$

1 **foreach** $(z, q) \in list(z)$ **do**   /* $(z, q) \in E : color(q) = $ red */
2   **if** $D(z) + w_{zq} < D(q)$ **then**
3          $D(q) := D(z) + w_{zq}$;
4          $P(q) := z$;
5   Heap_Improve$(Q, \langle q, D(q) \rangle)$;
6 **foreach** $(z, q) \in ownership(z)$ **do**   /* Check by ownership */
7   Update $F_q(z)$;

Fig. 4. Procedure $processRed$

To bound the cost of Phase 3, observe that the cycle at line 18 requires a time proportional to the number of **red** vertices and that the execution of the procedure *checkBestNonRed* performs both a scan by ownership and a scan by priority. The scan by ownership requires at most $O(k)$ time per **red** vertex while all the edges scanned by priority but the last are such that both their endpoints are **red**. As there are at most $O(k)$ of these edges per colored vertex, the total time required by the procedure is $O(k \log n)$ per **red** vertex. The remaining time of Phase 3 is dominated by the cost of executing line 24 of Procedure `decrementalBatch` and the whole Procedure *processRed*. Clearly the cost of line 24 of Procedure `decrementalBatch` is $O(\log n)$ per **red** vertex, while the cost of Procedure *processRed* can be bounded by noticing that all the edges considered are such that both their endpoints are **red**, that the cost of Line 5 of *processRed* is $O(\log n)$, and that updating the structures F_q requires at most $O(\log n)$ time per operation. It follows that the total cost of Phase 3 is $O(k \log n)$ per **red** vertex and that the overall cost of algorithm `decrementalBatch` is $O(|\beta| \log n + |\hat{\delta}(G, \beta)| \cdot k \log n)$. $\qquad\square$

4 Incremental Batch Algorithm

In this section we provide an algorithm, named `incrementalBatch`, for the incremental batch single-source shortest path problem. Algorithm `incrementalBatch`, reported in Figure 5, computes the updated shortest path tree $T'(s)$ and distance function d' of G', where G' is the result of the application of an incremental batch $\beta = (\mu_1, \ldots, \mu_h)$ on G. We use the following notion of coloring. Given a vertex $q \in V$, *color*(q) is: **white** if q changes neither the distance from s nor the parent in $T(s)$; **blue** if q changes its distance from s.

The algorithm works in two phases, as follows. In the first phase, all the operations in the incremental batch are considered, one by one and all the data structures are updated. If an operation in the batch involving an edge (z, q) induces a decrease in the distance from s of one of the two endpoints $x \in \{z, q\}$, that is $d'(x) < d(x)$, then the procedure sets *color*(x) to **blue** and inserts it into a min-heap Q with priority equal to the new induced distance. Note that, if x is already in Q, the algorithm simply updates the priority.

In the second phase, the algorithm processes the vertices in Q: it extracts the min-priority vertex x from Q and, for each $y \in N(x)$, it performs a relaxing step: if a path from s to y passing through x is discovered in G' that is shorter than the shortest path from s to y in G, the color of y is set **blue**, its distance is updated and it is inserted into Q with priority equal to the new distance.

Lemma 1. *Given a graph $G = (V, E)$ and a set of additional edges F, let $G' = (V, E \cup F)$ and let k^* be the minimum integer such that a k^*-baf for G' exists. If a k-baf for G is known then it is possible to compute, in $O(|F|)$ time, a k'-baf for G' such that $k' \le \max\{2k, 4k^*\}$.*

Proof. Consider the graph $H = (V', F)$ where V' is the set of vertices adjacent to some vertex in F. Compute an accounting function for H using the linear-time

---

Procedure: incrementalBatch(β)
/* Phase 1   */
1   $F := \emptyset$;
2   **foreach** $\mu \in \beta : \mu$ *is an insertion of edge* (x, y) **do**
3   Add (x, y) to G with weight w_{xy};
4      $F := F \cup (x, y)$;
5   Recompute an ownership function for G using the set F, as shown by Lemma 1;
6   **foreach** $\mu \in \beta$ **do**
7   Let (x, y) the edge involved by μ;
8   **if** $D(y) < D(x)$ **then** Swap x and y
9   **if** $owner(x, y) = x$ **then** Update $B_y(x)$;   /* or insert */
10   **else** Update $B_x(y)$;   /* or insert */
11   **if** $D(x) + w_{xy} < D(y)$ **then**
12        $D(y) := D(x) + w_{xy}$;
13        $P(y) := x$;
14        $color(y) := $ **blue**;
15   Enqueue($Q, \langle y, D(y) \rangle$);   /* or update */
    /* Phase 2   */
16 **while** Q *is not empty* **do**
17     $\langle z, D(z) \rangle := $ extractMin(Q);
18   **foreach** $v \in N(z)$ **do**
19   **if** $(z, v) \in ownership(z)$ **then**
20   Update $B_v(z)$;
21   **if** $D(v) > D(z) + w_{zv}$ **then**
22           $D(v) := D(z) + w_{zv}$;
23           $P(v) := z$;
24           $color(v) := $ **blue**;
25   Enqueue($Q, \langle v, D(v) \rangle$);   /* or update */
26   Reinsert all the extracted vertices, with the same priority, into B_z;
27 Restore **white** color for all colored vertices

Fig. 5. Procedure incrementalBatch

2-approximation algorithm shown in [13] for finding a k-baf with the minimum value of k. Let $h \leq 2k^*$ be the value returned by this algorithm. By combining this function with the k-baf for G we get a $k' = (k + h)$-baf for G'. If $k \geq h$ then we have $k' \leq 2k$. Otherwise, if $k < h$ we have $k' \leq 2h \leq 2 \cdot 2k^*$.   □

Theorem 2. *If* G *has a* k-baf, *then* incrementalBatch *requires* $O(|\beta| + |\hat{\delta}(G, \beta)| \cdot \max\{k, k^*\} \cdot \log n)$ *worst case time to process an incremental batch* β, *where* k^* *is the minimum integer such that a* k^*-baf *exists for the resulting graph.*

Proof. Let F be the set of the newly inserted edges in β. Phase 1 requires time $O(|\beta|)$ as each operation of the batch β is processed in constant time and the ownership function can be updated in $O(|F|)$ time as shown in Lemma 1. Let $k' \leq \max\{2k, 4k^*\}$ be the maximum number of edges owned by a vertex with respect to the new ownership function.

To bound the complexity of Phase 2, we note that a vertex is in Q if and only if it decreases its distance from s. Moreover, once a vertex x has been extracted from Q its distance is $D(x) = d'(x)$ and it never changes again. As every vertex x is inserted in Q only if a better path from s to x is discovered this implies that vertices are inserted in Q at most once.

The above considerations imply that the number of processed vertices in Phase 2 is $|\hat{\delta}(G, \beta)|$, i.e. the number of blue vertices. Processing a vertex z takes at most $O(k' \log n)$ time: first the structures associated with vertices $v \in ownership(z)$ are updated in $O(k')$ total time, then the neighbors of z are examined. This is done by ownership, which requires at most $O(k')$ time, and by priority: vertices are extracted from $B_z(v)$ until they do not improve their distance any more. All (but the last) extracted vertices are blue and each extraction costs $O(\log n)$. We bound this cost by charging the owner of the edge each time an edge is processed by priority. As each vertex owns at most k' edges, the overall time needed is $O(k' \log n)$ per affected vertex. $\square$

5   Sequences of Incremental and Decremental Batches

In this section we analyze the case in which the input graph G is subject to arbitrary sequences of incremental and decremental batches. We remark that for any sequence of such updates that has either no insertion of new edges or has a constant number of insertions it is possible to repeatedly apply decrementalBatch and incrementalBatch when needed, to obtain the same bounds of Sections 3 and 4. The only modification needed in both algorithms is the following: every time F_x is updated for some vertex x, also B_x has to be updated, and viceversa.

In the more general case in which the number of insert operation is not constant, we need to apply a further modification to the algorithms. Before reverting the colored vertices to white, we perform an additional scan by ownership from each vertex in order to construct a list of edges that are incident to the affected vertices. Then, for each edge of this list, we change its owner. This means that the structures associated with both endpoints need to be updated. This costs $O(\log n)$ per affected vertex. During the execution of the incremental algorithm we no longer need to recompute an accounting function for newly inserted edges and we can just choose an arbitrary endpoint as its owner. In this case we can state the following theorem whose proof will be given in the full paper.

Theorem 3. *Given $G = (V, E)$ and a sequence $B = (\beta_1, \dots, \beta_h)$ of incremental and decremental batches on G, let $G_i = (V, E_i)$ be the graph obtained by applying the first i batches of B to G. If there exists an accounting function f for $G_B = (V, \bigcup_{i=1}^{h} E_i)$ such that f is k-bounded for each G_i, then it is possible to:*

- *preprocess G in order to compute a shortest path tree $T(s)$ and to initialize all the necessary data structures in $O(m + n \log n)$ worst case time.*
- *process B in $O(|B| \log n + |\hat{\Delta}(G, B)| \cdot k \log n)$ overall worst case time, that gives $O(|\beta_i| \log n + |\hat{\delta}(G_{i-1}, \beta_i)| \cdot k \log n)$ amortized time per batch β_i, in such a way that $T(s)$ and all distances from s are correctly updated at the end of each batch in B.*

References

1. Abraham, I., Delling, D., Goldberg, A.V., Werneck, R.F.: Hierarchical hub labelings for shortest paths. In: Epstein, L., Ferragina, P. (eds.) ESA 2012. LNCS, vol. 7501, pp. 24–35. Springer, Heidelberg (2012)
2. Bauer, R., Delling, D., Sanders, P., Schieferdecker, D., Schultes, D., Wagner, D.: Combining hierarchical and goal-directed speed-up techniques for dijkstra's algorithm. ACM Journal on Experimental Algorithms 15, Article 2.3 (2010)
3. Bauer, R., Wagner, D.: Batch dynamic single-source shortest-path algorithms: An experimental study. In: Vahrenhold, J. (ed.) SEA 2009. LNCS, vol. 5526, pp. 51–62. Springer, Heidelberg (2009)
4. Bernstein, A.: Maintaining shortest paths under deletions in weighted directed graphs. In: Proceedings of 45th ACM STOC, pp. 725–734. ACM (2013)
5. Brodal, G.S.: Worst-case efficient priority queues. In: Proceedings seventh ACM-SIAM Symposium on Discrete algorithms, pp. 52–58. SIAM (1996)
6. Bruera, F., Cicerone, S., D'Angelo, G., Di Stefano, G., Frigioni, D.: Dynamic multilevel overlay graphs for shortest paths. Mathematics in Computer Science 1(4), 709–736 (2008)
7. Buriol, L.S., Resende, M.G.C., Thorup, M.: Speeding up dynamic shortest-path algorithms. INFORMS Journal on Computing 20(2), 191–204 (2008)
8. Chan, E.P.F., Yang, Y.: Shortest path tree computation in dynamic graphs. IEEE Transactions on Computers 4(58), 541–557 (2009)
9. Delling, D., Goldberg, A.V., Pajor, T., Werneck, R.F.: Customizable route planning. In: Pardalos, P.M., Rebennack, S. (eds.) SEA 2011. LNCS, vol. 6630, pp. 376–387. Springer, Heidelberg (2011)
10. Dijkstra, E.W.: A note on two problems in connexion with graphs. Numerische Mathematik 1, 269–271 (1959)
11. Frigioni, D., Marchetti-Spaccamela, A., Nanni, U.: Semidynamic algorithms for maintaining single source shortest paths trees. Algorithmica 22(3), 250–274 (1998)
12. Frigioni, D., Marchetti-Spaccamela, A., Nanni, U.: Fully dynamic algorithms for maintaining shortest paths trees. J. of Algorithms 34(2), 251–281 (2000)
13. Frigioni, D., Marchetti-Spaccamela, A., Nanni, U.: Fully dynamic shortest paths in digraphs with arbitrary arc weights. J. of Algorithms 49(1), 86–113 (2003)
14. Geisberger, R., Sanders, P., Schultes, D., Vetter, C.: Exact routing in large road networks using contraction hierarchies. Transportation Sc. 46(3), 388–404 (2012)
15. Narváez, P., Siu, K.Y., Tzeng, H.Y.: New dynamic algorithms for shortest path tree computation. IEEE/ACM Transactions on Networking 8(6), 734–746 (2000)
16. Ramalingam, G., Reps, T.W.: An incremental algorithm for a generalization of the shortest paths problem. Journal of Algorithms 21, 267–305 (1996)
17. Ramalingam, G., Reps, T.W.: On the computational complexity of dynamic graph problems. Theor. Comput. Sci. 158(1&2), 233–277 (1996)
18. Roditty, L., Zwick, U.: Dynamic approximate all-pairs shortest paths in undirected graphs. SIAM J. on Computing 41(3), 670–683 (2012)

Simultaneous Consensus vs Set Agreement:
A Message-Passing-Sensitive Hierarchy
of Agreement Problems

Michel Raynal[1,2] and Julien Stainer[1]

[1] Institut Universitaire de France
[2] IRISA, Campus de Beaulieu, 35042 Rennes Cedex, France

Abstract. This paper investigates the relation linking the s-simultaneous consensus problem and the k-set agreement problem in wait-free message-passing systems. To this end, it first defines the (s, k)-SSA problem which captures jointly both problems: each process proposes a value, executes s simultaneous instances of a k-set agreement algorithm, and has to decide a value so that no more than sk different values are decided. The paper introduces then a new failure detector class denoted $Z_{s,k}$, which is made up of two components, one focused on the "shared memory object" that allows the processes to cooperate, and the other focused on the liveness of (s, k)-SSA algorithms. A novelty of this failure detector lies in the fact that the definition of its two components are intimately related. Then, the paper presents a $Z_{s,k}$-based algorithm that solves the (s, k)-SSA problem, and shows that the "shared memory"-oriented part of $Z_{s,k}$ is necessary to solve the (s, k)-SSA problem (this generalizes and refines a previous result that showed that the generalized quorum failure detector Σ_k is necessary to solve k-set agreement). Finally, the paper investigates the structure of the family of (s, k)-SSA problems and introduces generalized (asymmetric) simultaneous set agreement problems in which the parameter k can differ in each underlying k-set agreement instance. Among other points, it shows that, for $s, k > 1$, (a) the $(sk, 1)$-SSA problem is strictly stronger that the (s, k)-SSA problem which is itself strictly stronger than the $(1, ks)$-SSA problem, and (b) there are pairs (s_1, k_1) and (s_2, k_2) such that $s_1 k_1 = s_2 k_2$ and (s_1, k_1)-SSA and (s_2, k_2)-SSA are incomparable.

Keywords: Asynchronous system, Distributed computing, Distributed computability, Failure detector, Fault tolerance, Message-passing system, Quorum, Reduction, k-Set agreement, Simultaneous consensus, Wait-freedom.

1 Introduction

The k-set agreement problem The k-set agreement problem is a paradigm of coordination problems. Defined in the setting of systems made up of processes prone to crash failures, it is a simple generalization of the consensus problem (that corresponds to the case $k = 1$). The aim of this problem, introduced by Chaudhuri [9], was to investigate how the number of choices (k) allowed to the processes is related to the maximum number of processes t that can crash. The problem is defined as follows. Each process

T. Moscibroda and A.A. Rescigno (Eds.): SIROCCO 2013, LNCS 8179, pp. 298–309, 2013.

proposes an input value, and any process that does not crash must decide a value (termination), such that a decided value is a proposed value (validity), and no more than k distinct values are decided (agreement).

While it can be solved in synchronous systems prone to any number of process crashes (see [20] for a survey), the main result associated with k-set agreement is the impossibility to solve it in presence of both asynchrony and process crashes when $t \geq k$ [4,15,25].

A way to circumvent this impossibility consists in enriching the underlying pure asynchronous system with a failure detector [7,22,24]. A failure detector is a device that provides processes with information on failures. According to the type and the quality of this information, several failure detectors have been proposed (see [21] for a survey of failure detectors suited to k-set agreement). It has been shown that the failure detector $\overline{\Omega}_k$ (anti-omega-k) [19,26] is the weakest failure detector that allow k-set agreement to be solved despite any number of process crashes in asynchronous read/write systems [13].

The situation is different in asynchronous crash-prone *message-passing* system. More precisely, (a) while weakest failure detectors are known only for the cases $k = 1$ and $k = n - 1$ [8,11,12], (b) it has been shown that the generalized quorum failure detector denoted Σ_k is necessary [3]. Several k-set agreement algorithms based on failure detectors stronger than Σ_k can found in the literature (e.g., [3,5,10,16,17,18]).

The s-simultaneous consensus problem This problem has been introduced in [1]. Each of the n processes proposes the same value to s independent instances of the consensus problem, denoted $1, ..., s$. Each correct process has to decide a pair (c, v) (termination), where $c \in \{1, ..., s\}$ is a consensus instance and v is a proposed value (validity). Moreover, if (c, v) and (c, v') are decided we have $v = v'$ (agreement). (This is the scalar form of the problem: each process proposes the same value to each consensus instance. In the vector form, a process proposes a vector of s values, one value to each consensus instance. It is shown in [1] that both forms have the same computational power).

It is shown in [1] that the x-simultaneous consensus problem and the x-set agreement problem are computationally equivalent in asynchronous read/write systems where up to $t = n - 1$ processes may crash. It follows that in these systems, the failure detector $\overline{\Omega}_x$ is both necessary and sufficient to solve x-simultaneous consensus.

As far as asynchronous message-passing systems are concerned, it is shown in [6] that, for $x > 1$ and $t > \frac{n+x-2}{2}$, x-simultaneous consensus is strictly stronger than x-set agreement. This means that, differently from what can be done in asynchronous read/write systems, it is not possible to solve x-simultaneous consensus from a black box solving x-set agreement.

Content of the paper The aim of this paper is to (a) better understand the relations linking s-simultaneous consensus and k-set agreement, and (b) become closer to the weakest failure detector that allows k-set agreement to be solved in crash-prone asynchronous message-passing system.

To this end, the paper introduces first a problem that generalizes both s-simultaneous consensus and k-set agreement. This problem, denoted (s, k)-SSA (for s-Simultaneous k-Set Agreement) consists in s independent instances of the k-set agreement problem (hence, $(s, 1)$-SSA is x-simultaneous consensus, while $(1, k)$-SSA is k-set agreement).

Then, the paper introduces a new failure detector, denoted $Z_{s,k}$, that allows (s, k)-SSA to be solved in the asynchronous message-passing communication model, despite any number of process crashes. This failure detector is captured by an array of size s each entry of which is made up of two components. The first, which is nothing else than the *quorum* failure detector Σ_k, addresses the data sharing needed to correctly coordinate the processes. The second component states a leader-based property that allows the correct processes to always decide a value. When considering the $(1, k)$-SSA problem, it appears that $Z_{1,k}$ is a weaker failure detector than all the failure detectors proposed so far to solve k-set agreement. A noteworthy feature of $Z_{s,k}$ lies in the fact that these two components are not defined independently one from the other (e.g., as done in the pair (Σ, Ω) [11]), namely, the definition of the *leader* component of *some* entry of the array is intimately related to the associated *quorum* component.

The paper presents then a $Z_{s,k}$-based algorithm that solves the (s, k)-SSA problem, and shows that the quorum part of $Z_{s,k}$ is necessary to solve the (s, k)-SSA problem (this proof generalizes the proof given in [3] that shows that Σ_k captures information on process crashes that is necessary to solve k-set agreement).

Last but not least, the paper considers the family of asymmetric $\{k_1, ..., k_s\}$-SSA problems, defined by s simultaneous instances of the k_x-set agreement problem where $k_x = k_1, ..., k_s$. It shows that these problems define a strong hierarchy from a computability point of view. It follows from this hierarchy that (as indicated in the abstract) for $s, k > 1$, (a) the $(sk, 1)$-SSA problem is strictly stronger that the (s, k)-SSA problem which is itself strictly stronger than the $(1, ks)$-SSA problem, and (b) there are pairs (s_1, k_1) and (s_2, k_2) such that $s_1 k_1 = s_2 k_2$ and (s_1, k_1)-SSA and (s_2, k_2)-SSA are incomparable problems. More generally, given K, the paper shows that the structure of the set of symmetric (s, k)-SSA problems (where $sk = K$) is a lattice where an arrow from A to B means that B can be solved from a block box solving A, but not vice-versa. The paper associates also with each such pair a failure detector that is necessary to solve A and a failure detector that is sufficient to solve B.

Roadmap. The paper is made up of 5 sections. Section 2 defines the computation model, the (s, k)-SSA problem, and the Failure Detector class $Z_{s,k}$. Section 3 extends our previous results to the (s, k)-SSA problem, namely, it presents a simple $Z_{s,k}$-based algorithm that solves the (s, k)-SSA problem, and proves that the safety part of $Z_{s,k}$ is necessary when one wants to solve the (s, k)-SSA problem (from information on failures). Section 4 investigates the graph structure of the family of asymmetric SSA problems and shows that these problems define a strong hierarchy. Finally Section 5 concludes the paper. Due to page limitation, all the proofs are missing but can be found in [23].

2 Computation Model, (s, k)-SSA Problem, and the Failure Detector $Z_{s,k}$

2.1 Computation Model

Process model The system is made up of n asynchronous sequential processes denoted $\Pi = \{p_1, \ldots, p_n\}$ (to simplify notations, we sometimes consider that Π is the set $\{1, \ldots, n\}$). Each process is a Turing machine enriched with two operations, which allows it to send and receive messages. "Asynchronous" means that there is no assumption on the speed of processes: each process proceeds at its own speed, which may arbitrarily vary and is unknown from the other processes.

A process behaves correctly until it possibly crashes (a crash is an unanticipated premature stop). Up to $(n - 1)$ processes may crash (*wait-free* model). A process that crashes in a run is said to be *faulty* in that run, otherwise, it *correct*. Given a run, C denotes the set of processes which are correct in that run.

Communication model. Each pair of processes is connected by a bidirectional channel. The channels are failure-free (no creation, duplication, alteration, or loss of messages), and asynchronous. "Asynchronous" means that, while each message is received, there is no bound on message transfer delays.

Timing model. The underlying timing model is the set of natural integers $\mathbb{N}$. As the system is asynchronous, this time notion remains unknown to the processes. It is only used, from an external observer point of view, to state or prove properties. Time instants are denoted τ, τ', etc.

Notation. The previous wait-free message-passing model is denoted $\mathcal{AMP}_{n,n-1}[\emptyset]$.

2.2 The s-Simultaneous k-Set Agreement $-(s, k)$-SSA– Problem

As indicated in the introduction, the s-simultaneous k-set agreement problem (in short (s, k)-SSA) consists in the simultaneous execution of s instances of the k-set agreement problem. Moreover, each process proposes the same value to each instance of the k-set agreement problem. The (s, k)-SSA problem is defined by the three following properties.

- Termination. Every correct process decides.
- Validity. A decided value is a pair (c, v) where $1 \leq c \leq s$ and v is a value proposed by a process.
- Agreement. For any $c \in \{1, ..., s\}$, there are at most k different values v such that (c, v) is decided.

It is easy to see that at most $K = sk$ different values v are decided, and consequently, any algorithm solving the (s, k)-SSA problem solves the K-set agreement problem. Moreover, $(1, k)$-SSA is k-set agreement, while $(s, 1)$-SSA is s-simultaneous consensus.

2.3 The Failure Detector Class $Z_{s,k}$

Definition A failure detector of the class $Z_{s,k}$ provides each process p_i with two arrays denoted $qr_i[1..s]$ and $\ell d_i[1..s]$. Intuitively, $qr_i[z]$ and $\ell d_i[z]$, $1 \leq z \leq s$, denote, with respect to the index z, the current quorum and the current leader of p_i, respectively. $Z_{s,k}$ is defined by the following properties, where $qr_i^\tau[z]$ and $\ell d_i^\tau[z]$ denote the value of $qr_i[z]$ and $\ell d_i[z]$ at time τ.

- Safety property. $\forall\, z \in [1..s]$:
    - Quorum intersection property (QI).
      $$\forall\, i_1, ..., i_{k+1} \in \Pi,\ \forall\, \tau_1, ..., \tau_{k+1} : \exists h, \ell \in [1..k+1]:$$
      $$(h \neq \ell) \wedge (qr_{i_h}^{\tau_h}[z] \cap qr_{i_\ell}^{\tau_\ell}[z] \neq \emptyset).$$
    - Leader validity property(LV). $\forall\, \tau,\ \forall\, i : \ell d_i^\tau[z] \in \Pi$.
- Liveness property. $\exists z \in [1..s]$:
    - Quorum liveness property (QL). $\forall\, i \in C : \exists\, \tau : \forall\, \tau' \geq \tau : qr_i^{\tau'}[z] \subseteq C$.
    - Eventual leadership property (EL). $\exists \ell \in C : \forall i \in C :$
      $$\left[\forall\, \tau : \exists\, \tau', \tau'' \geq \tau : (qr_i^{\tau'}[z] \cap qr_\ell^{\tau''}[z] \neq \emptyset)\right]$$
      $$\Rightarrow \left[\exists\, \tau : \forall\, \tau' \geq \tau : (\ell d_i^{\tau'}[z] = \ell)\right].$$

The quorum intersection property states that, for any $z \in \{1, ..., s\}$, there are two quorum values that intersect in any set of $k + 1$ quorum values, each taken at any time. The leader validity property states that the leader domain is the set of processes.

While the safety properties concern all the entries of the arrays $qr_i[1..s]$ and $\ell d_i[1..s]$, the liveness properties are only on a single of these entries, say z. The quorum liveness property states that there is a finite time after which all quorum values (appearing in $qr_i[z]$ for every $i \in C$) contain only correct processes. The eventual leader liveness property involves only the quorum values taken by the entries $qr_i[z]$, for every $i \in C$. Hence, it relates these quorum values with the eventual leader values in the local variables $\ell d_i[z]$ at each correct process p_i. More precisely, it states that there is a correct process p_ℓ such that, for any correct process p_i whose quorum $qr_i[z]$ intersects infinitely often with the quorum $qr_\ell[z]$ of p_ℓ (left part of the implication), p_ℓ becomes eventually the permanent leader of p_i (saved in $\ell d_i[z]$, right part of the implication).

The generality of $Z_{s,k}$ wrt other failure detectors is investigated in [23].

Notation. Let $Z(Q)_{s,k}$ denote the quorum part of $Z_{s,k}$ (defined by the properties QI and QL). Similarly, let $Z(L)_{s,k}$ denote the leader part of $Z_{s,k}$ (defined by the properties LV and EL where the quorum part brings no information on failures, which means that we have then $\forall\, i, \forall\, z,\ \forall \tau : qr_i^\tau[z] = \Pi$).

Let FD be a failure detector class. $\mathcal{AMP}_{n,n-1}[FD]$ denotes the wait-free message-passing model enriched with a failure detector of the class FD. Sometimes FD is also used to denote a failure detector of the class FD.

3 Extending Two Previous Results

This section extends two of our previous results from k-set agreement to (s, k)-SSA.

3.1 A $Z_{s,k}$-Based Algorithm for the (s, k)-SSA Problem

An algorithm solving the (s, k)-SSA problem can be easily obtained by launching s concurrent instances of the previous k-set algorithm, the zth instance ($1 \leq z \leq s$) relying, at each process p_i, on the components $qr_i[z]$ and $\ell d_i[z]$ of $\mathcal{AMP}_{n,n-1}[Z_{s,k}]$. A process decides the value returned by the first of the s instances that locally terminates. Hence, it decides the pair (c, v) where c is its first deciding instance and v the value it decides in that instance. As there are s instances of the base k-set algorithm and at most k values can be decided in each of them, it follows that at most $K = sk$ different values can be decided. Moreover, as there is at least one instance z such that the failure detector outputs $\ell d_i[z]$ at each correct process p_i converge to the same correct process, it follows that the correct processes decide (if not done before) in at least one of the s underlying k-set agreement instances.

An underlying k-set agreement algorithm can be easily designed from an underlying abstraction (object) called alpha$_k$ (this object has been introduced in [18], which is a generalization of the alpha objects introduced in [14,24]).

Due to page limitation, both the underlying k-set agreement algorithm and the associated abstraction alpha$_k$ can be found in [23].

3.2 $Z(Q)_{s,k}$ is Necessary to Solve the (s, k)-SSA Problem

This section shows that $Z(Q)_{s,k}$ is necessary to solve the (s, k)-SSA problem as soon as we are looking for a failure detector-based solution. To that end, given a failure detector FD and an algorithm A that solves the (s, k)-SSA problem in $\mathcal{AMP}_{n,n-1}[FD]$, this section presents an algorithm that emulates the output of $Z(Q)_{s,k}$, namely an array $qr_i[1..s]$ at each process p_i, which satisfies the properties QI and QL. This means that it is possible to build $Z(Q)_{s,k}$ from any failure detector FD that can solve the (s, k)-SSA problem.

According to the usual terminology, $Z(Q)_{s,k}$ is *extracted* from the FD-based algorithm A. This extraction is a generalization of the algorithm introduced in [3], which extracts Σ_k from any failure detector-based algorithm solving k-set agreement.

The extraction algorithm. Each process p_i participates in several executions of the algorithm A. S being a set of processes, A^S denotes the execution of A in which exactly the processes of S participate. In this execution, each process of S either decides, blocks forever, or crashes. So the execution of the extraction algorithm is composed of $2^n - 1$ executions of A.

The behavior of each process p_i is described in algorithm 1. The internal statements of the tasks $T1$ and $T5$, and the tasks $T2$-$T4$ are locally executed in mutual exclusion. The local array $Q_i[1..s]$ is initialized to $[\Pi, \ldots, \Pi]$. The aim of $Q_i[c]$ is to contain all the sets S such that a value has been decided in the cth instance of the k-set agreement of the execution of A^S.

Initially, each process p_i proposes its identity i to all the instances of A in which it participates. To that end it invokes A^S.ssa_propose$_{s,k}(i)$ for each set S such that $i \in S$ (ssa_propose$_{s,k}()$ is the operation associated with each instance of the (s, k)-SSA problem). When it decides in the cth k-set agreement of A^S (task $T3$), p_i adds the set S to $Q_i[c]$ and informs each other process p_j, which includes S in $Q_j[c]$ when it learns it (task $T4$).

Each alive process p_i sends periodically messages ALIVE(i) (task $T1$) to inform the other processes that it is alive. When it receives a message ALIVE(j) (task $T2$), a process p_i moves j to the head of its local queue (denoted $queue_i$) which always contains all process identities. It follows that the identities of all the correct processes

---

Init: $Q_i[1, \ldots, s] \leftarrow [\Pi, \ldots, \Pi]$; $queue_i \leftarrow \langle 1, \ldots, n \rangle$;
    for each $S \subseteq \Pi$ **such that**$(i \in S)$ **do** A^S.ssa_propose$_{s,k}(i)$ **end for**; activate $T1$ to $T5$.

Task $T1$: **repeat periodically** send ALIVE(i) to each p_j such that $j \in \Pi \setminus \{i\}$ **end repeat**.

Task $T2$: **when** ALIVE(j) **is received**: move j at the head of $queue_i$.

Task $T3$: **when** $(c, -)$ **is decided by** p_i **in the** cth k-set agreement instance of A^S:
    $Q_i[c] \leftarrow Q_i[c] \cup \{S\}$; send DECISION$(c, S)$ to each p_j such that $j \in \Pi \setminus \{i\}$.

Task $T4$: **when** DECISION(c, S) **is received**: $Q_i[c] \leftarrow Q_i[c] \cup \{S\}$.

Task $T5$: **repeat forever**
    for each $c \in \{1, \ldots, s\}$ **do**
        $min\_rank_i \leftarrow \min\{\max\{rank(queue_i, j), j \in S\}, S \in Q_i[c]\}$;
        $qr_i[c] \leftarrow$ any $S_{min} \in Q_i[c]$ s.t. $\max\{rank(queue_i, j), j \in S_{min}\} = min\_rank_i$
    end for;
    end repeat.

Algorithm 1: Extracting $Z(Q)_{s,k}$ from a FD-based algorithm A solving (s, k)-SSA [3]

eventually precede in this queue the identities of all the faulty processes. (Initially, each queue $queue_i$ contains all process identities, in any order.)

$T5$ is a task whose aim is to repeatedly compute the current value of $qr_i[1..s]$. It uses the function $rank(queue_i, j)$ which returns the current rank of p_j in the queue $queue_i$. The value of $qr_i[c]$ is computed as follows. It is the "first set of $Q_i[c]$ with respect to $queue_i$" (i.e., with respect to the processes which are currently seen as being alive). This is captured with the help of the local variable $minrank_i$. As an example, let $Q_i[c] = \{\{3, 4, 9\}, \{2, 3, 8\}, \{4, 7\}, \{1, 2, 3, 4, 5, 6, 7, 8, 9\}\}$, and $queue_i = \langle 4, 8, 3, 2, 7, 5, 9, 1, 6 \rangle$. We have then $minrank = 4$, and $S_{min} = \{2, 3, 8\}$. This set of identities is the first set of $Q_i[c]$ with respect to $queue_i$ because each of the other sets $\{3, 4, 9\}$, $\{4, 7\}$, or $\{1, 2, 3, 4, 5, 6, 7, 8, 9\}$, includes an element (9, 7, and 6, respectively) that appears in $queue_i$ after all the elements of $\{2, 3, 8\}$ (in case several sets are "first", any of them can be selected).

Theorem 1. *Given any algorithm A that solves the (s, k)-SSA problem in the system model $\mathcal{AMP}_{n,n-1}[FD]$, the extraction algorithm described in Figure 1 is a wait-free construction of a failure detector $Z(Q)_{s,k}$.* (Proof in [23].)

4   The Structure of Generalized (s, k)-SSA Problems

This section studies the mathematical structure of the family of (s, k)-SSA problems for $sk = K$. To that end, it first introduces a straightforward generalization of this family and then shows that this generalized family can be represented by a directed graph where an arrow from A to B means that the problem B can be solved from a black box solving the problem A, while the opposite is impossible. To attain this goal, this section associates a pair of failure detectors with each pair of problems (A, B), such that one of these failure detectors is necessary to solve A while the other is sufficient to solve B.

4.1 The Generalized Asymmetric $\{k_1, ..., k_s\}$-SSA Problem

While the (s, k)-SSA problem is a symmetric problem which consists in s simultaneous instances of the k-set agreement problem, a simple generalization consists in considering an asymmetric version made up of s simultaneous instances of possibly different set agreement problems, namely the k_1-set agreement problem, the k_2-set agreement problem, etc., and the k_s-set agreement problem. Hence, among the proposed values, at most $K = \Sigma_{x=1}^{s} k_x$ different values are decided.

This asymmetric version is denoted $\{k_1, ..., k_s\}$-SSA where $\{k_1, ..., k_s\}$ is a multi-set[1]. The particular instance where $k_1 = \cdots = k_s = k$ is the symmetric (s, k)-SSA problem. As permuting the integers k_x does not change the problem, we consider the canonical notation where $k_1 \geq k_2 \geq ... \geq k_s \geq 1$.

4.2 Associating a Graph with a Family of Generalized $\{k_1, ..., k_s\}$-SSA Problems

Graph definition Given an integer K and starting from the source vertex labeled with the multiset $\{1, ..., 1\}$ (K times the integer 1), let us define a graph denoted $G(K)$ as follows. Given a vertex labeled $\{k_1, ..., k_s\}$ (initially, $s = K$ and $k_1 = \cdots = k_K = 1$), we add all possible vertices of $s - 1$ elements labeled $\{k'_1, ..., k'_{s-1}\}$ and directed edges from $\{k_1, ..., k_s\}$ to each vertex $\{k'_1, ..., k'_{s-1}\}$ defined as follows. Any pair of elements k_x, k_y of the multiset $\{k_1, ..., k_s\}$ gives rise to a vertex labeled by the multiset $\{k'_1, ..., k'_{s-1}\}$ such that

$$\{k'_1, ..., k'_{s-1}\} = \{k_1, ..., k_s\} \setminus \{k_x, k_y\} \cup \{k_x + k_y\}.$$

Then, the construction process is recursively repeated until we arrive at a sink node composed of a single element labeled $\{K\}$.

An example of graph for $K = 6$ is given on the right. The labels corresponding to symmetric instances $((s, k)$-SSA problems) are underlined. The graph (lattice) on the right side of the figure considers only the symmetric problem instances.

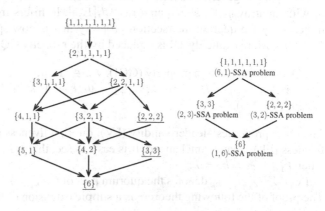

---

[1] The set notation is used to represent a multiset. A multiset is a set in which several elements can have the same value. As an example, $\{1, 2, 1, 1, 3\}$ is a multiset of 5 elements. Hence, the multisets $\{1, 2, 1, 1, 3\}$ and $\{2, 1, 3\}$ are different (while $\{1, 2, 1, 1, 3\} = \{2, 1, 3\}$ from a set point of view).

Meaning of the graph As we will see in Section 4.4, given an integer K, this graph describes the computability hierarchy linking all the $\{k_1, k_2, ...\}$-SSA agreement problems such that $k_1 + k_2 + \cdots = K$. Let the label A of a vertex denotes both the vertex itself and the associated agreement problem. An edge from a vertex A to a vertex B means that (a) given an algorithm that solves the problem A it is possible to solve the problem B, while (b) the opposite is impossible.

Lemma 1. $G(K)$ *is cycle-free.* (Proof in [23].)

As we will see in Lemma 2, the following predicate P characterizes (with the existence of a function f) the pairs of vertices connected by a path in $G(K)$. Given two nodes $\{k_1, ..., k_s\}$ and $\{k'_1, ..., k'_{s'}\}$ of $G(K)$, this function f maps each index of $\{k_1, ..., k_s\}$ to an index of $\{k'_1, ..., k'_{s'}\}$.

Definition Let $\{k_1, ..., k_s\}$ and $\{k'_1, ..., k'_{s'}\}$ be any pair of vertices of $G(K)$.
$$P(\{k_1, ..., k_s\}, \{k'_1, ..., k'_{s'}\}) \overset{def}{=}$$
$$\exists f : \{1, ..., s\} \to \{1, ..., s'\} \text{ s.t. } \forall y \in \{1, ..., s'\} : k_y = \sum_{x \in f^{-1}(y)} k_x.$$

Lemma 2. $(\exists \text{ a path: } \{k_1, ..., k_s\} \text{ to } \{k'_1, ..., k'_{s'}\}) \Leftrightarrow (P(\{k_1, ..., k_s\}, \{k'_1, ..., k'_{s'}\})).$ (Proof in [23].)

Theorem 2. *The transitive closure of $G(K)$ is a partial order.* (Proof in [23].)

4.3 Associated Generalized Failure Detector $GZ_{k_1,...,k_s}$

The failure detector $Z_{s,k}$ is implicitly tailored for the symmetric (s, k)-SSA problem. A simple generalization allows to extend it to obtain an "equivalent" failure detector suited to asymmetric problems.

As $Z_{s,k}$, this generalized failure detector, denoted $GZ_{k_1,...,k_s}$, provides each process p_i with an array $qr_i[1..s]$ and an array $\ell d_i[1..s]$. It differs from $Z_{s,k}$ in the constraint imposed by the quorum intersection property that is now specific to each entry $z \in \{1, ..., s\}$. More explicitly, QI is replaced by the property GQI defined as follows

- Quorum intersection property (GQI). $\forall z \in [1..s]$:
  $\forall i_1, ..., i_{k_z+1} \in \Pi, \forall \tau_1, ..., \tau_{k_z+1} : \exists h, \ell \in [1..k_z + 1] : (h \neq \ell) \wedge (qr_{i_h}^{\tau_h}[z] \cap qr_{i_\ell}^{\tau_\ell}[z] \neq \emptyset).$

The other properties –leader validity (LV), quorum liveness (QL), and eventual leader liveness (EL)– remain unchanged. It is easy to see, that $GZ_{k_1,...,k_s}$ boils down to $Z_{s,k}$ when $k_1 = \cdots = k_s = k$.

Let $GZ(Q)_{k_1,...,k_s}$ denotes the quorum part of $GZ_{k_1,...,k_s}$ (properties GQI and QL). The proof of the following theorem is a simple extension of the proof of Theorem 1.

Theorem 3. *Given any algorithm A that solves the $\{k_1, ..., k_s\}$-SSA problem in the system model $\mathcal{AMP}_{n,n-1}[FD]$, the extraction algorithm described in Figure 1 is a wait-free construction of a failure detector $GZ(Q)_{k_1,...,k_s}$.*

4.4 A Hierarchy of Agreement Problems

Problem hierarchy $\mathcal{AMP}_{n,n-1}[X]$ denotes the asynchronous message-passing model in which any number of processes may crash ($\mathcal{AMP}_{n,n-1}[\emptyset]$) enriched with an algorithm that solves the problem X.

Given the the message-passing model $\mathcal{AMP}_{n,n-1}[\emptyset]$, a problem A is *stronger* than a problem B (denoted $A \succeq B$) if B can be solved in $\mathcal{AMP}_{n,n-1}[A]$ (we also say that B is *weaker* than A, denoted $B \preceq A$). Moreover, A is *strictly stronger* than B (denoted $A \succ B$) if $A \succeq B$ and $\neg(B \succeq A)$ (A cannot be solved in $\mathcal{AMP}_{n,n-1}[B]$).

Lemma 3. $P(\{k_1, ..., k_s\}, \{k'_1, ..., k'_{s'}\})$ $\Rightarrow$ $(\{k_1, ..., k_s\}\text{-SSA} \succeq \{k'_1, ..., k'_{s'}\}$ -SSA). (Proof in [23].)

Lemma 4. *Let* $n > K \geq 2$.
$(\{k_1, ..., k_s\}\text{-SSA} \succeq \{k'_1, ..., k'_{s'}\}\text{-SSA}) \Rightarrow P(\{k_1, ..., k_s\}, \{k'_1, ..., k'_{s'}\})$.
(Proof in [23]; this proof is the most technical of this paper.)

Theorem 4. *Let* $n > K \geq 2$.
$(\{k_1, ..., k_s\}\text{-SSA} \succeq \{k'_1, ..., k'_{s'}\}\text{-SSA}) \Leftrightarrow P(\{k_1, ..., k_s\}, \{k'_1, ..., k'_{s'}\})$.
(Proof in [23].)

Theorem 5. *The relation* $\succ$ *on generalized*-SSA *problems is a partial order.*
(Proof in [23].)

The next corollary follows from the observation that, for any $K > 1$, the K-set agreement problem is a sink vertex in the directed graph $G(K)$.

Corollary 1. *The weakest failure detector for the K-set agreement problem does not allow to solve any* $\{k_1, ..., k_s\}$-SSA *problem such that* $s > 1$ *and* $k_1 + \cdots + k_s = K$.

4.5 The Lattice of Symmetric SSA Problems

As seen before, a symmetric vertex is a vertex $\{k_1, ..., k_s\}$ such that $k_1 = ... = k_s = k$. Let $SG(K)$ denote the graph whose vertices are the symmetric vertices of $G(K)$, and there is an edge from (s_x, k_x) to (s_y, k_y) iff there is a path in $G(K)$ from the vertex $\{k_x, ..., k_x\}$ (k_x appearing s_x times) to the vertex $\{k_y, ..., k_y\}$ (k_y appearing s_y times) and no path connecting these vertices passes through a symmetric vertex. As an example, $SG(6)$ is given in Section 4.2.

Theorem 6. *For any K, $SG(K)$ is a lattice.* (Proof in [23].)

The next corollary follows from the previous theorem.

Corollary 2. *Let* (s_1, k_1) *and* (s_2, k_2) *be two different pairs of integers such that* $s_1 k_1 = s_2 k_2$, *and none of k_1 and k_2 divides the other one. The symmetric* (s_1, k_1)-SSA *and* (s_2, k_2)-SSA *problems are incomparable in* $\mathcal{AMP}_{n,n-1}[\emptyset]$.

As far as agreement problems are concerned, this shows a strong difference between the message-passing model and the read/write model. In the read/write model, (s_1, k_1)-SSA and (s_2, k_2)-SSA are the same problem (they are both equivalent to the K-simultaneous problem which is itself equivalent to the K-set agreement problem, where $K = s_1 k_1 = s_2 k_2$).

5 Conclusion

This paper has investigated the comparative power of simultaneous agreement and set agreement in asynchronous message-passing systems prone to any number of process crashes. This study was initially motivated by the lasting (and difficult) quest for the weakest failure detector for the k-set agreement problem in message-passing systems.

While k-simultaneous consensus and k-set agreement are equivalent problems in asynchronous read/write systems prone to any number of process crashes [1], the paper has introduced a general formulation of agreement problems, namely the family of (s, k)-SSA (s-simultaneous k-set agreement) problems, and has shown that these agreement problems define a strong hierarchy, thereby showing that their shared memory equivalence is no longer true in message-passing systems. Hence, this study contributes to a better understanding of the relation (equivalence/difference) between the send/receive model and the read/write model (equivalence when a majority of processes are correct [2], and difference –from a problem ranking point of view– in the other cases).

Finally, it follows from the results of this paper that the (yet unknown) weakest failure detector for k-set agreement in asynchronous message-passing systems is not powerful enough to solve the generalized $\{k_1, ..., k_s\}$-SSA problem where $s > 1$.

Acknowledgment. This work has been partially supported by the French ANR project DISPLEXITY devoted to computability and complexity in distributed computing.

References

1. Afek, Y., Gafni, E., Rajsbaum, S., Raynal, M., Travers, C.: The k-Simultaneous Consensus Problem. Distributed Computing 22, 185–195 (2010)
2. Attiya, H., Bar-Noy, A., Dolev, D.: Sharing Memory Robustly in Message Passing Systems. Journal of the ACM 42(1), 121–132 (1995)
3. Bonnet, F., Raynal, M.: On the Road to the Weakest Failure Detector for k-Set Agreement in Message-passing Systems. Theoretical Computer Science 412(33), 4273–4284 (2011)
4. Borowsky, E., Gafni, E.: Generalized FLP Impossibility Impossibility Results for t-Resilient Asynchronous Computations. In: Proc. 25th ACM Symposium on Theory of Computation (STOC 1993), pp. 91–100 (1993)
5. Bouzid, Z., Travers, C.: (anti$-\Omega^x \times \Sigma_z$)-Based k-Set Agreement Algorithms. In: Lu, C., Masuzawa, T., Mosbah, M. (eds.) OPODIS 2010. LNCS, vol. 6490, pp. 189–204. Springer, Heidelberg (2010)
6. Bouzid, Z., Travers, C.: Simultaneous Consensus is Harder than Set Agreement in Message-Passing. In: Proc. 33rd Int'l IEEE Conference on Distributed Computing Systems (ICDCS 2013). IEEE Press (2013)
7. Chandra, T., Toueg, S.: Unreliable Failure Detectors for Reliable Distributed Systems. Journal of the ACM 43(2), 225–267 (1996)
8. Chandra, T., Hadzilacos, V., Toueg, S.: The Weakest Failure Detector for Solving Consensus. Journal of the ACM 43(4), 685–722 (1996)
9. Chaudhuri, S.: More Choices Allow More Faults: Set Consensus Problems in Totally Asynchronous Systems. Information and Computation 105, 132–158 (1993)

10. Chen, W., Zhang, J., Chen, Y., Liu, X.: Weakening Failure Detectors for k-Set Agreement Via the Partition Approach. In: Pelc, A. (ed.) DISC 2007. LNCS, vol. 4731, pp. 123–138. Springer, Heidelberg (2007)

11. Delporte-Gallet, C., Fauconnier, H., Guerraoui, R.: Tight Failure Detection Bounds on Atomic Object Implementations. Journal of the ACM 57(4), Article 22 (2010)

12. Delporte-Gallet, C., Fauconnier, H., Guerraoui, R., Tielmann, A.: The Weakest Failure Detector for Message Passing Set-Agreement. In: Taubenfeld, G. (ed.) DISC 2008. LNCS, vol. 5218, pp. 109–120. Springer, Heidelberg (2008)

13. Gafni, E., Kuznetsov, P.: On Set Consensus Numbers. Distributed Computing 24(3-4), 149–163 (2011)

14. Guerraoui, R., Raynal, M.: The Alpha of Indulgent Consensus. The Computer Journal 50(1), 53–67 (2007)

15. Herlihy, M.P., Shavit, N.: The Topological Structure of Asynchronous Computability. Journal of the ACM 46(6), 858–923 (1999)

16. Mostéfaoui, A., Raynal, M.: k-Set Agreement with Limited Accuracy Failure Detectors. In: Proc. 19th ACM Symposium on Principles of Distributed Computing (PODC 2000), pp. 143–152. ACM Press (2000)

17. Mostéfaoui, A., Raynal, M., Stainer, J.: Relations Linking Failure Detectors Associated with k-Set Agreement in Message-Passing Systems. In: Défago, X., Petit, F., Villain, V. (eds.) SSS 2011. LNCS, vol. 6976, pp. 341–355. Springer, Heidelberg (2011)

18. Mostéfaoui, A., Raynal, M., Stainer, J.: Chasing the Weakest Failure Detector for k-Set Agreement in Message-Passing Systems. In: Proc. 11th IEEE Int'l Symp. on Network Computing and Applications (NCA 2012), pp. 44–51. IEEE Press (2012)

19. Raynal, M.: K-anti-Omega. In: Rump Session at 26th ACM Symposium on Principles of Distributed Computing, PODC 2007 (2007)

20. Raynal, M.: Fault-Tolerant Agreement in Synchronous Message-Passing Systems, 165 pages. Morgan & Claypool Publishers (2010) ISBN: 978-1-60845-525-6

21. Raynal, M.: Failure Detectors to Solve Asynchronous k-Set Agreement: a Glimpse of Recent Results. Bulletin of the EATCS 103, 74–95 (2011)

22. Raynal, M.: Concurrent Programming: Algorithms, Principles, and Foundations, 515 pages. Springer (2013) ISBN 978-3-642-32027-9

23. Raynal, M., Stainer, J.: Simultaneous Consensus vs Set Agreement: a Message-passing Sensitive Hierarchy of Agreement Problems. Tech Report PI 2003, IRISA, University of Rennes (F) (2013), http://hal.inria.fr/hal-00787992

24. Raynal, M., Travers, C.: In Search of the Holy Grail: Looking for the Weakest Failure Detector for Wait-Free Set Agreement. In: Shvartsman, M.M.A.A. (ed.) OPODIS 2006. LNCS, vol. 4305, pp. 3–19. Springer, Heidelberg (2006)

25. Saks, M., Zaharoglou, F.: Wait-Free k-Set Agreement is Impossible: The Topology of Public Knowledge. SIAM Journal on Computing 29(5), 1449–1483 (2000)

26. Zielinski, P.: Anti-Omega: the Weakest Failure Detector for Set Agreement. In: Proc. 27th ACM Symp. on Principles of Distributed Computing (PODC 2008), pp. 55–64. ACM Press (2008)

Steiner Problems with Limited Number of Branching Nodes

Dimitri Watel[1,3], Marc-Antoine Weisser[1],
Cédric Bentz[2], and Dominique Barth[3]

[1] SUPELEC System Sciences, Computer Science Dpt., 91192 Gif sur Yvette, France
{dimitri.watel,marc-antoine.weisser}@supelec.fr
[2] CEDRIC-CNAM 292 rue Saint-Martin 75141 Paris, France
cedric.bentz@cnam.fr
[3] University of Versailles, 45 avenue des Etats-Unis, 78035, France
dominique.barth@prism.uvsq.fr

Abstract. Given an *undirected* weighted graph G with n nodes, the k-*Undirected Steiner Tree* problem is to find a minimum cost tree spanning a specified set of k nodes. If this problem and its directed version have several applications in multicast routing in packet switching networks, the modeling is not adapted anymore in networks based upon the circuit switching principle in which not all nodes are able to duplicate packets. In such networks, the number of branching nodes (with outdegree > 1) in the multicast tree must be limited.

We introduce the $(k, p)-$*Steiner Tree with Limited Number of Branching nodes* problems where the goal is to find an optimal Steiner tree with at most p branching nodes. We study, when p is fixed, its complexity depending on two criteria: the graph topology and the parameter k. In particular, we propose a polynomial algorithm when the input graph is acyclic and an other algorithm when k is fixed in an input graph of bounded treewidth. Moreover, in directed graphs where $p \leq k - 2$, or in planar graphs, we provide an n^{ϵ}-inapproximability proof, for any $\epsilon < 1$.

Keywords: graph algorithm, parameterized complexity, Steiner tree.

1 Introduction

The k-*Undirected Steiner Tree* problem (min-k-UST) consists, given an undirected weighted graph and k nodes called *terminals*, in the search of a minimum cost tree spanning the terminals. In the directed version (min-k-DST), the minimum cost directed tree must be rooted at a specific node r.

Those problems are known to have applications in multicast routing where one wants to minimize the bandwidth consumption [1–3]. Recent work emphasizes the fact that in optical networks, some nodes are not able to duplicate packets [4–6]. If such a node needs to transmit the same information to d neighbours, the root has to send the same message d times to these neighbours individually. As a consequence the bandwidth consumption considerably increases between the

T. Moscibroda and A.A. Rescigno (Eds.): SIROCCO 2013, LNCS 8179, pp. 310–321, 2013.
© Springer International Publishing Switzerland 2013

root and that node. Fortunately, there exist two different routers, the *branching routers*, capable of copying packets. The *opto-electronical router* receives an optical packet, translates it into a temporary electronical message, and creates for each neighbour a copy of the optical packet. The *splitter router* uses a set of mirrors to split the signal into several copies. However, those routers introduce supplementary cost to the network due to expensive maintenance, and when splitting packets, they introduce supplementary delay to the transmission (in the opto-electronical case) or an attenuation of the optical signal from the root to the terminals (in the splitter case). Therefore they have to be limited in the solution. It is why we will especially focus on the degree of the solution nodes, and see if it is possible to find an optimal solution with a limited number of branching routers and no repetition of the same message over the network.

Consequently, we modify the Steiner problems definition in order to consider this new element and introduce the (k, p)-*Steiner Tree with Limited number of Branching nodes* problems in an undirected or a directed graph ((k, p)-USTLB and (k, p)-DSTLB).

Definition 1. *In a undirected (resp. directed) tree, a* branching node *is a node whose degree (resp. outdegree) is strictly greater than 2 (resp. 1).*

Problem 1. min-(k, p)-USTLB: Given an undirected graph $G = (V, E)$ with n nodes and a non negative cost function ω on its edges, a set $X \subset V$ of k terminals, determine, if it exists, a minimum cost tree T^* spanning all the nodes of X and containing at most p branching nodes.

Problem 2. min-(k, p)-DSTLB: Given a directed graph $G = (V, E)$ with n nodes and a non negative cost function ω on its arcs, a node r and a set $X \subset V$ of k terminals, determine, if it exists, a minimum cost directed tree T^* rooted at r, spanning all the nodes of X and containing at most p branching nodes.

When $p = k-1$, the min-$(k, k-1)$-DSTLB problem is the min-k-DST problem because an optimal solution of this problem cannot have more than $k-1$ branching nodes. Similar observation holds for the undirected case (with $p = k - 2$). When $k = n$, the problem is equivalent to the minimum spanning tree problem with few branch vertices [6]. When $p = 0$, the problem searches for a path going through each terminal, which is equivalent to the Steiner Cycle problem [7]. It was shown to be difficult in directed graphs when $k \geq 2$ [8]. We extend those results for any $p \leq k - 2$.

We define as (k, p)-DSTLB and (k, p)-USTLB the problems of finding a feasible solution in a min-(k, p)-DSTLB or min-(k, p)-USTLB instance. Finally we define as min-$(*, p)$-DSTLB, min-$(*, p)$-USTLB, $(*, p)$-DSTLB and $(*, p)$-USTLB the previous problems where the parameter k is not fixed.

Table 1 summarizes the results of this article and the remaining open problems. We first prove the general directed case with a fixed number of terminals to be NP-Complete, even if we just look for a feasible solution. The proof is a reduction from the two vertex-disjoint paths problem, which consists in connecting with two node-disjoint paths the nodes of two distinct couples of nodes.

This problem is NP-Complete in directed graphs [9]. However, this is not true in the general undirected case [10], in planar graphs [11] or in the acyclic directed case [9], and the reduction is not applicable. Consequently, the other results are dedicated to determining the complexity class of those cases. We finally study the undirected case in bounded treewidth graph as it is to our knowledge the only relevant case where one can find a solution to the k-vertex disjoint paths problem at minimum total cost with an FPT algorithm in the treewidth of the graph [12]. Indeed, a parameterized algorithm in p and k for USTLB follows from that property.

Table 1. Results for the USTLB and DSTLB problems

Problem	Conditions	Find a feasible solution	Minimization
(k,p)-DSTLB	General graph, $p \leq k - 2$	NPC (Th. 1)	n^ϵ-Inappr. (Th. 2)
(k,p)-USTLB	General graph, $p \leq k - 3$	P (Th. 8)	OPEN
(k,p)-USTLB	Bounded treewidth graph	P (Th. 9)	P (Th. 9)
$(*,p)$-USTLB		OPEN	OPEN
(k,p)-DSTLB		P (Th. 7)	OPEN
$(*,p)$-DSTLB	Planar graph	NPC (Th. 3)	n^ϵ-Inappr. (Th. 4)
(k,p)-USTLB		P (Th. 8)	OPEN
$(*,p)$-USTLB		NPC (Th. 3)	n^ϵ-Inappr. (Th. 4)
$(*,p)$-DSTLB	Acyclic graph	W[2]-hard in p (Th. 5)	P (Th. 6)

In the following section, we present a survey of some results related to the Steiner problem. We then divide our study into three parts. In Section 3, we prove the hardness results of Table 1. In Section 4, we propose a polynomial algorithm for the min-$(*,p)$-DSTLB case where we impose the graph to be acyclic. In Section 5, we prove the last polynomial results of Table 1. We lastly give the conclusions and perspectives for further works.

2   Related Work

The Undirected Steiner Tree problem is NP-hard and can be approximated within constant ratio [13–15]. For a general survey, see [16].

The Directed Steiner Tree was first studied in acyclic graphs [17]. A paper then developed a non trivial series of approximation within k^ϵ in the general case [18], which is currently the best known ratio although the algorithm itself was improved [19] or the approximation ratio was rediscovered by other methods [20]. As a generalization of the Set Cover problem, it was known to be inapproximable within a $O(\log(k))$ ratio unless $NP \subseteq DTIME[n^{O(\log \log n)}]$ (see [21]).

Given an undirected graph $G = (V, E)$ and a list of k subsets of nodes called *groups*, the k-Group Steiner Tree problem (min-k-GST) consists in finding a minimum cost tree spanning at least one node in each group. We can reduce the

GST problem to the DST problem. The GST problem can not be approximated within a $O(\log(k)^2)$ ratio [22] and so does the DST problem. In addition, it can be randomly approximated within a $O(\log(n)\log(k))$ ratio if the input graph is a tree and within a $O(\log^3(n)\log(k))$ ratio in the general case [23].

The min-k-GST, min-k-UST and min-k-DST problems are FPT with respect to the parameter k as it exists an exact algorithm in time $O(3^k n + 2^k(k + \log(n))n + n^2)$ and in space $O(2^k n)$ [24].

In this paper, by reducing the number of branching nodes, we modify the set of feasible solutions and cannot easily adapt those previous results in order to solve or approximate the USTLB and DSTLB problems.

An other generalization of the Steiner problems, the rooted connectivity problem, asks for a subgraph which has a specified number of openly (node or arc) disjoint paths from a root to each terminal. The undirected version with requirements for node-disjoint paths and all the directed versions of the problem cannot be approximated within $O(k^\epsilon)$ [25]. As we will see for the (k,p)-DSTLB problem, the introduction of constraints implying to find disjoint paths makes the problem harder to approximate.

3   Hardness Results

In this section, we will prove seven results of Table 1: the NP-hardness and inapproximability of the problem in general digraphs when both parameters k and p are fixed, and in planar graphs when only the parameter p is fixed. We will finally prove a parameterized complexity hardness result in the acyclic case.

3.1   The General Directed Case with Fixed k and p

This part extends previous results of [8] about the k-Directed Steiner Cycle problem (equivalent to min-$(k,0)$-DSTLB) to any $p \leq k - 2$.

NP-Completeness of Finding a Feasible Solution to (k,p)-DSTLB

Theorem 1. *Let the two parameters k and p satisfy $k \geq 2$ and $p \leq k - 2$. (k,p)-DSTLB is NP-Complete.*

In order to prove Theorem 1, we first define a reduction from the *2 Vertex Disjoint Paths* problem (2VDP).

Problem 3. **2VDP**: Given a directed graph and four distinct nodes s_1, s_1', s_2 and s_2', find two node-disjoint paths from s_1 to s_1' and from s_2 to s_2'.

The 2VDP problem is known to be NP-Complete [9]

Let $\mathcal{I} = (G = (V, E), s_1, s_1', s_2, s_2')$ be an instance of the 2VDP problem. We will construct an instance $\mathcal{I}' = (G', r, X, \omega)$ of the (k,p)-DSTLB problem as follows. G is a subgraph included in G'. We add a node r, p nodes $\{b^1, b^2, \dots b^p\}$ and k nodes $X = \{t_1, t_2^1, t_2^2 \dots t_2^{k-1}\}$ into V'. If $p \neq 0$, we add into E' the arcs

(r, s_1), (s'_1, t_1), (t_1, s_2), (s'_2, b^1), (b^i, t^i_2) for all $i \le p$, (b^i, b^{i+1}) for all $i \le p-1$ and (b^p, t^i_2) for all $i \in [\![p; k-1]\!]$. An example is shown in Figure 1. If $p = 0$, we add the arcs (r, s_1), (s'_1, t_1), (t_1, s_2), (s'_2, t^1_2) and (t^{i-1}_2, t^i_2) for all $2 \le i \le k-1$. This last case is not treated in the following proof, but is similar to the previous case. We define the instance $\mathcal{I}'$ by setting r as the root, X as the terminals, the unit function as the cost function ω.

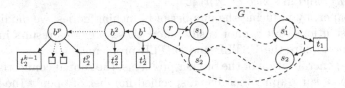

Fig. 1. The reduction from 2VDP to (k, p)-DSTLB. The dashed ellipse represents the copy of G in G'. The dashed lines depict potential paths from s_1 to s'_1 and s_2 to s'_2.

Lemma 1. *If there exists a solution for $\mathcal{I}$, there exists a solution for $\mathcal{I}'$.*

Proof. Let ps be the set of arcs contained in the two disjoint paths of $\mathcal{I}$ and T' the graph induced by $ps \cup (E' - E)$. This solution is a directed tree because the paths of ps are node-disjoint and $E' - E$ is a forest. T' is rooted at r and spans X. Lastly, T' contains only p branching nodes (b^1, b^2 ... b^p). Thus, T' is a feasible solution for $\mathcal{I}'$. ☐

Lemma 2. *If there exists a solution for $\mathcal{I}'$, there exists a solution for $\mathcal{I}$.*

Proof. Let T' be the solution of $\mathcal{I}'$. Since for each $i \in [\![1; p]\!]$ T' spans t^i_2, it also spans its only predecessor b^i. Consequently, b^1, b^2 ... b^p are the p branching nodes of T'. In the same way, it contains s'_1 as it is the only predecessor of terminal t_1.

Let P be the path from r to b^1 in T'. No branching node can be in G so P is an elementary path containing in that order s_1, s'_1, s_2, s'_2. Thus, the subpath of P going from s_1 to s'_1 and the subpath of P going from s_2 to s'_2 are disjoint. ☐

Proof (Theorem 1). (k, p)-DSTLB is in the complexity class NP as we can polynomially decide if a set of arcs is a directed tree rooted at a specific node r, spanning specific nodes X with at most p branching nodes.

Furthermore, Lemmas 1 and 2 prove that there exists a polynomial-time reduction from 2VDP to (k, p)-DSTLB. So (k, p)-DSTLB is NP-Complete. ☐

Inapproximability of the Problem. The previous reduction tells us that computing a feasible solution may be hard. In order to work with an optimization problem where the feasible solution set is not empty, we add to the previous instance a long path starting from the root and going through each terminal. This long path is a feasible solution with no branching node. If this path is long enough between each terminals, it will not be allowed to be part of any approximate solution. We now prove the following theorem.

Theorem 2. *Let the two parameters k and p satisfy $k \geq 2$ and $p \leq k - 2$. Let $\epsilon < 1$ be a real number. If $P \neq NP$, the min-(k,p)-DSTLB problem with unit costs cannot be approximated within a factor of $\mathcal{N}^\epsilon$ where $\mathcal{N}$ is the number of nodes in the instance, even if a feasible solution is given.*

Proof. Let $\mathcal{I} = (G = (V, E), s_1, s'_1, s_2, s'_2)$ be an instance of the 2VDP problem with n nodes such that $n \geq k > p$. We construct a min-(k,p)-DSTLB instance $\mathcal{I}'$ like we did in the last section except that we also add for each terminal a long path with h arcs. The first path links r to t_1, the second links t_1 to t_2^1, and the others link t_2^i to t_2^{i+1} for $i \in [\![1; k - 1]\!]$. We will fix the value of h later.

The number of nodes $\mathcal{N}$ in G' is $k \cdot h + n + p + 1$.

Let T^* be an optimal solution of $\mathcal{I}'$. It exists because the new paths are a solution with no branching node. Let $\epsilon < 1$, and suppose it exists a $\mathcal{N}^\epsilon$-approximation algorithm for min-(k,p)-DSTLB.

If there exist two node-disjoint paths in $\mathcal{I}$, T^* contains at most $n + p + k + 1$ nodes (for example, the n nodes of G, the k terminals, $b^1 \ldots b^p$ and r), thus at most $n + k + p$ arcs. So the approximate solution has a cost $c_{\text{YES}} \leq (n + k + p) \cdot \mathcal{N}^\epsilon$.

If there are not two node-disjoint paths in $\mathcal{I}$, the previous section proves that without one of the long paths we cannot build a feasible solution. So the approximate solution uses at least one long path and has a cost $c_{\text{NO}} > h$.

If $c_{\text{NO}} > h > c_{\text{YES}}$, then the approximation algorithm can decide whether there are two node-disjoint paths in $\mathcal{I}$ or not.

Let h satisfies $h = 6^{\frac{1}{1-\epsilon}} n^{\frac{1+\epsilon}{1-\epsilon}} + 1$. Notice that $h > 2$ for all $\epsilon < 1$, $n \geq k > p$.

$$h > 6^{\frac{1}{1-\epsilon}} n^{\frac{1+\epsilon}{1-\epsilon}} \tag{1}$$

$$h^{1-\epsilon} > 6n^{1+\epsilon} \tag{2}$$

$$h > 3 \cdot 2n^{1+\epsilon}h^\epsilon > 3 \cdot 2^\epsilon n^{1+\epsilon}h^\epsilon \tag{3}$$

$$h > 3n^{1+\epsilon}(2h)^\epsilon > 3n^{1+\epsilon}(2 + h)^\epsilon \tag{4}$$

$$h > 3n(2 + h)^\epsilon \cdot n^\epsilon = 3n(2n + n \cdot h)^\epsilon \tag{5}$$

$$h > (n + p + k)(n + p + 1 + k \cdot h)^\epsilon \tag{6}$$

$$c_{\text{NO}} > h > c_{\text{YES}} \tag{7}$$

As a consequence, if $P \neq NP$, such an algorithm does not exist.   □

3.2   The Undirected and Directed Planar Cases with Fixed p

The previous reduction is not applicable in a planar graph as the 2VDP problem is polynomial in this case [11]. However a similar result holds if k is not fixed.

Theorem 3. *$(*, p)$-DSTLB and $(*, p)$-USTLB are NP-Complete, even if G is planar.*

Proof. The $(k, 0)$-DSTLB problem is equivalent to the Hamiltonian Path problem in a graph with k nodes. Thus $(*, 0)$-DSTLB is NP-Complete even if G is planar. We now extend this reduction to any fixed parameter p.

Let p be a positive integer and $G = (V, E)$ be a directed planar graph. Let $\mathcal{B}$ be a binary arborescence rooted at r with p internal nodes and $p + 1$ leaves $\mathcal{L}$. Let G_v be the graph obtained when adding $\mathcal{B}$ to G by linking one leaf of $\mathcal{L}$ to a node $v \in V$. We denote r as the root and $X = \mathcal{L} \cup V$ as the terminals. By solving the $(*, p)$-DSTLB problem on G_v for all $v \in V$ we can decide whether or not G contains an hamiltonian path.

A similar proof holds for undirected planar graphs.   □

Like in Section 3.1, we now add an expensive feasible solution to prove an inapproximability result. However, this is more complex than in the previous section because we have to keep a the graph planar while doing this. Due to lack of space, we do not detail the proof in this paper but provide it in [26].

Theorem 4. *Let $\epsilon < 1$ be a real number. If $P \neq NP$, the min-$(*, p)$-DSTLB and the min-$(*, p)$-USTLB problems in planar graphs with unit costs and $\mathcal{N}$ nodes cannot be approximated within a factor of $\mathcal{N}^\epsilon$, even if a feasible solution is given.*

3.3   The Directed Acyclic Case with Fixed p

In a directed acyclic graph, the reduction given in Section 3.1 is not applicable because 2VDP is polynomial [9]. However, when k is not fixed, we are able to prove that finding a feasible solution is W[2]-hard in p.

Theorem 5. *The $(*, p)$-DSTLB problem is W[2]-hard with respect to the parameter p even if the graph is acyclic.*

Proof. We use a variant of the classic Directed Steiner Tree reduction from the Set Cover problem. Given a set of elements U, a set S of subsets of U and an integer N, the set cover problem is to find N or less sets of S covering all the elements in U. This problem is W[2]-complete in N [27].

We define a fixed-parameter reduction from this parameterized problem to min-$(*, p)$-DSTLB on acyclic graphs. We construct an instance $\mathcal{I} = (G = (V, A), r, X)$ of $(*, p)$-DSTLB. For each set in S we add a *set node* s in V, a terminal t_s in X_S and link s to it. For each element in U, we add an *element terminal* in X_e. We set $X = X_S \cup X_e$. Finally we add a root r to V. We link r to each set node, and link a set node to an element terminal if the associated element is in the associated set in the set cover instance. We set the parameter p to $N + 1$. An example is shown in Figure 2.

As a feasible tree T covers every terminals, it covers X_S and thus contains all the set nodes, which outdegree is at least 1. The root is consequently a branching node in T, and each set node father of an element node in T is a branching node. If there exists a cover $c \subset S$ with $|c| \leq N$, the arborescence using each set node associated with a set of c to cover X_e has $|c| + 1 \leq p$ branching nodes. If there exists an arborescence T using the set nodes $c \subset V$ to cover X_e, the corresponding subset of S covers every elements of U with $|c| \leq N$ sets. This FPT reduction proves the $(*, p)$-DSTLB problem to be W[2]-hard in p.   □

Although Theorem 5 proves min-$(*, p)$-DSTLB to be W[2]-hard with respect to the parameter p, the next section shows it remains in the XP class.

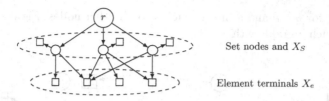

Set nodes and X_S

Element terminals X_e

Fig. 2. Example of reduction from a Set Cover instance with $U = \{x_1, x_2, x_3, x_4\}$ and $S = \{\{x_1, x_2\}, \{x_2, x_3, x_4\}, \{x_2, x_4\}\}$

4   A Polynomial Algorithm for Solving the Directed Acyclic Case with Fixed p

This section provides an algorithm running in polynomial time when p is fixed to compute an optimal solution when the directed graph G contains no circuit.

Theorem 6. *The min-$(*, p)$-DSTLB problem can be solved in $O(n^{p+2} m^3 \log(n))$ time where n and m respectively are the number of nodes and arcs in the graph.*

The remaining of this section is dedicated to the proof of Theorem 6. The main idea is to try every possible set of branching nodes. We compute afterwards in polynomial time the minimum cost tree whose branching nodes are among the set and then try another set. In the end, we return the minimum cost tree.

We now describe the polynomial algorithm which finds the minimum cost tree using a given subset of V as branching nodes. We use a *Minimum Cost Flow* instance where an integral flow joins the sink if and only if there exists a rooted tree with the branching nodes constraint. We get the minimum-cost rooted tree by computing an integral minimum-cost flow.

We first recall the definition of the *Minimum Cost Flow* problem and then define the instance which solves our problem:

Problem 4. **MCF**: Given a directed graph G with a cost function ω on its arcs, a capacity function c on its nodes and with a source node S, a sink node S' and an integer l, send l units of integral flow from S to S' at minimum cost.

MCF is polynomial [28]. Let $\mathcal{I} = (G = (V, E), r, X, \omega)$ be a min-$(*, p)$-DSTLB instance, $j \le p$ an integer and $\kappa = \{v_1, v_2, ..., v_j\}$ j distinct nodes of V. We construct an instance $\mathcal{F}_\kappa$ of the MCF problem. Let $\mathcal{D}$ be the set $\kappa \cup X$. Consider now the graph G where each node v of $\mathcal{D}$ is replaced by two nodes v^- and v^+. Let $\mathcal{D}^-$ and $\mathcal{D}^+$ be respectively the images of $\mathcal{D}$ by the functions v^- and v^+. Finally let d be the function which satisfies $d(v^-) = v$ and $d(v^+) = v$.

Each arc entering (resp. leaving) a node v of $\mathcal{D}$ is replaced by an arc with same cost entering v^- (resp. leaving v^+). We add to $\mathcal{F}_\kappa$ a source S and a sink S'. We link S to each node of $\mathcal{D}^+$ and r, and each node of $\mathcal{D}^-$ to S', by arcs with cost 0. As a consequence, the flow leaving the source goes into $\mathcal{D}^+ \cup \{r\}$ at first and later $\mathcal{D}^-$ before entering the sink. Notice that if $r \in \mathcal{D}$, the graph contains the arc (S, r^-) instead of (S, r). At last, we give an infinite capacity to

S, S' and v^+ for $v \in \kappa$ and a unit capacity to all other nodes. Figure 3 gives an example of such a graph with $j = 1$.

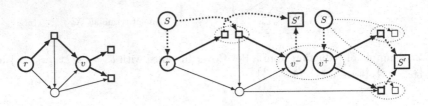

Fig. 3. Graph $\mathcal{F}_{\{v\}}$ (on the right) obtained from a graph G (on the left) with 3 terminals (square nodes). Only S, S' and v have infinite capacity. The source S and the sink S' have been duplicated for readability. Bold elements are one feasible solution.

If $\mathcal{I}$ contains a feasible solution with the set κ as branching nodes, we can send $|\mathcal{D}|$ flow units from S to S' using the same arcs in $\mathcal{F}_\kappa$, as shown with the bold arcs and nodes in Figure 3. Consider now an integral solution f of $\mathcal{F}_\kappa$. We define G_κ as the subgraph of G induced by r and by the arcs where f is not null.

Lemma 3. *The flow f goes through each node of $\mathcal{D}^-$.*

Proof. The only $|\mathcal{D}|$ nodes preceding the sink with a unit capacity are in $\mathcal{D}^-$. □

Lemma 4. G_κ *is a feasible solution for $\mathcal{I}$ and its branching nodes are in κ.*

Proof. Since from Lemma 3 the flow f goes through every node of $\mathcal{D}^-$, G_κ contains $\mathcal{D} = \kappa \cup X$ and it contains r by definition. As $\mathcal{D}^+$ contains the only nodes with a non unit capacity, all the branching nodes of G_κ are in κ.

G_κ has only one connected component. Otherwise there would be a connected component with a root node different from r. Thus, there would be a node w of $\mathcal{F}_\kappa$ through which the flow goes with no other predecessor than the source S. So w would have to belong to $\mathcal{D}^+$. However f goes through every node of $\mathcal{D}^-$ and particularly $d(w)^-$, which would have a predecessor and so would $d(w)$. Therefore, $d(w)$ could not be a root in a connected component.

Finally, G_κ is a rooted tree, otherwise it would contain a cycle or a circuit. There is no circuit as G is acyclic. There is no cycle as only the nodes of $\mathcal{D}^+$ have a capacity $c > 1$, as their only predecessor is S, and as the flow is integral. □

Proof (Theorem 6). A subgraph of G and a flow f using the same arcs in $\mathcal{F}_\kappa$ have the same cost because the output node of each arc in $\mathcal{F}_\kappa$ has a unit capacity except the arcs with cost 0. Let κ^* be the set for which G_{κ^*} has minimum cost. Let $T^\diamond$ be an optimal rooted tree of $\mathcal{I}$ and $\kappa^\diamond$ its branching nodes. We can send in $\mathcal{F}_{\kappa^\diamond}$ a flow $f^\diamond$ using the arcs of $T^\diamond$. By definition of $G_{\kappa^\diamond}$, the cost of $f^\diamond$ is greater than or equal to the cost of $G_{\kappa^\diamond}$, which is greater than or equal to the cost of G_{κ^*}. By Lemma 4 G_{κ^*} is an optimal solution for $\mathcal{I}$.

We try $\binom{n}{1} + \binom{n}{2} + \dots + \binom{n}{p}$ sets κ. A minimum cost flow instance can be solved in time $O(n^2 m^3 \log(n))$ [28]. The running time of this algorithm follows. □

5 Polynomial-Time Algorithm for Fixed k and p

The parameters k and p are now fixed. We provide a polynomial algorithm for solving (k,p)-USTLB in general graphs, (k,p)-DSTLB in planar digraphs and min-(k,p)-USTLB in bounded treewidth graphs.

The main idea of the algorithm is to define and enumerate some feasible solution classes, called *patterns*, whose number depends on k and p, such that if the instance has no feasible solution, each pattern is empty, and on the other hand, if it has a solution, one of the pattern contains the optimal solution.

5.1 Patterns

Assuming it exists a feasible solution T of a (k,p)-DSTLB instance, let V_P be a set containing its branching nodes, the root and the terminals. By contracting into a single edge each path of T having both endpoints in V_P and no other node in V_P, we get a smaller tree P with V_P as nodes. P still contains k terminals, at most p branching nodes among V and the root. P is called the *pattern* of T and we write $T \to P$. An example is shown in Figure 4.

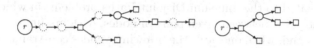

Fig. 4. An example of trees T and P with $T \to P$. Dashed nodes are contracted.

Definition 2. *Let $\mathcal{I}$ be a (k,p)-DSTLB instance. We denote as $\Pi_{\mathcal{I}}$ the set of patterns of $\mathcal{I}$. For $P \in \Pi_{\mathcal{I}}$, $S(P)$ is the feasible solutions T satisfying $T \to P$.*

If $S(P)$ is empty for every P, the instance has no feasible solution. As a pattern is a rooted tree with the k terminals and at most p nodes among V, the set $|\Pi_{\mathcal{I}}| = O(n^p)$ and thus is polynomial in n.

5.2 Algorithm

Lemma 5. *Let $\mathcal{I}$ be a (k,p)-DSTLB instance in a planar graph. Let $P = (V_P, A_P)$ be a pattern of $\Pi_{\mathcal{I}}$. We can decide in polynomial time if $S(P) = \emptyset$. If not, we can return a directed tree of $S(P)$.*

Proof. Assume $S(P)$ is not empty. In any tree T such that $T \to P$, each arc (u,v) of A_P corresponds to a path p_{uv} of T, and the paths of the set $\{p_{uv}, (u,v) \in A_P\}$ are pairwise vertex-disjoint, except, possibly, at their endpoints. There are at most $k + p$ such paths as there are at most $k + p$ arcs in P.

In a directed planar graph, we can find in polynomial time a fixed number of vertex disjoint paths [11]. If there are no such path, $S(P)$ is empty.

On the other hand, if we find $k + p$ vertex disjoint paths linking nodes u to nodes v for $(u,v) \in A_P$, the union of the paths form a directed tree rooted at r with at most p branching nodes among V, covering X. □

By iterating on every pattern P of $\Pi_\mathcal{I}$ whose size is polynomial in n, we check if $S(P)$ is empty or not. There is no feasible solution if all sets are empty. If not, by Lemma 5, we can find a solution. This proves the following theorem:

Theorem 7. *The (k,p)-DSTLB problem, in planar graphs, is polynomial.*

We can similarly prove the following theorems, knowing that the Undirected Vertex Disjoint Paths problem (UVDP) is FPT in the number of paths [10], and the min-sum UVDP (in which one wants to find vertex disjoint paths at minimum total cost) is FPT in the treewidth of the graph [12].

Theorem 8. *The (k,p)-USTLB problem is polynomial.*

Theorem 9. *The min-(k,p)-USTLB problem, restricted to bounded treewidth graphs, is polynomial.*

6   Concluding Remarks and Open Problems

The results we obtained and the main open questions concerning the problems studied in this paper are summarized in Table 1. It appears that DSTLB and USTLB are related to the min-sum Disjoint Paths problem (in which one wants to find disjoint paths between given pairs of nodes, with minimum total length).

On the one hand, when one adds the branching nodes constraint to the steiner problem, any solution must contain an elementary path between each couple of successive branching nodes. That property allows us to build a reduction from a Disjoint Paths problem, and make the general directed case hard to approximate. New hardness results on the min-sum Disjoint Paths problem (especially in planar graphs) may enable us to complete the results of Table 1.

On the other hand, we gave a parameterized algorithm for the (k,p)-DSTLB problem in planar graphs, the (k,p)-USTLB problem, and the min-(k,p)-USTLB problem in bounded treewidth graphs, based on Disjoint Paths polynomial algorithms. However, some minimization cases and the parameterized complexity class of those problems with respect to couple of parameters (k,p) remain open. New polynomial results on the min-sum Disjoint Paths problem may also enable us to complete the results of Table 1.

References

1. Cheng, X., Du, D.Z.: Steiner trees in industry, vol. 11. Kluwer (2001)
2. Voß, S.: Steiner tree problems in telecommunications, pp. 459–492 (January 2006)
3. Rugeli, J., Novak, R.: Steiner tree algorithms for multicast protocols (1995)
4. Reinhard, V., Tomasik, J., Barth, D., Weisser, M.-A.: Bandwidth Optimization for Multicast Transmissions in Virtual Circuit Networks. In: Fratta, L., Schulzrinne, H., Takahashi, Y., Spaniol, O. (eds.) NETWORKING 2009. LNCS, vol. 5550, pp. 859–870. Springer, Heidelberg (2009)
5. Reinhard, V., Cohen, J., Tomasik, J., Barth, D., Weisser, M.A.: Optimal configuration of an optical network providing predefined multicast transmissions. Comput. Netw. 56(8), 2097–2106 (2012)

6. Gargano, L., Hell, P., Stacho, L., Vaccaro, U.: Spanning trees with bounded number of branch vertices. In: Widmayer, P., Triguero, F., Morales, R., Hennessy, M., Eidenbenz, S., Conejo, R. (eds.) ICALP 2002. LNCS, vol. 2380, pp. 355–365. Springer, Heidelberg (2002)
7. Salazar-González, J.J.: The Steiner cycle polytope. EJOR 147(3), 671–679 (2003)
8. Steinová, M.: Approximability of the Minimum Steiner Cycle Problem. Computing and Informatics 29(6+), 1349–1357 (2010)
9. Fortune, S., Hopcroft, J., Wyllie, J.: The directed subgraph homeomorphism problem. Theoretical Computer Science 10(111), 111–121 (1980)
10. Robertson, N., Seymour, P.: The disjoint paths problem. Journal of Combinatorial Theory, Series B, 65–110 (1995)
11. Schrijver, A.: Finding k disjoint paths in a directed planar graph. SIAM Journal on Computing, 1–10 (1994)
12. Scheffler, P.: A Practical Linear Time Algorithm for Disjoint Paths in Graphs with Bounded Tree Width. Technical Report 396/1994, Fachbereich Mathematik (1994)
13. Kou, L., Markowsky, G., Berman, L.: A fast algorithm for steiner trees. Acta informatica 15(2), 141–145 (1981)
14. Zelikovsky, A.: An 11/6-approximation algorithm for the network steiner problem. Algorithmica 9(5), 463–470 (1993)
15. Hougardy, S., Prömel, H.: A 1.598 approximation algorithm for the Steiner problem in graphs. In: Proc. SODA, pp. 448–453 (1999)
16. Du, D., Lu, B., Ngo, H., Pardalos, P.: Steiner tree problems. Encyclopedia of Optimization 5, 227–290 (2000)
17. Hsu, T.S., Tsai, K., Wang, D., Lee, D.: Steiner problems on directed acyclic graphs. Computing and Combinatorics, 21–30 (1996)
18. Charikar, M., et al.: Approximation algorithms for directed steiner problems. In: Proc. SODA, pp. 192–200 (1998)
19. Ming-IHsieh, E., Tsai, M.: Fasterdsp: A faster approximation algorithm for directed steiner tree problem. JISE 22, 1409–1425 (2006)
20. Rothvoß, T.: Directed steiner tree and the lasserre hierarchy. CoRR abs/1111.5473 (2011)
21. Feige, U.: A threshold of ln n for approximating set cover. J. of the ACM 45(4), 634–652 (1998)
22. Halperin, E., Krauthgamer, R.: Polylogarithmic inapproximability. In: Proc. STOC, pp. 585–594. ACM (2003)
23. Garg, N., Konjevod, G., Ravi, R.: A polylogarithmic approximation algorithm for the group steiner tree problem. In: Proc. SODA, pp. 253–259 (1998)
24. Ding, B., Yu, J.X., Wang, S., Qin, L., Zhang, X., Lin, X.: Finding top-k min-cost connected trees in databases. In: Chirkova, R., Dogac, A., Özsu, M.T., Sellis, T.K. (eds.) ICDE, pp. 836–845. IEEE (2007)
25. Cheriyan, J., Laekhanukit, B., Naves, G., Vetta, A.: Approximating rooted steiner networks. In: Proc. SODA, pp. 1499–1511 (2012)
26. Watel, D., Weisser, M.A., Bentz, C.: Inapproximability proof of DSTLB and USTLB in planar graphs,
    http://hal-supelec.archives-ouvertes.fr/hal-00793424
27. Downey, R.G., Fellows, M.R.: Parameterized complexity, vol. 3. Springer (1999)
28. Goldberg, A.V., Tarjan, R.E.: Finding minimum-cost circulations by canceling negative cycles. Journal of the ACM 36(4), 873–886 (1989)

Exact and Approximate Algorithms for Movement Problems on (Special Classes of) Graphs*

Davide Bilò[1], Luciano Gualà[2], Stefano Leucci[3], and Guido Proietti[3,4]

[1] Dipartimento di Scienze Umanistiche e Sociali, University of Sassari, Italy
[2] Dipartimento di Ingegneria dell'Impresa, University of Rome "Tor Vergata", Italy
[3] Dipartimento di Ingegneria e Scienze dell'Informazione e Matematica,
University of L'Aquila, Italy
[4] Istituto di Analisi dei Sistemi ed Informatica, CNR, Rome, Italy
davidebilo@uniss.it, guala@mat.uniroma2.it,
{stefano.leucci,guido.proietti}@univaq.it

Abstract. When a large collection of objects (e.g., robots, sensors, etc.) has to be deployed in a given environment, it is often required to plan a coordinated motion of the objects from their initial position to a final configuration enjoying some global property. In such a scenario, the problem of minimizing the distance travelled, and therefore energy consumption, is of vital importance. In this paper we study several motion planning problems that arise when the objects must be moved on a *network*, in order to reach certain goals which are of interest for several network applications. Among the others, these goals include broadcasting messages and forming connected or interference-free networks. We study these problems with the aim to minimize a number of natural measures such as the average/overall distance travelled, the maximum distance travelled, or the number of objects that need to be moved. To this respect, we provide approximability and inapproximability results, most of which are tight.

1 Introduction

In many practical applications a number of objects need to be moved in a given environment in order to complete some task. Problems of this kind often occur in robot motion planning where we seek to move a set of robots from their starting position to a set of ending positions such that a certain property is satisfied. For example, if the robots are equipped with a short range communication device we might want to move them so that a message originating from one of the robots can be routed to all the others. If the robots' goal is to monitor a certain area we might want to move them so that they are not too close to each other. Other interesting problems include gathering (placing robots next to each other), monitoring of

---

* Research partially supported by the Research Grant PRIN 2010 "ARS TechnoMedia" from the Italian Ministry of University and Research.

T. Moscibroda and A.A. Rescigno (Eds.): SIROCCO 2013, LNCS 8179, pp. 322–333, 2013.

traffic between two locations, building interference-free networks, and so on. To make things harder, objects to be moved are often equipped with a limited supply of energy. Preserving energy is a critical problem in ad-hoc networking, and movements are expensive. To prolong the lifetime of the objects we seek to minimize the energy consumed during movements and thus the distance travelled. Sometimes, instead, movements are cheap but before and/or after an object moves it needs to perform expensive operations. In this scenario we might be interested in moving the minimum number of objects needed to reach the goal.

In this paper, we assume the underlying environment is actually a *network*, which can be modelled as an undirected graph G, and the moving objects are centrally controlled *pebbles* that are initially placed on vertices of G, and that can be moved to other vertices by traversing the graph edges. To this respect, we study several movement planning problems that arise by various combinations of final positioning goals and movement optimization measures. In particular, we focus our study on the scenarios where we want the pebbles to be moved to a *connected subgraph* (CON), an *independent set* (IND), or a *clique* (CLIQUE) of G, while minimizing either the *overall movement* (SUM), the *maximum movement* (MAX), or the *number of moved pebbles* (NUM). We also give some preliminary results on the problem of moving the pebbles to a *set of vertices whose removal separates two given vertices* (s-t-CUT) while minimizing the above measures.

We will denote each of the above problems with ψ-c, where ψ represents the goal to be achieved and c the measure to be minimized. For a more rigorous definition of the problems we refer the reader to Section 2.

Related work. Although movement problems were deeply investigated in a distributed setting (see [12] for a survey), quite surprisingly the centralized counterpart has received attention from the scientific community only in the last few years.

The first paper which defines and studies these problems in this latter setting is [4]. In their work, the authors study the problem of moving the pebbles on a graph G of n vertices so that their final positions form a connected component, a path (directed or undirected) between two specified nodes, an independent set, or a matching (two pebbles are matched together if their distance is exactly 1). Regarding connectivity problems, the authors show that all the variants are hard and that the approximation ratio of CON-MAX is between 2 and $\mathcal{O}(1 + \sqrt{k/c^*})$, where k is the number of pebbles and c^* denotes the measure of an optimal solution. This result has been improved in [2], where the authors show that CON-MAX can be approximated within a constant factor.

In [4] it is also shown that CON-SUM and CON-NUM are not approximable within $\mathcal{O}(n^{1-\epsilon})$ (for any positive ϵ) and $o(\log n)$, respectively, while they admit approximation algorithms with ratios of $\mathcal{O}(\min\{n \log n, k\})$ (where n is the number of vertices of G) and $\mathcal{O}(k^\epsilon)$, respectively. Moreover, they also provide an exact polynomial-time algorithm for CON-MAX on trees.

Concerning independency problems, in [4] the authors remark that it is NP-hard even to find any feasible solution on general graphs since it would require to find an independent set of size at least k. This clearly holds for all three objective functions. For this reason, they study an Euclidean variant of these problems where pebbles have to be moved on a plane so that their pairwise distances are strictly greater than 1. In this case, the authors provide an approximation algorithm that guarantees an additive error of at most $1 + 1/\sqrt{3}$ for IND-MAX, and a polynomial time approximation scheme for IND-NUM.

More recently, in [7], a variant of the classical facility location problem has been studied. This variant, called *mobile facility location*, can be modelled as a movement and is approximable within $(3 + \epsilon)$ (for any constant $\epsilon > 0$) if we seek to minimize the total movement [1]. A variant where the maximum movement has to be minimized admits a tight 2-approximation [1,4].

As it is frequent, in practice, to have a small number of pebbles compared to the size of the environment (i.e., the vertices of the graph), the authors of [5] turn to study fixed-parameter tractability. They show a relation between the complexity of the problems and their *minimal configurations* (sets of final positions of the pebbles that correspond to feasible solutions, such that any removal of an edge makes them unacceptable).

Our results. We study independency motion problems on graphs where a maximum independent set (and thus a feasible solution for the corresponding motion problem) can be computed in polynomial time. This class of graphs includes, for example, perfect and claw-free graphs. More precisely, we show that IND-MAX and IND-SUM are NP-hard even on bipartite graphs (which are known to be perfect graphs [3]). Moreover we devise a polynomial-time approximation algorithm for IND-MAX that computes solutions where the maximum movement is at most the value of the optimum plus 1. This result is clearly tight.

Concerning the problem of moving pebbles towards a clique of a general graph, we prove that all the three variants are NP-hard. Then, we provide an approximation algorithm for CLIQUE-MAX which is optimal unless an additive term of 1 (this result is clearly tight). We show that both CLIQUE-SUM and CLIQUE-NUM are approximable within a factor of 2 but they are not approximable within a factor better than $10\sqrt{5} - 21 > 1.3606$. If the *unique game conjecture* [10] is true, then both problems are not approximable within a factor better than 2 and the provided approximation algorithms are tight. We also show that an exact solution for CLIQUE-NUM can be computed in polynomial time on every class of graphs which finding a maximum weight clique requires polynomial time (these classes of graphs also include perfect and claw-free graphs).

Finally, we present a strong inapproximability results of $\Omega(n^{1-\epsilon})$ (for any $\epsilon > 0$) for s-t-CUT-MAX and s-t-CUT-SUM along with two approximation algorithms. The approximation algorithm for s-t-CUT-MAX is essentially tight, while we show that any constant-factor approximation for s-t-CUT-NUM would imply a tight approximation for s-t-CUT-SUM.

We were also able to devise five exact polynomial-time algorithms for solving IND-MAX on paths and IND-SUM, IND-NUM, CON-SUM and CON-NUM on trees, respectively.[1] The latter two algorithms complement the already known polynomial-time algorithm for CON-MAX on trees, shown in [4].

---

[1] These results and some of the proofs can be found in the full version of the paper.

The state of the art of the studied problems, along with the results of this paper, is summarized in Table 1.

Table 1. Known and new (in bold) results for the various motion problems on general graphs (G), bipartite graphs (B), graphs on which a maximum independent set or a maximum weight clique can be computed in polynomial time (IS, MWC), trees (T), and paths (P). n and d denote the number of vertices and the diameter of G, respectively. k is the number of pebbles, ρ denotes the best approximation ratio for the corresponding problem, c^* is the measure of an optimal solution. For independency problems on general graphs it is NP-hard even to find any feasible solution. All the inapproximability results hold under the assumption that $P \neq NP$.

	MAX		SUM		NUM	
CON	G: $2 \leq \rho = \mathcal{O}(1)$ T: polynomial	[2,4] [4]	G: $\rho = \Omega(n^{1-\epsilon})$ $\rho = \mathcal{O}(\min\{n,k\})$ T: polynomial	[4] [4]	G: $\rho = \Omega(\log n)$ $\rho = \mathcal{O}(k^2)$ T: polynomial	[4] [4]
IND	G: NP-hard **IS: $c^* + 1$, $\rho \leq 2$** **B: $\rho \geq 2$** **P: polynomial**	[4]	G: NP-hard **B: NP-hard** **T: polynomial**	[4]	G: NP-hard **T: polynomial**	[4]
CLIQUE	**G: NP-hard** **$c^* + 1$**		**G: NP-hard** **$10\sqrt{5} - 21 \leq \rho \leq 2$** **MWC: polynomial**		**G: NP-hard** **$10\sqrt{5} - 21 \leq \rho \leq 2$**	
s-t-CUT	**G: $\rho = \Omega(n^{1-\epsilon})$** **$\rho \leq d$**		**G: $\rho = \Omega(n^{1-\epsilon})$** **$\rho \leq k \cdot d$**		**G: ρ-apx $\implies (\rho \cdot d)$-apx** **for s-t-CUT-SUM**	

2   Formal Definitions

A pebble motion problem, denoted by ψ-c, is an optimization problem whose instances consist of a loop-free connected undirected graph G on n nodes, a set $P = [k] = \{1, \ldots, k\}$ of *pebbles*, a function $\sigma : P \to V(G)$ that assigns each pebble to a *start vertex* of G and a boolean predicate $\psi : 2^{V(G)} \to \{\top, \bot\}$ that assigns a truth values to every possible subset of vertices of G.

A (feasible) solution is a function $\mu : P \to V(G)$ that maps each pebble to an *end vertex* such that $\psi(\mu[P])$ is true, where $\mu[P]$ denotes the image of P under μ. Notice that, in general, it is not required for σ or μ to be injective.

Finally, $c(\mu) \in \mathbb{N}_0$ is a measure function that assigns a non-negative integer to each feasible solution. A solution μ^* that minimizes c is said to be *optimal*.

In the following, we will study some of the movement problems that arise from the different choices of predicates and measures. We will assume that a pebble moving from a vertex u to a vertex v always uses a shortest path in G between u and v. Moreover we denote by $d_G(u, v)$ the length of such a path. In particular, we will consider the following predicates:

Independency: $\text{IND}(U)$ is true if and only if U is an independent set of size k ($|U| = k$) for G, i.e., there is at most one pebble per vertex and no two pebbles are on adjacent vertices;

Clique: $\text{CLIQUE}(U)$ is true if and only if U is a clique of G, i.e., for each pair u, v of distinct vertices in U there exists the edge $(u, v) \in E(G)$;

s-t-**Cut:** Given $s, t \in V(G)$ with $s \neq t$. s-t-CUT(U) is true if and only if $s \notin U$, $t \notin U$ and U is a cut that disconnects s from t, i.e., there exists no path between s and t in the graph induced by the vertices in $V(G) - U$;

and the following measures:

Overall movement: The sum of the distances travelled by pebbles has to be minimized: every pebble $p \in P$ moves from its starting vertex $\sigma(p)$ to his end vertex $\mu(p)$, so the overall distance is: $\text{SUM}(\mu) = \sum_{p \in P} d_G(\sigma(p), \mu(p))$;

Maximum movement: We seek for a solution that minimizes the maximum distance travelled by a pebble. This is captured by the measure $\text{MAX}(\mu) = \max_{p \in P} d_G(\sigma(p), \mu(p))$;

Number of moved pebbles: We aim to minimize the number of pebbles that need to be moved from their starting positions. The associated measure is $\text{NUM}(\mu) = |\{p \in P : \sigma(p) \neq \mu(p)\}|$.

3   Independency Motion Problems

In this section we focus on independency motion problems for which we provide both positive and negative results. Since if $k \geq n$ there is no feasible solution, we will consider only instances where $k < n$.

As we already pointed out, for independency problems on general graphs it is NP-hard even to find any feasible solution since it would require to find an independent set of size at least k. Nevertheless, one may wonder whether independency motion problems are tractable on instances on which a maximum independent set can be found in polynomial time. We provide a negative answer to this question, by showing that IND-MAX and IND-SUM are NP-hard even if the input graph is bipartite. Moreover, we show that on these graphs IND-MAX is not approximable within a factor better than 2. This is tight, as we provide an optimal solution, unless an additive term of 1, to IND-MAX on any class of graphs where a maximum independent set can be found in polynomial time.

We start by stating the following result:

Theorem 1. IND-MAX *and* IND-SUM *are* NP-*hard on bipartite graphs.*

Actually, IND-MAX is hard already when the cost of an optimal solution is 1 and this immediately implies the following:

Corollary 1. IND-MAX *is not approximable in polynomial time within a factor of* $2 - \epsilon$ *for any positive* ϵ, *unless* P = NP. *This also holds for bipartite graphs.*

We now show how to find a solution for IND-MAX such that its cost is at most the optimum cost plus one. This can be done in polynomial time for every class of graphs where a maximum independent set can be found in polynomial time. These include perfect graphs and claw-free graphs.

Given a graph H and a subset of vertices $A \subseteq V(H)$ we will denote the open neighbourhood of A by $N_H(A) = \{v \in V(H) : \exists u \in A \text{ s.t. } (u, v) \in E(H)\}$. Moreover we will denote the closed neighbourhood of A by $N_H[A] = A \cup N_H(A)$.

Let U^* be a maximum independent set of G, the following lemma holds:

Lemma 1. *For each independent set U of G it is true that $|U^* \cap N_G[U]| \geq |U|$.*

Proof. By contradiction, let $|U^* \cap N_G[U]| < |U|$ then $U' = (U^* \setminus N_G[U]) \cup U = (U^* \setminus (U^* \cap N_G[U])) \cup U$ is an independent set of G and $|U'| > |U^*|$. □

To prove the following lemma we use the following well known result [8]:

Theorem 2 (Hall's Matching Theorem). *Let $H = (V_1 + V_2, E)$ be a bipartite graph. There exists a matching of size $|V_1|$ on H iff $|A| \leq |N_H(A)|, \forall A \subseteq V_1$.*

Lemma 2. *For each independent set U of G, there exists an injective function $f : U \to U^*$ such that $d_G(u, f(u)) \leq 1$.*

Proof. Construct the bipartite graph $H = (U + U^*, E)$ where all the vertices of U are considered to be distinct from the ones in U^* and $E = \{(u, v) : u \in U \wedge v \in U^* \wedge v \in N_G[\{u\}]\}$.

Notice that, by construction, if two vertices u, v are adjacent in H either they are the same vertex or they are adjacent in G, i.e., $d_G(u, v) \leq 1$.

Lemma 1 shows that, for every $A \subseteq U$, we have $N(A) = |U^* \cap N_G[A]| \geq |A|$. By Hall's Matching Theorem, the above implies the existence of a matching of size U on H (and thus the existence of the function f). □

Theorem 3. *There exists a polynomial-time algorithm for IND-MAX which, for every class of graphs where the maximum independent set can be found in polynomial time, computes a solution $\tilde{\mu}$ such that $c(\tilde{\mu}) \leq c^* + 1$ where c^* is the measure of an optimal solution.*

Proof. The algorithm works as follows: first it computes a maximum independent set U^* of G then, for every value of z from 0 to $n - 1$, it computes a solution S_z for maximum matching on the bipartite graph $H = (P + U^*, E)$ where $(p, v) \in E$ if and only if $d(\sigma(p), v) \leq z$.

Let $\tilde{z}$ be the first value of z such that $|S_{\tilde{z}}| = k$, i.e., all the pebbles have been matched. Set $\tilde{\mu}(p) = v$ where v is the only vertex such that $(p, v) \in S_{\tilde{z}}$ and return $\tilde{\mu}$. Clearly $c(\tilde{\mu}) = \tilde{z}$. Let μ^* be an optimal solution to IND-MAX and let $U = \mu^*[P]$. By Lemma 2 there exists an injective function f that maps every vertex of the independent set U on an adjacent vertex of U^*.

For every $p \in P$ we have $d_G(\sigma(p), \mu^*(p)) + d_G(\mu^*(p), f(\mu^*(p))) \leq c^* + 1$, therefore there exists a way to place all the pebbles on vertices of U^* while travelling a maximum distance of at most $c^* + 1$. This implies $\tilde{z} \leq c^* + 1$. □

Concerning the independency motion problems on trees and paths, we are able to prove the following results:

Theorem 4. IND-SUM *on trees can be solved in $\mathcal{O}(n^2 \cdot k^2)$ time.*

Theorem 5. IND-NUM *on trees can be solved in $\mathcal{O}(n^2 \cdot k^2)$ time.*

Theorem 6. IND-MAX *on paths can be solved in $\mathcal{O}(n + k \log n)$ time.*

We end this section by mentioning that it remains open to establish whether IND-MAX on trees can be solved in polynomial time.

4   Clique Motion Problems

In this section we prove that the problems CLIQUE-MAX, CLIQUE-SUM and CLIQUE-NUM are NP-hard. Then, we give a tight approximation algorithm for CLIQUE-MAX that computes a solution that costs at most one more than the optimal solution. As of CLIQUE-SUM and CLIQUE-NUM, we show that the problems are not approximable within any factor smaller than $10\sqrt{5} - 21$ and we devise two 2-approximation algorithms.

Actually, we will show that any approximation for CLIQUE-SUM or CLIQUE-NUM implies an approximation with the same ratio for *minimum vertex cover*, which is known to be not approximable under $10\sqrt{5} - 21$ [6]. Moreover, if the unique game conjecture [10] is true, then both approximation algorithms are tight as the corresponding problems are not approximable within any constant factor better than 2 [11].

Finally, for classes of graphs where we can find a maximum weight clique in polynomial time, we can also solve CLIQUE-NUM in polynomial time.

4.1   Approximability of CLIQUE-MAX

We prove the following:

Theorem 7. CLIQUE-MAX *is* NP-*hard.*

Proof. We show a reduction from the problem of determining if there exists a *dominating clique* in a graph H, which is known to be NP-Complete [9]. A dominating clique is a subset of vertices $C \subseteq V(H)$ such that C is both a clique and a dominating set for H.

We construct an instance of CLIQUE-MAX by setting $G = H$ and placing a pebble on each vertex of G. We claim that there exists a dominating clique in H if and only if the optimal solution for CLIQUE-MAX has measure c^* at most 1.

Suppose that there exists a dominating clique C in H. By definition, C is also a dominating set of G. We define μ so that every pebble initially placed on a vertex $u \notin C$ is moved to a vertex $v \in C$ such that $(u, v) \in E(G)$ (notice that such a vertex always exists). After their movement the pebbles are placed on a clique of G and each pebble has travelled a distance of at most 1.

Now suppose that there exists a solution μ for CLIQUE-MAX such that $c(\mu) \leq 1$. Clearly $\mu[P]$ is a clique, to show that it is also a dominating set note that, for each vertex $u \notin \mu[P]$, there exists a vertex $v \in \mu[P]$ such that $(u, v) \in G$.   □

Theorem 8. *It is possible to compute a solution* $\tilde{\mu}$ *for* CLIQUE-MAX *such that* $c(\tilde{\mu}) \leq c^* + 1$ *in polynomial time, where* c^* *is the measure of an optimal solution.*

Proof. Consider the following algorithm: for each vertex $u \in V(G)$ construct a solution μ_u that moves all the pebbles to u (i.e., set $\mu_u(p) = u$, $\forall p \in P$) and compute $c(\mu)$. Among the n possible solutions choose the one of minimum cost and call it $\tilde{\mu}$. Let μ^* be an optimal solution.

Recall that $c(\mu^*) = \max_{p \in P} \{d_G(\sigma(p), \mu^*(p))\}$. Now call $C = \mu^*[P]$ the clique where pebbles have been placed by μ^* and notice that when the above algorithm considers a node $u \in C$ we have: $d_G(\sigma(p), \mu_u(p)) \leq d_G(\sigma(p), \mu^*(p)) + 1 \; \forall p \in P$, thus $c(\tilde{\mu}) \leq c(\mu_u) \leq c(\mu^*) + 1$. □

4.2   Approximability of CLIQUE-NUM

We prove the following:

Theorem 9. CLIQUE-NUM *is 2-approximable and it is not approximable within any constant factor smaller than* $10\sqrt{5} - 21$, *unless* P = NP.

Proof. Let $\langle G, P, \sigma \rangle$ be an instance of CLIQUE-NUM and let $\varphi(u) = |\{p \in P : \sigma(p) = u\}|$ be the number of pebbles that are initially placed on vertex $u \in V$.

Let us assume that $\varphi(u) \leq 1$ for every $u \in V$, i.e., no two pebbles are placed on the same vertex. We will show later that this assumption is not restrictive.

Call H the graph induced by the vertices u with $\varphi(u) = 1$. Let $\bar{H}$ be the complement graph of H w.r.t. the edge set, that is the graph such that $V(\bar{H}) = V(H)$ and $E(\bar{H}) = \{(u, v) : u, v \in V(H) \land u \neq v \land (u, v) \notin E(H)\}$.

We will show that there exists a vertex cover C for $\bar{H}$ if and only if there exists a solution for the instance $\langle G, P, \sigma \rangle$ of CLIQUE-NUM of cost $|C|$.

Let C be a vertex cover for $\bar{H}$, this implies that $Q = V(H) - C$ is an independent set for $\bar{H}$ and therefore a clique for H and G. We construct a solution μ for CLIQUE-NUM by moving the $|C|$ pebbles that are not yet placed on vertices in Q to one of such vertices.

Now let μ be a solution for the instance $\langle G, P, \sigma \rangle$ of CLIQUE-NUM and let $Q = \{u \in V(H) : \exists p \in P \text{ s.t. } \sigma(p) = \mu(p) = u\}$. Notice that the cost of μ is exactly $k - |Q| = |V(H)| - Q|$. Clearly $Q \subseteq \mu[P]$ is a clique for G and H, therefore it is also an independent set for $\bar{H}$. This implies that $C = V(H) - Q$ is a vertex cover for $\bar{H}$. From the above it follows that the cost of an optimal solution is equal to the size of the minimum vertex cover for $\bar{H}$.

To approximate CLIQUE-NUM we construct the graph $\bar{H}$, compute a 2-approximate minimum vertex cover $\tilde{C}$ and reconstruct the solution μ.

If the previous assumption is not met, i.e., there exists at least a vertex on which two or more pebbles are placed by σ, a slight modification to the instance is needed before we can apply the previous approach. We modify the graph G by replacing each vertex u such that $\varphi(u) > 1$ with a clique of size $\varphi(u)$. Each edge e incident to u is replaced by $\varphi(u)$ edges connecting every vertex of the clique to the other endpoint of e. Then, we modify the function σ so that the $\varphi(u)$ pebbles that were placed on u are assigned to each of the $\varphi(u)$ vertices of the corresponding clique. After the modifications, the cost of an optimal solution has not changed, moreover every solution for the modified instance can be easily reconverted to a solution for the original instance without increasing its cost.

We prove the inapproximability result by contradiction: suppose that there exists an algorithm that approximates CLIQUE-NUM with a ratio better than $10\sqrt{5} - 21$, this would allow to approximate minimum vertex cover with the same approximation ratio. Let $\bar{H}$ be the an instance of minimum vertex cover, call G the complement of $\bar{H}$ w.r.t. the edge set, let $P = [|V(G)|]$, and let σ be a function that places a single pebble on each vertex of G. We can now compute an approximate solution for the instance $\langle G, P, \sigma \rangle$ of CLIQUE-NUM and reconstruct a solution with the same cost for minimum vertex cover, as shown before.   □

By considering a weighted variant of the graph G where each vertex v has a weight equal to the number of pebbles starting on v, we can prove the following:

Theorem 10. *An exact solution to* CLIQUE-SUM *can be found in polynomial time on every class of graphs where a maximum weight clique can be found in polynomial time.*

4.3   Approximability of CLIQUE-SUM

We prove the following:

Theorem 11. CLIQUE-SUM *is 2-approximable and is not approximable within any constant factor better than* $10\sqrt{5} - 21$, *unless* P = NP.

Proof. Let μ^* be an optimal solution to CLIQUE-SUM. If μ^* moves all the pebbles (i.e., $\sigma(p) \neq \mu^*(p)$, $\forall p \in P$) then we can compute a 2-approximate solution $\tilde{\mu}$ by guessing a vertex $u \in \mu^*[P]$ and moving all the pebbles to u (i.e., setting $\tilde{\mu}(p) = u$, $\forall p \in P$). Indeed, we have:

$$\frac{c(\tilde{\mu})}{c(\mu^*)} = \frac{\sum_{p \in P} d_G(\sigma(p), \tilde{\mu}(p))}{\sum_{p \in P} d_G(\sigma(p), \mu^*(p))} \leq \frac{|P| + \sum_{p \in P} d_G(\sigma(p), \mu^*(p))}{\sum_{p \in P} d_G(\sigma(p), \mu^*(p))} \leq 2$$

where we used the fact that $d_G(\sigma(p), \tilde{\mu}(p)) \leq d_G(\sigma(p), \mu^*(p)) + 1$ and that $d_G(\sigma(p), \mu^*(p)) \geq 1$, for every pebble $p \in P$.

On the other hand, if there exists at least one pebble $p' \in P$ such that $\sigma(p') = \mu^*(p')$, then we guess its starting vertex $u = \sigma(p')$. We call P_0 the set of pebbles whose starting vertex is u, P_1 the set of pebbles whose starting vertex is adjacent to u, and P_2 the set of pebbles that are initially placed on a vertex at distance 2 or more from u. We set $\tilde{\mu}(p) = u$ if $p \in P_0$ or $p \in P_2$. With a reasoning similar to the one of the previous case we can show that:

$$\sum_{p \in P_2} d_G(\sigma(p), \tilde{\mu}(p)) \leq |P_2| + \sum_{p \in P_2} d_G(\sigma(p), \mu^*(p)) \leq 2 \sum_{p \in P_2} d_G(\sigma(p), \mu^*(p)).$$

Concerning P_1, assume $P_1 \neq \emptyset$, and so we need to compute $\tilde{\mu}$ for the pebbles in P_1. To do that, consider the instance $\langle H, P_1, \sigma \rangle$ of CLIQUE-NUM where H is the subgraph of G induced by the vertices initially occupied by pebbles in $P_1 \cup \{p'\}$, and compute a 2-approximate solution μ' as shown in Theorem 9. Set $\tilde{\mu}(p) = \sigma(p)$ for every pebble $p \in P_1$ such that $\mu'(p) = \sigma(p)$ and set $\tilde{\mu}(p) = u$ for the remaining pebbles in P_1.

Clearly $\tilde{\mu}[P]$ is a clique for G as the vertices in $\mu'[P_1]$ are a clique for G, u is adjacent to every vertex in $\mu'[P_1]$, and $\tilde{\mu}[P] \subseteq \mu'[P_1] \cup \{u\}$.

Notice that the cost of moving the pebbles in P_1 w.r.t. μ^* is greater than or equal to the cost of the optimal solution for the instance $\langle H, P_1, \sigma \rangle$ of CLIQUE-NUM. Moreover the cost of moving the pebbles in P_1 w.r.t. $\tilde{\mu}$ is equal to the cost of μ'. From the above it follows that:

$$\sum_{p \in P_1} d_G(\sigma(p), \tilde{\mu}(p)) \leq 2 \sum_{p \in P_1} d_G(\sigma(p), \mu^*(p)).$$

Therefore, the overall cost of this approximated solution is:

$$c(\tilde{\mu}) \leq 2 \sum_{p \in P_1} d_G(\sigma(p), \mu^*(p)) + 2 \sum_{p \in P_2} d_G(\sigma(p), \mu^*(p)) \leq 2\,c(\mu^*).$$

To prove the inapproximability result, take a graph $H = (V, E)$ and construct the graph G by complementing H w.r.t. the edge set and adding an additional vertex v_0 adjacent to every other vertex. Let $P = [|V(H)|]$ and let σ be a function that places a pebble to each vertex of G except v_0. We will show that, given any solution for CLIQUE-SUM, it is possible to construct a vertex cover of H of the same cost, and vice versa. This implies that any approximation algorithm for IND-SUM translates into an approximation algorithm for minimum vertex cover with the same approximation ratio, therefore no approximation algorithm with an approximation ratio less than $10\sqrt{5} - 21$ can exist for IND-SUM [6].

Let C be a vertex cover for H, then $V(H) - C$ is an independent set for H and a clique for G. The solution that moves all the pebble of C to v_0 and leaves the other on their starting position is feasible and costs C.

Now let μ be a solution for the instance IND-SUM. Let $Q = \{u \in V(H) : \exists p \in P \text{ s.t. } \sigma(p) = \mu(p) = u\}$ (notice that $v_0 \notin Q$) and let $C = V(H) - Q$. The cost of μ is $|V(H)| - |Q| = |C|$, and $Q \subseteq \mu[P]$ is a clique for G. Therefore Q is also an independent set for H and C is a vertex cover for H. $\qquad\square$

5 s-t-CUT Motion Problems

In this section we discuss (in)approximability results for s-t-CUT-MAX and s-t-CUT-SUM. Among the others, we provide an essentially tight approximation algorithm for s-t-CUT-MAX. Regarding s-t-CUT-NUM, establishing its tractability remains open, but we will show that approximating such a problem can be useful to approximate s-t-CUT-SUM, as well. We start by proving the following:

Theorem 12. s-t-CUT-MAX *and* s-t-CUT-SUM *are* NP-*hard even if* G *is a bipartite graph.*

Proof. Due to space limitations we only sketch the proof. The reduction is from the decisional version of 3-SAT: given a C.N.F. boolean formula f decide whether there is a truth assignment to the variables such that f is satisfied.

Given a formula f, we construct an instance of s-t-CUT-MAX (resp., s-t-CUT-SUM) similar to the one shown in Figure 1 where, for each variable in x_i of f, there is a gadget composed of three nodes $u_i, x_i, \bar{x}_i$, and, for each clause c_j of

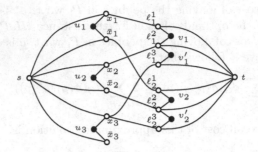

Fig. 1. Instance of s-t-CUT-MAX corresponding to the formula $(x_1 \vee x_2 \vee x_3) \wedge (\bar{x}_1 \vee \bar{x}_2 \vee x_3)$. Pebbles are placed on black vertices. All the edges not incident to a black vertex represent paths of length h between their endpoints.

f, there is a gadget of five nodes $\ell_1^j, v_j, \ell_2^j, v_j', \ell_3^j$. We connect all the vertices u_i to s and all the vertices v_j, v_j' to t. Moreover when a literal appears in a clause we connect the corresponding literal to an ℓ-vertex of the clause using a path of length $h > k$. We place a pebble on each of the vertices u_i, v_j and v_j'.

The claim follows as there exists a truth assignment satisfying f iff an optimal solution for s-t-CUT-MAX (resp., s-t-CUT-SUM) has cost at most 1 (resp., k). $\qquad\square$

We now show that s-t-CUT-MAX and s-t-CUT-SUM are actually very hard to approximate:

Theorem 13. *s-t-CUT-MAX and s-t-CUT-SUM are not approximable within a factor of $n^{1-\epsilon}$ for every $\epsilon > 0$, unless* P = NP. *This also holds for bipartite graphs.*

Proof. As shown in the proof of Theorem 12, it is possible to construct instances of s-t-CUT-MAX (and s-t-CUT-SUM) such that the optimal solution has measure at most 1 (resp., $k = \tau + 2m$) if and only if a boolean C.N.F. formula with three literals per clause, τ variables and m clauses is satisfiable.

Moreover any solution with cost z such that $\tau + 2m < z < h$ (recall that h is the length of the "long" paths) can be easily transformed into a solution of cost at most $\tau + 2m$ by moving each pebble that has been placed on the long paths to an appropriate endpoint (i.e., the one adjacent to its starting position).

This implies that, if f is satisfiable, the measure z^* of an optimal solution is at most k for both the problems, while, if f is not satisfiable, z^* is at least h.

Take a formula f, construct an instance for s-t-CUT-MAX (and s-t-CUT-SUM) and suppose that there exists a polynomial time algorithm that approximates the optimal solution within a factor of $n^{1-\epsilon}$, for some positive $\epsilon \leq 1$. Let $\tilde{c}$ be the measure of an approximate solution.

It is possible to upper bound the number of vertices n of the instance with the quantity $2 + 3\tau + 5m + 2\tau h + 6mh \leq 11h(\tau + m) \leq 11hk$.

If f is satisfiable, we have $\tilde{c} \leq kn^{1-\epsilon}$, while if f is not satisfiable we have $\tilde{c} \geq h$. If $kn^{1-\epsilon} < h$ holds then it is possible to decide 3-SAT by running the approximation algorithm and looking the measure of the approximate solution. This can be guaranteed by choosing $h > (11k)^{2/\epsilon}$ as we have: $h > (11k)^{2/\epsilon} \implies h^\epsilon > 11k^2 \implies h^\epsilon > 11^{1-\epsilon}k^{2-\epsilon} \implies h > k(11hk)^{1-\epsilon} \implies h > kn^{1-\epsilon}$. $\square$

By moving all the pebbles on a minimum s-t-cut of G we can show that the inapproximability result provided above is tight for s-t-CUT-MAX:

Theorem 14. s-t-CUT-MAX *is d-approximable in polynomial time, where* $d < n$ *is the diameter of* G. s-t-CUT-SUM *is* $(k \cdot d)$-*approximable in polynomial time.*

We close this section stating a theorem which is useful in linking the approximability of s-t-CUT-NUM to that of s-t-CUT-SUM:

Theorem 15. *If there exists a* ρ-*approximation algorithm for* s-t-CUT-NUM *then there exists a* $(\rho \cdot d)$-*approximation algorithm for* s-t-CUT-SUM, *where* $d < n$ *is the diameter of* G.

References

1. Ahmadian, S., Friggstad, Z., Swamy, C.: Local-search based approximation algorithms for mobile facility location problems. arXiv preprint arXiv:1301.4478 (2013)
2. Berman, P., Demaine, E.D., Zadimoghaddam, M.: $O(1)$-approximations for maximum movement problems. In: Goldberg, L.A., Jansen, K., Ravi, R., Rolim, J.D.P. (eds.) APPROX/RANDOM 2011. LNCS, vol. 6845, pp. 62–74. Springer, Heidelberg (2011)
3. Bollobás, B.: Modern graph theory, vol. 184. Springer (1998)
4. Demaine, E.D., Hajiaghayi, M., Mahini, H., Sayedi-Roshkhar, A.S., Oveisgharan, S., Zadimoghaddam, M.: Minimizing movement. In: Proceedings of the Eighteenth Annual ACM-SIAM Symposium on Discrete Algorithms, pp. 258–267. Society for Industrial and Applied Mathematics (2007)
5. Demaine, E.D., Hajiaghayi, M., Marx, D.: Minimizing movement: Fixed-parameter tractability. In: Fiat, A., Sanders, P. (eds.) ESA 2009. LNCS, vol. 5757, pp. 718–729. Springer, Heidelberg (2009)
6. Dinur, I., Safra, S.: On the hardness of approximating minimum vertex cover. Annals of Mathematics pp. 439–485 (2005)
7. Friggstad, Z., Salavatipour, M.R.: Minimizing movement in mobile facility location problems. ACM Transactions on Algorithms (TALG) 7(3), 28 (2011)
8. Hall, P.: On representatives of subsets. J. London Math. Soc. 10(1), 26–30 (1935)
9. Haynes, T.W., Hedetniemi, S.T., Slater, P.J.: Domination in graphs: advanced topics, vol. 40. Marcel Dekker, New York (1998)
10. Khot, S.: On the power of unique 2-prover 1-round games. In: Proceedings of the Thiry-fourth Annual ACM Symposium on Theory of Computing, pp. 767–775. ACM (2002)
11. Khot, S., Regev, O.: Vertex cover might be hard to approximate to within 2- ε. Journal of Computer and System Sciences 74(3), 335–349 (2008)
12. Prencipe, G., Santoro, N.: Distributed algorithms for autonomous mobile robots. In: Navarro, G., Bertossi, L., Kohayakwa, Y. (eds.) IFIP TCS 2006. IFIP, vol. 209, pp. 47–62. Springer, Bostan (2006)

Maximum Distance Separable Codes
Based on Circulant Cauchy Matrices

Christian Schindelhauer and Christian Ortolf

Department of Computer Science, University of Freiburg, Germany
{schindel,ortolf}@informatik.uni-freiburg.de
http://cone.informatik.uni-freiburg.de/

Abstract. We present a maximum-separable-distance (MDS) code suitable for computing erasure resilient codes for large word lengths. Given n data blocks (words) of any even bit length w the Circulant Cauchy Codes compute $m \leq w+1$ code blocks of bit length w using XOR-operations, such that every combination of n data words and code words can reconstruct all data words. The number of XOR bit operations is at most $3nmw$ for encoding all check blocks. The main contribution is the small bit complexity for the reconstruction of $u \leq m$ missing data blocks with at most $9nuw$ XOR operations.

We show the correctness for word lengths of form $w = p - 1$ where p is a prime number for which two is a primitive root. We call such primes Artin numbers. We use efficiently invertible Cauchy matrices in a finite field $GF[2^p]$ for computing the code blocks To generalize these codes for all even word lengths w we use ℓ independent encodings by partitioning each block into sub-blocks of size $p_i - 1$, i.e. $w = \sum_{i=1}^{\ell} p_i - \ell$ for Artin numbers p_i. While it is not known whether infinitely many Artin numbers exist we enumerate all Circulant Cauchy Codes for $w \leq 10^5$ yielding MDS codes for all $m + n \leq \frac{10}{62}w$.

Keywords: RAID, erasure codes, storage, fault-tolerance.

1 Introduction

Computer systems are prone to data loss in many situations, ranging from the failure of a hard disk, communication errors in the physical layer, to the incomplete transmission of large files in overlay networks. Data is often stored in fixed block sizes and many systems rely on creating extra code blocks, from which one can recover the original data. A wide-spread example of such codes are RAID storage systems [14], where the parity of data blocks are stored on extra hard disks to recover the data. If the number of code blocks and data blocks necessary to restore a message equals the original number of data blocks, say n, such coding schemes are called *maximum distance separable (MDS)*.

We consider the input data partitioned into n blocks of w bits. We generate m additional check blocks with w bits each. MDS codes allow by definition to retrieve all n data blocks from any combination of the $n + m$ data and check blocks. MDS codes have been studied extensively and the standard method is Reed-Solomon codes [17]. These codes can be represented as a matrix multiplication over a finite field $\mathbb{F}[2^w]$. Addition in a Galois field is a bitwise XOR operation, while multiplication is a product of polynomials modulo a polynomial (and modulo base 2). The multiplication operation is the

T. Moscibroda and A.A. Rescigno (Eds.): SIROCCO 2013, LNCS 8179, pp. 334–345, 2013.

most time consuming part of such codes and therefore there have been research efforts to establish MDS codes only based on XOR operations.

It should be noted that Luby has found more efficient codes [11] based on XOR operations, spawning a lot of work in this area. However, these LT codes and successors are not MDS codes, and are thus not relevant for many uses, e.g. in RAID systems.

$$
\begin{pmatrix}
I & 0 & 0 \\
0 & I & 0 \\
0 & 0 & I \\
H_{11} & H_{12} & H_{13} \\
H_{21} & H_{22} & H_{23}
\end{pmatrix}
\begin{pmatrix}
d_1 \\
d_2 \\
d_3
\end{pmatrix}
=
\begin{pmatrix}
d_1 \\
d_2 \\
d_3 \\
c_1 \\
c_2
\end{pmatrix}
\tag{1}
$$

1.1 State of the Art

For RAID systems, MDS codings are considered where the original data is part of the encoding and the check blocks are generated by a matrix multiplication. Such MDS codes use a $nw \times (n+m)w$ *systematic* generator matrix G where the first nw columns are occupied with the identity matrix I, see (1). For the complexity of such an approach Blaum has proved in Proposition 3.4 [5] that at least $nw(m+1)$ nonzero entries are in the generator matrix, where n is the number of original blocks, m is the number of check blocks and w is the word length. For $\mathbb{F}_2$ he improves his bound in Proposition 5.2 to a minimum of $(m+1)nw + \frac{1}{2}\frac{mn}{1-1/n}$ entries of value 1 in the generator matrix. This corresponds to $nmw + \frac{1}{2}\frac{mn}{1-1/n}$ XOR operations for computing the check blocks, but does not imply a lower bound, since a small number of XOR operations can construct full matrices. Note that sparser matrices exist, if one drops the systematic matrix property, see [19].

In [4] MDS codes based on Galois-fields $\mathbb{F}[2^p]$ with generator polynomials $1 + x + \ldots + x^p$, where p is a prime number and 2 is a generator, i.e. $\{2^0, 2^1 \bmod p, \ldots, 2^{p-1} \bmod p\} = \{1, \ldots, p-1\}$ are shown. The authors use this approach to establish efficient coding for up to 8 check blocks. In this paper, we use the same generator polynomial and extend it to general block sizes and number of check blocks.

Many MDS codes like Even-Odd [3], Row-Diagonal Parity (RDP) [7], and Liberation Codes [15] construct only two parities and optimize on the number of XOR operation for computing the code blocks. In [2] RDP was generalized to compute up to eight check blocks while maintaining the optimal encoding complexity of RDP. An MDS code optimizing the reconstruction complexity has been presented in [12].

Blömer et al. [6] use Cauchy Reed-Solomon matrices to construct check matrices for systematic MDS codes. It turns out that the number of operations over the Galois field for reconstructing n data words is bounded by $O(nm)$ operations of the Galois field. In general, a multiplication in a Galois field can be performed in time $w \log w 2^{O(\log^* w)}$ using Fürer's integer multiplication algorithm [10]. For small word lengths, such multiplication can be performed on modern computers in constant time using table lookups with table size $O(2^w)$ and corresponding precomputing time. In [6] the authors recommend to precompute the factors for coding and decoding and word-parallel computation which reduces the complexity of computing m check blocks to $O(nmw)$ word-wise

XOR operations assuming the processor has word size w. Based on the approach of [6] Plank uses exhaustive search techniques to reduce the XOR complexity of such Cauchy Reed-Solomon matrices [16].

In [8] the usage of circular Boolean matrices helped to find an efficient MDS code for tolerating three disk failures. They showed that encoding takes $3wn$ XOR operations and decoding can be done within $3wn + 9(w+1)$ XOR operations. Using these Cauchy Reed-Solomon matrices (aka. Rabin Codes) the same authors generalized this result in [9] to a MDS code with complexity of $9wn$ XOR operations for encoding and $O(w^3 m^4)$ XOR operations for decoding when m disks fail (and $(9n + 95)(w + 1)$ for $m = 4$). The approach presented here has a similar coding complexity of $9mn$, but reduces the decoding complexity to $9nuw$ XOR operations.

1.2   Contribution

We present MDS codes, called Circulant Cauchy Codes, with an asymptotic optimal bit complexity for computing the check blocks and reconstructing data blocks. Bit complexity denotes the overall number of bit operations used in the calculation. In particular, we have an encoding bit complexity of $3nmw$ for computing m check blocks from n data blocks of word length w. The reconstruction of u data blocks from any n data or check blocks needs $9unw$ XOR bit operations. We prove that Circulant Cauchy Codes are defined for any even word length w.

However the maximum number of check blocks m depends on n and w. We prove that any number of check blocks can be generated if a conjecture of Emil Artin is true. It states that there are infinitely many prime numbers with two as primitive root. This conjecture does not give a good bound on the number of check blocks unless these prime numbers $A = \{3, 5, 11, 13, \ldots\}$ are dense. At least we can show that for all even $w \leq 10^5$ we can generate at least $m = \frac{10}{62}w - n$ check blocks. If w is a power of two we have evaluated that at least $\frac{29}{128}w - n$ parity check blocks can be generated for all $w \leq 2^{20}$. We conjecture that if w tends to infinity the number of possibly parity check blocks approaches $\frac{1}{4}w - n$.

An implementation can be downloaded from our website [18].

2   Circulant Boolean Matrices

The key to our efficient MDS codes are circulant Boolean matrices. A circulant matrix $n \times n$ matrix $A = \mathrm{Cir}_n(s_1, \cdots, s_n)$ has the following form

$$\mathrm{Cir}(s_0, \cdots, s_{n-1}) := (s_{i-j \mod n})_{i,j \in [n]} \tag{2}$$

We consider Boolean circulant matrices encoding bit-strings of length n. For $s_i \in \{0, 1\}$ we denote by $s = \sum_{i=0}^{p-1} s_i 2^i$ the matrix $\mathrm{Cir}_n(s) = \mathrm{Cir}(s_0, \ldots, s_{n-1})$ and operations are defined over $\mathbb{F}_2$ (XOR is addition and AND denotes the multiplication of bits). So, the cyclic shift by one position is denoted by a multiplication with $\mathrm{Cir}(2)$. $\mathrm{Cir}(1)$ is the neutral element for multiplication and $\mathrm{Cir}(0)$ is the neutral element for addition, see Fig. 1.

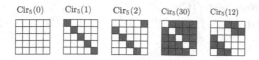

Fig. 1. Examples for $\text{Cir}_5(x)$

It is well known, that Boolean circulant matrices describe a semiring. $\text{Cir}(2)^i = \text{Cir}(2^i)$ describe cyclic rotations by i positions. Since the rank of all Boolean circulant matrices is at most $n - 1$, Boolean circulant matrices do not describe a finite field.

An alternative way to describe these semigroups is to consider them as monic polynomials $\sum_{i=0}^{n-1} s_i z^i$ modulo the monic polynomial $z^n + 1$ which is reducible, since $z^n + 1 = (z^{n-1} + \ldots + z + 1)(z + 1) \bmod 2$. Now, $z^{n-1} + \ldots + z + 1$ is irreducible if n is a prime number and 2 is a primitive root for p. Such finite fields have been already considered by Blaum et al. [4]. From now on we assume p to be such a prime with primitive root 2 and we call this set of numbers A.

Lemma 1. *For $p \in A$, i.e. a prime number where 2 is a primitive root modulo p, the set of all matrices $\mathcal{E}_p$ of $\text{Cir}_p(s)$ for binary strings $s \in \{0,1\}^p$ with even parity forms a finite field for addition and multiplication as described above.*

Proof. If p is a prime number and 2 is a primitive root for m, then the polynomial $1 + z + \ldots + z^{p-1}$ is irreducible according to [13].

Note that $(z + 1)(1 + z + \ldots + z^{p-1}) \equiv z^p + 1 \pmod{2}$. Operations modulo the polynomial $z^p + 1$ do not describe a field, since the polynomial is not irreducible. However, addition and multiplication modulo $z^p + 1$ correspond to the addition and multiplication of circulant Boolean matrices.

Given $\text{Cir}_p(a)$ with binary representation $a_0, \ldots, a_{p-1}$ we prove an operation preserving isomorphism described by the mapping $f : \mathcal{E}_p \to \{0,1\}^{p-1}$ where

$$f(\text{Cir}_p(a_0, \ldots, a_{p-1})) = (a_i + a_{p-1})_{i \in \{0,\ldots,p-2\}} \tag{3}$$

and the inverse function is

$$f^{-1}((a_i)_{i \in \{0,\ldots,p-2\}}) = \text{Cir}_p \left(\left(\sum_{j \in \{0,\ldots,p-2\} \setminus \{i\}} a_j \right)_{i \in \{0,\ldots,p-1\}} \right) \tag{4}$$

We prove that

$$f(\text{Cir}_p(a) + \text{Cir}_p(b)) \equiv f(a) + f(b) \pmod{1 + z + \ldots + z^{p-1}} \tag{5}$$
$$f(\text{Cir}_p(a) \cdot \text{Cir}_p(b)) \equiv f(a)f(b) \pmod{1 + z + \ldots + z^{p-1}} \tag{6}$$

Note that for $a, b \in \mathcal{E}_p$ we have $a + b \in \mathcal{E}_p$ and $a \cdot b \in \mathcal{E}_p$. From the irreducibility of $1 + z + \ldots + z^{p-1}$ the claim follows since we have an isomorphism to a finite field.

It remains to prove (5) and (6).

1. Addition

$$f(\text{Cir}_p(a) + \text{Cir}_p(b))$$
$$= (a_i + b_i + a_{p-1} + b_{p-1})_{i \in \{0,\ldots,p-2\}}$$
$$= f(\text{Cir}_p(a)) + f(\text{Cir}_p(b)) \tag{7}$$

2. Multiplication: Note that $f(\mathrm{Cir}_p(2^p - 3)) = (0,1,0,\ldots,0) = x$, $f(\mathrm{Cir}_p(2^p - 1)) = 1$, $f(\mathrm{Cir}_p(0)) = 0$, $f(\mathrm{Cir}_p(2^p - 3)^i) = f(\mathrm{Cir}_p(2^p - 1 - 2^{i \bmod p-2})) = x^i$. Furthermore,

$$f(\mathrm{Cir}_p(2^p - 3)) \cdot a = f(\mathrm{Cir}_p(a_{i-1 \bmod p})_{i \in \{0,\ldots,p-1\}}$$
$$= (a_{i-1 \bmod p} + a_{p-2})_{i \in \{0,\ldots,p-2\}}$$
$$= \sum_{i=0}^{p-2} (a_i + a_{p-1}) x^{i+1}$$
$$= f(\mathrm{Cir}_p(a)) \cdot x \tag{8}$$

since $x^{p-1} \equiv \sum_{i=0}^{p-2} x^i \pmod{1 + x + x^2 + \ldots + x^{p-1}}$.

So, for $\mathrm{Cir}_p(b) = \sum_{i=0}^{p-1} b_i \mathrm{Cir}_p(2)^i$ we have

$$f(\mathrm{Cir}_p(a) \cdot \mathrm{Cir}_p(b)) = f\left(\mathrm{Cir}_p(a) \cdot \sum_{i=0}^{p-1} b_i \mathrm{Cir}_p(2)^i\right)$$
$$= f\left(\sum_{i=0}^{p-1} b_i \mathrm{Cir}_p(a) \mathrm{Cir}_p(2^p - 3)^i\right)$$
$$= \sum_{i=0}^{p-1} b_i f(\mathrm{Cir}_p(a) \mathrm{Cir}_p(2^p - 3)^i))$$
$$= \sum_{i=0}^{p-1} f(\mathrm{Cir}_p(a)) b_i x^i$$
$$= \sum_{i=0}^{p-2} f(\mathrm{Cir}_p(a))(b_i + b_{p-2}) x^i$$
$$= f(\mathrm{Cir}_p(a)) \cdot f(\mathrm{Cir}_p(b)) \tag{9}$$

Using the observation that Circulant matrices with even parity form a finite field, one can reduce the number of XOR operations. For this, we use the complement of an element as an alternative representation. So, define for each element $a \in [0, 2^{p-1} - 1]$: $[a] := \{\mathrm{Cir}(a), \mathrm{Cir}(2^p - 1 - a)\}$. Define the addition over these sets $[a] + [b] = \{x + y \mid x \in [a], y \in [b]\}$ and similarly, $[a] \cdot [b] = \{x \cdot y \mid x \in [a], y \in [b]\}$. Now, the observation of Figures 2 and 3 can be generalized as follows.

$$[a + b] = [a] + [b], \tag{10}$$
$$[a \cdot b] = [a] \cdot [b]. \tag{11}$$

When coding or decoding all inputs $a \in \{0, 1\}^{p-1}$ of word size $p - 1$ are mapped to such sets $[a]$ of word size p at the beginning. All subsequent operations will be done on the increased word size p since in the representation fewer XOR operations are needed. Therefore, we do not distinguish between both representations $[a] = \{a, 2^p - 1 - a\}$. Eventually, we eliminate the ambiguity by a final computation where we choose the

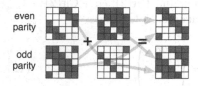

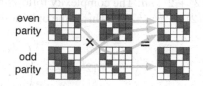

Fig. 2. Example for addition $[a + b]$ **Fig. 3.** Example for multiplication $[a] \cdot [b]$

element of which the least significant bit is 0. This final operation costs $p - 1$ XOR operations and reduces the output to the original word length $p - 1$. See Fig. 4. The transition from the data to one of its representation can be denoted by multiplication with matrix R, which is just a copy operation and its reverse operation with matrix L where $p - 1$ XORs are needed.

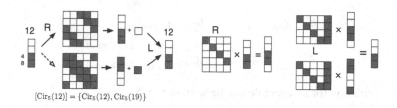

$[\mathrm{Cir}_5(12)] = \{\mathrm{Cir}_5(12), \mathrm{Cir}_5(19)\}$

Fig. 4. Example for representation of 12 in $[\mathrm{Cir}_5(12)]$

Lemma 2. *The basic operations in the table below need the given number of XOR operations for input word length* $w = p - 1$, *where* i *and* j *are constants and* a, b *input variables.*

Operation	XOR operations
$R \cdot a = \mathrm{Cir}_p(a)$	0
$L \cdot \mathrm{Cir}(a) = a$	w
$\mathrm{Cir}(a) + \mathrm{Cir}(b)$	$w + 1$
$\mathrm{Cir}(2)^i \cdot a$	0
$\mathrm{Cir}(2^i + 2^j) \cdot a$	$w + 1$
$[\mathrm{Cir}(2^i + 2^j)^{-1} \cdot a]$	$2w - 1$

Proof. 1. $a \mapsto \mathrm{Cir}_p(a)$: For this operation we simply append a constant 0 to the representation.

2. $\mathrm{Cir}(a) \mapsto (a_i + a_p)_{i \in \{0,\dots,p-1\}}$: This operation is necessary to produce the output and to transform the representation of a by Circulant Boolean matrices to the corresponding finite field element. Clearly, p XORs are sufficient to compute the result.

3. $\mathrm{Cir}(a) + \mathrm{Cir}(b)$: We compute pairwise XORs of the corresponding bits of a and b.

4. $\mathrm{Cir}(2)^i \cdot a$: For constant i this constitutes a clock-wise right shift by i steps. No XOR operations are necessary.

5. $\text{Cir}(2^i + 2^j) \cdot a$: The complexity follows by computing $\text{Cir}(2)^i \cdot a + \text{Cir}(2)^j \cdot a$.
6. $[\text{Cir}(2^k + 2^\ell)^{-1} \cdot a]$: This is the only non-trivial case. Note that

$$\text{Cir}(2^k + 2^\ell) \cdot \begin{pmatrix} b_0 \\ \vdots \\ b_{p-1} \end{pmatrix} = \begin{pmatrix} a_0 \\ \vdots \\ a_{p-1} \end{pmatrix}. \tag{12}$$

This is equivalent to

$$a_{i+\ell} + \sum_j a_j = b_{i \bmod p} + b_{i+(\ell-k) \bmod p} \tag{13}$$

$$\text{for all } i \in \{0, \ldots, p-1\}.$$

Since, any of the two elements of $[a] = \{a, 2^p - 1 - a\}$ is allowed as result we choose $b_0 = 0$. So we get for $i \in \{0, \ldots, p-2\}$

$$b_{(i+1)2(\ell-k) \bmod p} = b_{2i(\ell-k) \bmod p}$$
$$+ a_{2(i+1)(\ell-k)+k \bmod p}$$
$$+ a_{(2i+1)(\ell-k)+k \bmod p}. \tag{14}$$

The calculation for an example can be seen in Fig. 5

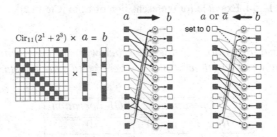

Fig. 5. Multiplication and division by $\text{Cir}_{11}(9)$

It is straightforward that the result is either b or $2^p - b$. The result might be inverted, but this is acceptable, as we have described above.

3   Circulant Cauchy Matrix

Cauchy Reed-Solomon matrices have been introduced for MDS codes by Blömer et al. [6]. For n data blocks and m check blocks of word length w we use a $m \times n$ Cauchy matrix where each entry is a Boolean Circulant matrix defined as

$$M_{ij} = \frac{1}{x_i + y_j} \tag{15}$$

for $x_i \neq y_j$ for all $i \in \{1, \ldots, m\}, j \in \{1, \ldots, n\}$. For the Circulant Cauchy matrix we choose $x_i = \mathrm{Cir}_p(2^{i-1})$ for $i \in \{1, \ldots, m\}$ and $y_1 = 0$, $y_j = \mathrm{Cir}_p(2^{p-j})$ for $j \in \{2, \ldots, n\}$. Therefore for $n + m \leq p + 1$ the sets $\{x_1, \ldots, x_n\}$ and $\{y_1, \ldots, y_m\}$ are distinct which is the prerequisite for the full rank property of the Cauchy matrix.

So the check blocks $c_1, \ldots, c_m \in \{0, 1\}^{p-1}$ are computed from the data block $d_1, \ldots, d_n \in \{0, 1\}^{p-1}$ by

$$
\begin{pmatrix} L & & 0 \\ & \ddots & \\ 0 & & L \end{pmatrix} \cdot M \cdot \begin{pmatrix} R & \cdots & 0 \\ \vdots & \ddots & \vdots \\ 0 & \cdots & R \end{pmatrix} \cdot \begin{pmatrix} d_1 \\ \vdots \\ d_n \end{pmatrix} = \begin{pmatrix} c_1 \\ \vdots \\ c_m \end{pmatrix} \tag{16}
$$

where L is a $(p-1) \times p$ matrix with $L_{ii} = 1$, $L_{i,p} = 1$ for $i \in \{1, \ldots, p-1\}$ and $L_{i,j} = 0$ elsewhere. R is a $p \times (p-1)$ matrix with $R_{ii} = 1$ and $R_{i,p} = 0$ elsewhere.

Theorem 1. *The computation of m check blocks of the Circulant Cauchy matrix from n data blocks of block size w can be computed with $(3m-2)nw + n - m = (3 + o(1))nmw$ XOR-operations.*

Proof. The transformation from d_i to a circulant matrix does not use any operations. The multiplication with the first row of the Circulant Cauchy matrix is a cyclic shift operation since $x_1 = \mathrm{Cir}_p(0)$ which does not involve any XOR bit operation. For the multiplication of the residual $m - 1$ operations we divide by terms of the form $\mathrm{Cir}_p(2^k + 2^\ell)$. Lemma 2 states that $2w - 1$ XOR operations are sufficient resulting in $(m-1)n(2w-1)$ XOR operations. For adding all results we need $(n-1)m(w+1)$ XOR operations. A multiplication with an L matrix needs w XOR operations resulting in mw operations.

So, the overall number of XOR operations is $(n-1)m(w+1) + n(m-1)(2w-1) + mw = 3mnw - 2nw - m + n$.

Theorem 2. *The Circulant Cauchy matrix is an MDS code, i.e. from every set of n data or check blocks the data can be recovered, if $n + m \leq p + 1$.*

Proof. The determinant of a $n \times n$ Circulant Cauchy matrix $M_{ij} = \left(\frac{1}{x_i + y_j} \right)$ is

$$
\det(M) = \frac{\prod_{i<j}(x_i + x_j)(y_i + y_j)}{\prod_{i,j} x_i + y_j} \tag{17}
$$

Let $C_{rs} = M_{rs}$ be the adjugate matrix of M (deleting row i and column j). Note that the adjugate matrix is again a Circulant Cauchy matrix.

$$
\det(C_{rs}) = \frac{\prod_{i<j,i,j \neq r}(x_i + x_j) \prod_{i<j,i,j \neq s}(y_i + y_j)}{\prod_{i \neq r, j \neq s} x_i + y_j} \tag{18}
$$

Now for the inverse matrix d_{ij} we have

$$
d_{ij} = \frac{\det(C_{ji})}{\det(M)} = \frac{\prod_u x_j + y_u \prod_u x_u + y_i}{\prod_{u \neq j}(x_u + x_j) \prod_{u \neq i}(y_u + y_i)} \frac{1}{(x_i + y_j)} \tag{19}
$$

For computing the data blocks, we choose the square submatrix B of the generator matrix corresponding to the indices of the u unknown data blocks and the k given check blocks. Note that B is a Circulant Cauchy matrix and thus invertible. The $k \times n - k$ submatrix A corresponds to the given data blocks and unknown check blocks. So, we can calculate the missing data blocks by

$$
\begin{pmatrix} d_{i_1} \\ \vdots \\ d_{i_u} \end{pmatrix} = B^{-1} \left(\begin{pmatrix} c_{j_1} \\ \vdots \\ c_{j_u} \end{pmatrix} + A \cdot \begin{pmatrix} d_{g_1} \\ \vdots \\ d_{g_{n-u}} \end{pmatrix} \right) \tag{20}
$$

where $i_1, \ldots, i_u$ are the indices of the unknown data blocks, $j_1, \ldots j_u$ the indices of the given check blocks, and $g_1, \ldots, g_{n-u}$ the indices of the given data blocks.

Our main contribution is the efficiency of this operation.

Theorem 3. *Given $n - u$ data blocks and u check blocks of a Circulant Cauchy Code the missing u data blocks can be computed with at most $3nuw + 6u^2w \leq 9nuw$ XOR bit operations.*

Proof. Using Equation 20 the given input $d_{g_1}, \ldots, d_{g_{n-k}}$ is multiplied with the submatrix $k \times (n-k)A$. From Theorem 1 this can be done with $3u(n-u)w - 2(n-u)w + n - 2u = 3nuw - 3u^2w - 2nw + 2uw - 2u$ XOR operations. With another uw XOR operations the result is added to the given u check blocks of word size w giving the intermediate result $y_1, \ldots, y_u \in \{0,1\}^{w+1}$.

Then, the Circulant Cauchy matrix is reduced to the $u \times u$ sub-matrix B corresponding to the check blocks which results in a smaller Circulant Cauchy matrix. Following the approach in [6] the inverse $(d_{ij})_{i,j \in 1, \ldots, n}$ of a Circulant Cauchy matrix $(\frac{1}{x_i + y_j})_{i,j}$ can be computed as follows.

$$
a_k = \prod_{i \neq k} (x_i + x_k) , \quad b_k = \prod_{i \neq k} (y_i + y_k) , \quad e_k = \prod_{i=1}^{n} (x_k + y_i) ,
$$
$$
f_k = \prod_{i=1}^{n} (x_i + y_k) , \quad d_{ij} = \frac{e_i f_j}{a_i b_j (x_i + y_j)} . \tag{21}
$$

We multiply the inverse with the intermediate result $y_1, \ldots, y_u$. We have to compute the data blocks $d_1, \ldots, d_u$ (WLOG we assume that the first u data blocks need to be restored) such that for all $j \in \{1, \ldots, u\}$

$$
d_i = \sum_{j=1}^{u} d_{ij} y_j = \sum_{j=1}^{u} \frac{e_i f_j}{a_i b_j (x_i + y_j)} y_j = \frac{e_i}{a_i} \sum_{j=1}^{u} \frac{1}{(x_i + y_j)} \frac{f_j}{b_j} y_j . \tag{22}
$$

First we compute for all $i \in \{1, \ldots, u\}$

$$
C'_j = \frac{f_j}{b_j} C_j = \frac{\prod_{i=1}^{n} (x_i + y_j)}{\prod_{i \neq j} (y_i + y_j)} C_j \tag{23}
$$

This results in u^2 multiplications with terms of the form $\mathrm{Cir}(2^\nu + 2^\eta)$ and $u(u-1)$ divisions by $\mathrm{Cir}(2^\nu + 2^\eta)$. Some factors or divisors may be of form $\mathrm{Cir}(2^\nu)$ since we choose $x_1 = 0$. This case only reduces the complexity and is from now on omitted for simplicity. The number of XOR operations is therefore at most $u(u-1)(w+1) + u^2(2w-1) = 3u^2w - uw - u$. Then, we compute all terms of the form $\frac{C'_j}{x_i + y_j}$ which takes at most $u^2(2w-1) = 2u^2w - u^2$ XOR operations.

The following sum costs $u(u-1)(w+1) = u^2w - uw + u^2 - u$ XOR operations. Finally each of the u sums need to be multiplied by

$$\frac{e_i}{a_i} = \frac{\prod_j (x_i + y_j)}{\prod_{j \neq i}^u (x_i + x_j)} \tag{24}$$

So, u^2 multiplications by terms of the form $\mathrm{Cir}(2^i + 2^j)$ are necessary ($w+1$ single XOR operations each) and $u(u-1)$ divisions by terms of form $\mathrm{Cir}(2^i + 2^j)$ with ($2w-1$ XOR operations). Hence, $u^2(w+1) + u(u-1)(2w-1) = 3u^2w - 2uw + u$ operations suffice.

Finally, the size of the u data bits of word length $w+1$ need to be reduced to size w adding another uw Xor operations.

The overall number of XOR bit operations is therefore at most $3nuw + 6u^2w - 2nw - 4u \leq 3nuw + 6u^2w \leq 9nuw$.

Because of the run-time of $3nmw$ for computing the check blocks and $9nuw$ for reconstructing the data, the length of w does not play any role. It is advisable to choose w as large as possible, since it increases the number of possible code words and does not change the run-time, e.g. if the input consist of N bits, then the complexity for computing all the m check blocks is $3\frac{N}{w}nw = 3Nm$ and similarly for reconstruction N data bits from u check blocks: $9\frac{N}{w}uw = 9Nu$. One might argue that usually the input size is not a multiple of w (since we have a restriction on $w+1$ being an Artin number). However, we overcome this problem in the next section where we show that w can be chosen to be any even number.

4 Generalizing the Word Length

While Artin numbers do not appear to be scarce, it is an open problem first conjectured by Emil Artin in 1927 [1] (p. 246) whether an infinite number of such prime numbers exist. On the positive side there are efficient methods to test the Artin number property.

Most notably, the run-time grows only linearly with the length of the words. Therefore, for the time complexity partitioning the data in word size $w = 4$ or $w = 1018$ does not make a difference for the run-time. Since larger word sizes allow more redundancy and the combination of more data blocks, it is advisable to increase it as large as the CPU cache and the block size of the data allows.

We now overcome the restriction that the word size has the form $w = p - 1$, where p is an Artin number.

Fact 1. *A partition of an integer w is valid if*

$$w = \sum_{i=1}^{\ell} (p_i - 1), \tag{25}$$

where p_i are Artin primes, i.e. 2 is a primitive root of p_i.

- *There exists a valid partition for every even number $w \leq 10^5$ such that the minimum term is at least $\frac{10}{62}w$.*
- *If $w \leq 2^{20}$ is a power of two the minimum term of a valid partition is at least $\frac{28}{128}w$.*

We have verified this fact using computer algebra programs and exhaustive testing.

To overcome the word size restriction we us a valid partition and apply the Circulant Cauchy codes for each of the sub-words separately. The MDS property is preserved and also for the run-time we get $3nm(w_1 + \ldots + w_k) = 3nmw$ XOR operations for encoding and $9nuw$ for decoding. Since, the number of reconstructable blocks is limited by $w_i + 1$ it is desirable to maximize the smallest term. While it is an open question whether infinitely many Artin primes exist, the above fact shows that the density for numbers up to 10^5 is high enough to guarantee good partitions.

Clearly, the most interesting word lengths are powers of two. The following valid partitions maximize the size of the smallest subword size.

$$
\begin{aligned}
8 &= 4 + 4 & 16 &= 4 + 12 \\
32 &= 10 + 10 + 12 & 64 &= 28 + 36 \\
128 &= 28 + 100 & 256 &= 60 + 196 \\
512 &= 196 + 316 & 1024 &= 372 + 652 \\
2048 &= 940 + 1108 & 4096 &= 2028 + 2068 \\
8192 &= 3796 + 4396 & 16384 &= 8116 + 8268
\end{aligned}
\tag{26}
$$

5   Conclusions

We have presented a new approach to MDS codes using long word lengths w which after adding a parity bit to a given word uses only cyclic shift operations and bit-wise XOR-operations. Our method allows the generation of arbitrarily many check blocks and arbitrarily large word sizes. It is based on the isomorphism between Boolean circulant matrices and finite fields where $w + 1$ is a prime number and 2 is a primitive root modulo $w + 1$. We show how this method can be applied to arbitrary word length by a partitioning of the words without an impact to the coding and decoding complexity. We use Cauchy Reed-Solomon codes and present the first MDS scheme with asymptotical optimal XOR complexity for computing the check blocks and recovering data blocks. Furthermore, the constant factors are small and these codes can be easily implemented on existing computer architectures.

References

1. Artin, E., Hasse, H., Frei, G., Roquette, P., Lemmermeyer, F.: Emil Artin und Helmut Hasse: Die Korrespondenz, 1923-1934. In: Einleitung in engl. Sprache, Universitätsverlag Göttingen (2008)
2. Blaum, M.: A family of mds array codes with minimal number of encoding operations. In: 2006 IEEE International Symposium on Information Theory, pp. 2784–2788 (July 2006)
3. Blaum, M., Brady, J., Bruck, J., Menon, J.: EVENODD: An efficient scheme for tolerating double disk failures in RAID architectures. IEEE Trans. Computers 44(2), 192–202 (1995)

4. Blaum, M., Bruck, J., Vardy, A.: MDS array codes with independent parity symbols. IEEE Transactions on Information Theory 42(2), 529–542 (1996)
5. Blaum, M., Roth, R.M.: On lowest density MDS codes. IEEE Transactions on Information Theory 45(1), 46–59 (1999)
6. Blömer, J., Kalfane, M., Karpinski, M., Karp, R.M., Luby, M., Zuckerman, D.: An XOR-based erasure-resilient coding scheme. ICSI Technical Report TR-95-048, ICSI (August 1995)
7. Corbett, P., English, B., Goel, A., Grcanac, T., Kleiman, S., Leong, J., Sankar, S.: Row-diagonal parity for double disk failure correction. In: Proceedings of the USENIX FAST 2004 Conference on File and Storage Technologies, pp. 1–14. Network Appliance, Inc., USENIX Association, San Francisco (2004)
8. Feng, G.-L., Deng, R.H., Bao, F., Shen, J.-C.: New efficient mds array codes for raid part i: Reed-solomon-like codes for tolerating three disk failures. IEEE Trans. Comput. 54(9), 1071–1080 (2005)
9. Feng, G.-L., Deng, R.H., Bao, F., Shen, J.-C.: New efficient mds array codes for raid part ii: Rabin-like codes for tolerating multiple (greater than or equal to 4) disk failures. IEEE Trans. Comput. 54(12), 1473–1483 (2005)
10. Fürer, M.: Faster integer multiplication. In: ACM (ed.) STOC 2007: Proceedings of the 39th Annual ACM Symposium on Theory of Computing, San Diego, California, USA, June 11-13, pages 57–66. ACM Press (2007); pub-ACM:adr
11. Luby, M.: L.T Codes. In: IEEE (ed.) Proceedings of the 43rd Annual IEEE Symposium on Foundations of Computer Science, FOCS 2002, Vancouver, BC, Canada, November 16-19, pp. 271–280. IEEE Computer Society Press (2002); pub-IEEE:adr
12. Luo, J., Xu, L.: Scan: An efficient decoding algorithm for raid-6 codes. In: 2011 10th IEEE International Symposium on Network Computing and Applications (NCA), pp. 91–98 (August 2011)
13. Menezes, A.J., Blake, I.F., Gao, X., Mullin, R.C., Vanstone, S.A., Yaghoobian, T. (eds.): Applications of Finite Fields. Kluwer Academic Publishers (1993)
14. Patterson, D.A., Gibson, G., Katz, R.H.: A Case for Redundant Arrays of Inexpensive Disks (RAID). In: Proceedings of the ACM Conference on Management of Data (SIGMOD), pp. 109–116 (1988)
15. Plank, J.S.: The RAID-6 liberation codes. In: Baker, M., Riedel, E. (eds.) 6th USENIX Conference on File and Storage Technologies, FAST 2008, pp. 97–110, February 26-29. USENIX, San Jose (2008)
16. Plank, J.S., Xu, L.: Optimizing cauchy reed-solomon codes for fault-tolerant network storage applications. In: Proceedings of the Fifth IEEE International Symposium on Network Computing and Applications, NCA 2006, pp. 173–180. IEEE Computer Society, Washington, DC (2006)
17. Reed, I.S., Solomon, G.: Polynomial codes over certain finite fields. Journal of the Society for Industrial and Applied Mathematics 8(2), 300–304 (1960)
18. Schindelhauer, C.: C implementation of our circulant cauchy based codes, http://archive.cone.informatik.uni-freiburg.de/pubs/CirculantCauchy.zip
19. Zaitsev, G.V., Zinovev, V.A., Semakov, N.V.: Minimum check-density codes for correcting bytes of errors, erasures, or defects. Probl. Inform. Transm. 19(3), 29–37 (1983)

Author Index